D0214071

Introduction to Topology

Second Edition

Theodore W. Gamelin
and
Robert Everist Greene

DOVER PUBLICATIONS, INC.
Mineola, New York

Copyright

Published in Canada by General Publishing Company, Ltd., 30 Lesmill Road, Don Mills, Toronto, Ontario.
Published in the United Kingdom by Constable and Company, Ltd., 3 The Lanchesters, 162–164 Fulham Palace Road, London W6 9ER.

Bibliographical Note

This Dover edition, first published in 1999, is an unabridged reprint of the first edition of *Introduction to Topology,* published by W B Saunders Company, Philadelphia, in 1983. The authors prepared a new addendum, Solutions to Selected Exercises, especially for this edition.

Library of Congress Cataloging-in-Publication Data

Gamelin, Theodore W.
Introduction to topology / Theodore W. Gamelin, Robert Everist Greene. — 2nd ed.
p. cm.
An unabridged reprint of the first edition published by W.B. Saunders Company, Philadelphia in 1983. A new addendum has been added.
Includes index.
ISBN 0-486-40680-6 (pbk.)
1. Topology. I. Greene, Robert Everist, 1943– . II. Title.
QA611.G35 1999
514—dc21 99-14612
CIP

Manufactured in the United States of America
Dover Publications, Inc., 31 East 2nd Street, Mineola, N.Y. 11501

To our parents

Frank and Ruth Gamelin

Lee and Dorothy Greene

PREFACE

One of the most important developments in mathematics in the twentieth century has been the formation of topology as an independent field of study and the subsequent systematic application of topological ideas to other fields of mathematics. Ideas that are topological in nature occurred explicitly in the nineteenth century and in embryonic form even earlier. The part of topology most relevant to analysis, that part usually called point-set topology or general topology, had its beginnings in the nineteenth-century works that first established calculus on a rigorous basis. The beginnings of the second large branch of topology, algebraic and geometric topology, are from an even earlier period; for instance, Euler's results in the eighteenth century on the combinatorics of polyhedral figures were already a clear precursor of contemporary ideas. However, the systematic investigation of topology as a separate field of mathematics, which began in the late nineteenth century and continues unabated today, has given the ideas of topology both new generality and new depth. By now, the ideas of point-set topology are a large part of the basic language and technique of analysis. The methods of algebraic topology play an important role in algebra as well as forming a large field of mathematics unto themselves. Also, the great growth of differential geometry in recent times is associated with a viewpoint involving topological concepts.

The by now well-established importance of topology has led naturally to the writing of many works on the subject, including a large number of introductory texts. The purposes of the present text are, however, somewhat different from those of most introductory topology texts. First, we have attempted early in the book to lead the reader through a number of nontrivial applications of metric space topology to analysis, so that the relevance of topology to analysis is apparent both with more immediacy and also on a deeper level than is commonly the case. Second, in the treatment of topics from elementary algebraic topology later in the book, we have concentrated upon results with concrete geometric meaning and have presented comparatively little algebraic formalism; at the same time, however, we have provided proofs of some highly nontrivial results (e.g., the noncontractibility of S^n). These goals have been accomplished by treating homotopy theory without considering homology theory. Thus the reader can immediately see important applications without undertaking the development of a large formal program. We hope that these applications, besides having intrinsic interest, will lead the reader toward the detailed study of algebraic topology with the feeling that putting its methods on a formal, general basis is well worthwhile. The metric space and point-set topology material occupies the first two chapters, the algebraic topological material the remaining two chapters.

This book arose from our experiences in teaching introductory courses in various aspects of topology to upper division undergraduate students and beginning graduate students. All the material was found accessible by the students, but it could not all be covered in a single one-term course. The book is arranged so that considerable flexibility in the choice of topics is possible. Chapter II depends on Chapter I only for motivation, and Chapters III and IV depend on only the most basic material of Chapter II. In fact, by restricting attention to homotopy groups of metric spaces, Chapters III and IV can be studied after only the basic material (e.g., compactness, continuity) of Chapter I. Thus a short course with strong analytical emphasis can easily be constructed by using Chapters I and II in their entirety and only a few topics from Chapters III and IV; a short course emphasizing the geometric and algebraic parts of the subject can be constructed by using little of Chapters I and II and then Chapters III and IV in toto. In all cases, starred sections, which are in general of somewhat greater difficulty than the other material, can be omitted without detriment to understanding of later portions of the text.

In the strict logical sense, the book is almost independent of prior mathematical knowledge. Only familiarity with the real numbers and with some basic set theory, such as countability and uncountability, is needed in this strict sense. Readers who are not to some extent familiar with the subject usually called "rigorous calculus" (e.g., $\varepsilon - \delta$ definitions of limits and continuity) may find the material a strain on their capacity to absorb abstraction without concrete motivation. Readers familiar with rigorous calculus will recognize easily that Chapter I and Chapter II are generalized expressions of phenomena that occur in concrete form in rigorous calculus. Throughout, readers are urged to construct for themselves many concrete illustrations of the general definitions and results. Topology is a subject that, in spite of its appearance of abstractness, is deeply rooted in the concrete and geometrically comprehensible world; the development of intuitive insight by consideration of these roots is a vital part of learning topology, but it is a part that one must do largely for oneself. In this connection also, we strongly urge the reader to do as many of the exercises as possible. Topology is as much a mode of thought as it is a body of information, and mastery of the mode of thought cannot be really well developed by passive reading alone. This remark applies to most mathematics, but it seems to us to apply especially strongly to the learning of topology. Thus the exercises are to us an integral part of the text, and readers should so treat them. In any case, it is more fun to play the game than just to learn the rules.

Cross references to lemmas, theorems and exercises are handled with notations such as "III.4.7" or "5.7." Thus "Theorem III.4.7" refers to Theorem 7 of Chapter III, Section 4, and "Exercise 5.7" refers to Exercise 7 of Section 5 of the Chapter in which the reference occurs. A list of special symbols is given at the end of the book, as are various bibliographical references.

We thank with pleasure the many students and colleagues at UCLA who gave us encouragement and valuable comments on the preliminary lecture notes from which this book evolved over the past eight years. We especially thank Laurie Beerman for her fine job typing the various revisions of the manuscript.

University of California
Los Angeles, California

THEODORE W. GAMELIN
ROBERT EVERIST GREENE

CONTENTS

Introduction to Topology

Second Edition

Metric Spaces

ONE

The ideas of "metric" and "metric space," which are the subject matter of this chapter, are abstractions of the concept of distance in Euclidean space. These abstractions have turned out to be particularly fundamental and useful in modern mathematics; in fact, the aspects of the Euclidean idea of distance retained in the abstract version are precisely those that are most useful in a wide range of mathematical activities. The determination of this usefulness was historically a matter of experience and experiment. By now, the reader can be assured, the mathematical utility of the metric-space information developed in this chapter entirely justifies its careful study.

Sections 1 through 6 of this chapter are devoted to the basic definitions and main theorems about metric spaces in general. Among the theorems established, two are especially substantial: the result called the Baire Category Theorem, in Section 2, and the equivalence of compactness and sequential compactness, in Section 5. The material in these first six sections is basic to modern analysis.

Sections 7 through 9 treat more specialized topics. Section 7 introduces some special classes of metric spaces—the normed linear spaces and Banach spaces—that are particularly important in applications. These spaces have not only the abstract idea of distance common to all metric spaces but also a vector-space structure that interacts with the distance idea in a desirable fashion. In Section 8, an important result about functions from metric spaces to themselves, the Contraction Mapping Theorem, is proved. This result is applied to obtain solutions to various specific problems of analysis: the solution of certain integral equations and the proof of the existence of solutions to certain differential equations (the Cauchy-Picard Theorem). In Section 9, the idea of differentiability for functions on normed linear spaces is introduced. This derivative concept, known as the Frechet derivative, is used to develop analogues for these general spaces of the standard inverse and implicit function theorems of vector calculus, which are themselves incidentally proved in the process.

Sections 7 through 9 are not essential to an understanding of the remainder of the text, and they can thus be omitted or deferred. Nevertheless, we urge the reader at least to glance at the material in these sections to gain some insight into the scope and power of the general metric-space methods.

Most of the ideas about metric spaces in general are motivated by geometric ideas about sets in $\mathbb{R}$ or $\mathbb{R}^n$, $n > 1$. Since this is true in general, explicit statement to this effect is usually omitted in the specific discussions that follow. The reader should nonetheless consider, each time a new idea occurs, what geometric meaning the idea has for sets in Euclidean space. The concrete pictures thus formed are very helpful in developing intuition about the general metric-space situation, even though some caution is, as always in abstract mathematics, necessary to ensure that intuition does not lead one astray.

1. OPEN AND CLOSED SETS

A *metric* on a set X is a real-valued function d on $X \times X$ that has the following properties:

$$d(x,y) \geq 0, \qquad x,y \in X, \tag{1.1}$$

$$d(x,y) = 0 \quad \text{if and only if} \quad x = y, \tag{1.2}$$

$$d(x,y) = d(y,x), \qquad x,y \in X, \tag{1.3}$$

$$d(x,z) \leq d(x,y) + d(y,z), \qquad x,y,z \in X. \tag{1.4}$$

The idea of a metric on a set X is an abstract formulation of the notion of distance in Euclidean space. The intuitive interpretation of property (1.4) is particularly suggestive. This property is the abstract formulation of the fact that the sum of the lengths of two sides of a triangle is greater than or equal to the length of the third side. Consequently (1.4) is referred to as the *triangle inequality*.

A *metric space* (X,d) is a set X equipped with a metric d on X. Sometimes we suppress mention of the metric d and refer to X itself as being a metric space.

If (X,d) is a metric space and Y is a subset of X, then the restriction d' of d to $Y \times Y$ is clearly a metric on Y. The metric space (Y,d') is called a *subspace* of (X,d).

The set of real numbers $\mathbb{R}$, with the usual distance function

$$d(x,y) = |x - y|, \qquad x,y \in \mathbb{R},$$

is a metric space since properties (1.1) through (1.4) all hold. More generally, the n-dimensional Euclidean space $\mathbb{R}^n$, consisting of all n-tuples $x = (x_1, \ldots, x_n)$ of real numbers, becomes a metric space when endowed with the metric

$$d(x,y) = \left[\sum_{j=1}^{n} (x_j - y_j)^2 \right]^{1/2}, \qquad x,y \in \mathbb{R}^n. \tag{1.5}$$

Actually, it is not immediately clear that (1.5) defines a metric. The verification of properties (1.1), (1.2), and (1.3) is straightforward, but the verification of (1.4) requires some effort, and a proof is outlined in Exercise 3. We shall be especially interested in the metric spaces $\mathbb{R}$ and $\mathbb{R}^n$ and in their subspaces.

Note that the cases $n = 2$ and $n = 3$ correspond to the spaces of Euclidean plane geometry and solid geometry, respectively. In this setting, our definition of distance is adopted from the statement of the Theorem of Pythagoras. In mathematics, many good theorems are eventually converted into definitions.

As another example of a metric space, let S be any set and let $B(S)$ denote the set of bounded real-valued functions on S. Endowed with the metric

$$d(f,g) = \sup\{|f(s) - g(s)| : s \in S\}, \tag{1.6}$$

$B(S)$ becomes a metric space. The verification of (1.1) through (1.4) in this case is left to the reader (Exercise 5).

Any set X can be made into a "discrete" metric space by associating with X the metric d defined by

$$d(x,y) = \begin{cases} 1, & x \neq y, \\ 0, & x = y. \end{cases} \tag{1.7}$$

The verification that d is indeed a metric is also left to the reader (Exercise 2).

Let (X,d) be any metric space. The *open ball* $B(x;r)$ with center $x \in X$ and radius $r > 0$ is defined by

$$B(x;r) = \{y \in X : d(x,y) < r\}.$$

The balls centered at x form a nested family of subsets of X that increase with r, that is, $B(x;r_1) \subseteq B(x;r_2)$ if $r_1 \leq r_2$. Furthermore,

$$\bigcup_{r>0} B(x;r) = X,$$

and, because of (1.2),

$$\bigcap_{r>0} B(x;r) = \{x\}.$$

Let Y be a subset of X. A point $x \in X$ is an *interior point* of Y if there exists $r > 0$ such that $B(x;r) \subset Y$. The set of interior points of y is the *interior* of Y, and it is denoted by int(Y). Note that every interior point of Y belongs to Y:

$$\text{int}(Y) \subseteq Y.$$

A subset Y of X is *open* if every point of Y is an interior point of Y, that is, if int(Y) $= Y$. In particular, the empty set $\varnothing$ and the entire space X are open subsets of X.

1.1 Theorem: Any open ball $B(x;r)$ in a metric space X is an open subset of X.

Proof: Let $y \in B(x;r)$. It suffices to find some open ball centered at y that is contained in $B(x;r)$. Let $s = r - d(x,y)$. Then $s > 0$. If $z \in B(y;s)$, i.e., $d(y,z) < s$, then $d(x,z)$

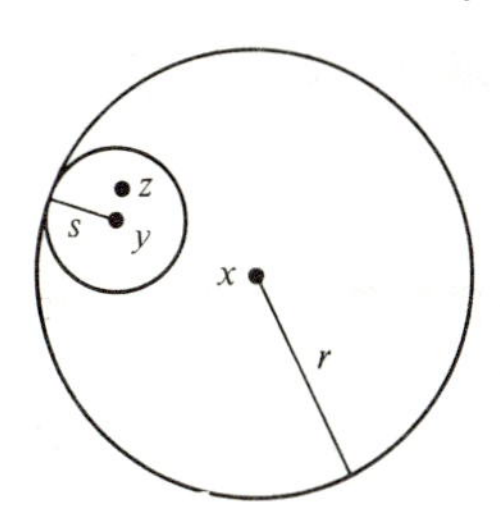

$\leq d(x,y) + d(y,z) < d(x,y) + s = r$, so that $z \in B(x;r)$. Consequently $B(y;s) \subseteq B(x;r)$. □

1.2 Theorem: Let Y be a subset of a metric space X. Then $\text{int}(Y)$ is an open subset of X. In other words, $\text{int}(\text{int}(Y)) = \text{int}(Y)$.

Proof: Let $x \in \text{int}(Y)$. Then there exists $r > 0$ such that $B(x;r) \subset Y$. It suffices to show that $B(x;r) \subset \text{int}(Y)$.

Since $B(x;r)$ is open, there is for each $y \in B(x;r)$ an open ball $B(y;s)$ contained in $B(x;r)$. In particular, each $B(y;s)$ is contained in Y, so that each $y \in B(x;r)$ belongs to $\text{int}(Y)$. □

1.3 Theorem: The union of a family of open subsets of a metric space X is an open subset of X.

Proof: Let $\{U_\alpha\}_{\alpha\in A}$ be a family of open subsets of X and let $U = \cup_{\alpha\in A} U_\alpha$. Suppose $x \in U$. Then there is some index α such that $x \in U_\alpha$. Since U_α is open, there exists some $r > 0$ such that $B(x;r) \subset U_\alpha$. Then $B(x;r) \subset U$, so that x is an interior point of U. Since this is true for all $x \in U$, U is open. □

1.4 Theorem: A subset U of a metric space X is open if and only if U is a union of open balls in X.

Proof: By Theorems 1.1 and 1.3, any set that is a union of open balls is open. On the other hand, suppose that U is an open subset of X. For each $x \in U$, there then exists $r(x) > 0$ such that $B(x;r(x)) \subset U$. Utilizing U as an index set for a union, we obtain an expression

$$U = \bigcup_{x\in U} B(x;r(x))$$

for U as a union of open balls. □

In general, the union in Theorem 1.4 will have to be infinite. For example, the set $\{(x_1,x_2) \in \mathbb{R}^2 : 0 < x_1 < 1, 0 < x_2 < 1\}$ is open but is not a finite union of open balls.

1.5 Theorem: The intersection of any finite number of open subsets of a metric space is open.

Proof: Let $U_1, \ldots, U_m$ be open subsets of X and let $U = U_1 \cap \ldots \cap U_m$. Let $y \in U$. Since each U_j is open, there exist $r_j > 0$ such that

$$B(y;r_j) \subset U_j, \qquad 1 \leq j \leq m.$$

Set $r = \min(r_1, \ldots, r_m)$. Then $r > 0$ (the minimum of a *finite* set of positive numbers is a positive number). Since $B(y;r) \subset U_j$, $1 \leq j \leq m$, we obtain $B(y;r) \subset U$. It follows that U is open. □

The finiteness assumption in the theorem just given is essential. In $\mathbb{R}^1$, the intersection of open balls $\bigcap_{i=1}^{\infty} B(0;1/i)$ is the set $\{0\}$ consisting of 0 only. This set is not open in $\mathbb{R}^1$.

A subset of X that is open in a subspace of X need not be open in X. For instance, an open interval (a,b) on the real line $\mathbb{R}$ is an open subset of $\mathbb{R}$. However, $\mathbb{R}$ can be regarded as a subspace of the plane $\mathbb{R}^2$ (by identifying $\mathbb{R}$ with the x-axis in $\mathbb{R}^2$), and then the interval is not an open subset of $\mathbb{R}^2$. What *is* true is the following.

1.6 Theorem: Let Y be a subspace of a metric space X. Then a subset U of Y is open in Y if and only if $U = V \cap Y$ for some open subset V of X.

Proof: Because the metric $d' : Y \times Y \to R$ is the restriction to $Y \times Y$ of the metric $d : X \times X \to R$ on X, the open ball in the metric space Y with center $y \in Y$ and radius $r > 0$ is just the intersection $B(y;r) \cap Y$ of Y and the open ball $B(y;r)$ in X. If V is open in X, then for each $y \in V \cap Y$, there exists $r > 0$ such that $B(y;r) \subset V$. Then the open ball in Y centered at y with radius r is contained in $V \cap Y = U$. Consequently each $y \in V \cap Y$ is an interior point of $V \cap Y$ in the subspace Y, so that $U = V \cap Y$ is open in Y.

Conversely, suppose that U is an open subset of the subspace Y. Let $y \in U$. Then there exists $r(y) > 0$ such that the open ball $B(y;r(y)) \cap Y$ in Y is contained in U. Then the open subset

$$V = \bigcup_{y \in U} B(y;r(y))$$

of X satisfies $V \cap Y \subset U$. Since each $y \in U$ belongs to V, we obtain $V \cap Y = U$. □

Let Y be a subset of a metric space X. A point $x \in X$ is *adherent* to Y if for all $r > 0$,

$$B(x;r) \cap Y \neq \varnothing.$$

The *closure* of Y, denoted by $\overline{Y}$, consists of all points in X that are adherent to Y. Evidently each point of Y is adherent to Y, so that

$$Y \subset \overline{Y}.$$

The subset Y is *closed* if $Y = \overline{Y}$. In particular, the empty set $\varnothing$ and the entire space X are closed subsets of X.

Intuitively speaking, a subset is closed if it contains all its boundary points. (For a precise version of this idea, see Exercise 14.) For example, the union of a circle in $\mathbb{R}^2$ and its inside is a closed set. If any points of the circle are deleted from this union, however, then the resulting set is not closed.

1.7 Theorem: If Y is a subset of a metric space X, then the closure of Y is closed, that is, $\overline{\overline{Y}} = \overline{Y}$.

Proof: Since $\overline{Y} \subset \overline{\overline{Y}}$, it suffices to obtain the reverse inclusion, $\overline{\overline{Y}} \subset \overline{Y}$.

Let $x \in \overline{\overline{Y}}$ and let $r > 0$. It suffices to show that $B(x;r) \cap Y \neq \varnothing$. Since x is adherent to $\overline{Y}$, there is a point $z \in B(x;r/2) \cap \overline{Y}$. Since z is adherent to Y, there exists a point $y \in B(z;r/2) \cap Y$. Then

$$d(x;y) \leq d(x,z) + d(z,y) < \frac{r}{2} + \frac{r}{2} = r,$$

so that $y \in B(x;r) \cap Y$ and the intersection is not empty. □

1.8 Theorem: A subset Y of a metric space X is closed if and only if the complement of Y is open.

Proof: The set-theoretic difference of two sets U and V consists of those points in U that do not belong to V. Here, and forevermore, the notation used for this difference is $U \backslash V$.

Suppose first that Y is closed. For each $x \in X \backslash Y$, there then exists $r > 0$ such that $B(x;r) \cap Y = \varnothing$. Then $B(x;r) \subset X \backslash Y$, so that x is an interior point of $X \backslash Y$, and $X \backslash Y$ is open.

Conversely, suppose that $X \backslash Y$ is open. For each $x \in X \backslash Y$, there then exists $r > 0$ such that $B(x;r) \subseteq X \backslash Y$. Hence $B(x;r) \cap Y = \varnothing$ and $x \notin \overline{Y}$. It follows that $Y = \overline{Y}$. □

Despite Theorem 1.8, it should be noted that being open and being closed are not opposite notions. The empty set is both open and closed, and so is the entire space X. In a discrete metric space, every set is both open and closed. On the other hand, the semiopen interval $(0,1]$, regarded as a subset of $\mathbb{R}$, is neither open nor closed.

1.9 Theorem: The intersection of any family of closed sets is closed. The union of any finite family of closed sets is closed.

Proof: The statement on intersections follows from Theorems 1.3 and 1.8, together with the identity

$$X \backslash \bigcap_{\alpha \in A} E_\alpha = \bigcup_{\alpha \in A} (X \backslash E_\alpha),$$

valid for any family $\{E_\alpha\}_{\alpha \in A}$ of subsets of X (Exercise 1). Indeed, if each E_α is closed, then each $X \backslash E_\alpha$ is open, so that the union of the $X \backslash E_\alpha$ is open and the intersection of the E_α is closed. The statement on unions follows from Theorems 1.5 and 1.8, together with the identity (Exercise 1)

$$X \backslash \bigcup_{\alpha \in A} E_\alpha = \bigcap_{\alpha \in A} (X \backslash E_\alpha). \ \square$$

A sequence $\{x_n\}_{n=1}^{\infty}$ in a metric space X *converges to* $x \in X$ if

$$\lim_{n \to \infty} d(x_n,x) = 0.$$

In this case, x is the *limit* of $\{x_n\}$ and we write $x_n \to x$, or

$$\lim_{n \to \infty} x_n = x.$$

1.10 Lemma: The limit of a convergent sequence in a metric space is unique.

Proof: Suppose that x and $y \in X$ are both limits of a sequence $\{x_n\}$ in X. Then for all n,

$$d(x,y) \leq d(x,x_n) + d(x_n,y).$$

As n tends to infinity, the right-hand side tends to 0, so that $d(x,y) = 0$. Consequently, $x = y$. □

1.11 Theorem: Let Y be a subset of the metric space X, Then $x \in X$ is adherent to Y if and only if there is a sequence in Y that converges to x.

Proof: If there is a sequence in Y that converges to x, then every open ball centered at x contains points of the sequence, so that x is adherent to Y. Conversely, suppose that x is adherent to Y. For each integer $n \geq 1$, there exists then some point $x_n \in B(x;1/n) \cap Y$. The sequence $\{x_n\}_{n=1}^{\infty}$ then satisfies $d(x,x_n) < 1/n \to 0$, so that x_n converges to x. □

EXERCISES

1. Let U, V, and W be subsets of some set. Recall that $U \backslash V$ consists of all points in U that do not belong to V.
   (a) Prove that $(U \cup V) \backslash W = (U \backslash W) \cup (V \backslash W)$.
   (b) Prove that $(U \cap V) \backslash W = (U \backslash W) \cap (V \backslash W)$.
   (c) Does $U \backslash (V \backslash W)$ coincide with $(U \backslash V) \backslash W$? Justify your answer by proof or counterexample.
   (d) Prove the two set-theoretic identities used in the proof of Theorem 1.9.
2. Show that (1.7) defines a metric on X. Show that every subset of the resulting metric space is both open and closed.
3. (a) Show that if $a,b,c \in \mathbb{R}$ are such that for all $\lambda \in \mathbb{R}$, $a\lambda^2 + b\lambda + c \geq 0$, then $b^2 - 4ac \leq 0$. *Hint:* Find the minimal value of the polynomial in λ.

(b) Show that for any $(x_1, \ldots, x_n), (y_1, \ldots, y_n) \in \mathbb{R}^n$,

$$\sum_{i=1}^{n} x_i y_i \le \left(\sum_{i=1}^{n} x_i^2\right)^{1/2} \left(\sum_{i=1}^{n} y_i^2\right)^{1/2}.$$

Remark: This is a version of the important *Cauchy-Schwarz inequality*. (Perhaps the most common spelling error in mathematics is to replace "Schwarz" by "Schwartz.") For the proof, apply part (a) to

$$\Sigma_{i=1}^{n} (x_i - \lambda y_i)^2.$$

(c) Using the Cauchy-Schwarz inequality, show that the function d defined by (1.5) satisfies the triangle inequality.

4. Show that the semiopen interval $(0,1]$ is neither open nor closed in $\mathbb{R}$.
5. Show that (1.6) defines a metric on the space $B(S)$ of bounded real-valued functions on a set S.

For the following exercises let (X,d) be a metric space.

6. Prove that the interior of a subset Y of X coincides with the union of all open subsets of X that are contained in Y. (Thus the interior of Y is the largest open set contained in Y.)
7. Prove that the closure of a subset Y of X coincides with the intersection of all closed subsets of X that contain Y. (Thus the closure of Y is the smallest closed set containing Y.)
8. A set of the form $\{y \in X : d(x,y) \le r\}$ is called a *closed ball*. Show that a closed ball is a closed set. Is the closed ball $\{y \in X : d(x,y) \le r\}$ always the closure of the open ball $B(x;r)$? What if $X = \mathbb{R}^n$? Prove your answers.
9. Let Y be a subspace of X and let S be a subset of Y. Show that the closure of S in Y coincides with $\bar{S} \cap Y$, where $\bar{S}$ is the closure of S in X.
10. A point $x \in X$ is a *limit point* of a subset S of X if every ball $B(x;r)$ contains infinitely many points of S. Show that x is a limit point of S if and only if there is a sequence $\{x_j\}_{j=1}^{\infty}$ in S such that $x_j \to x$ and $x_j \neq x$ for all j. Show that the set of limit points of S is closed.
11. A point $x \in S$ is an *isolated point* of S if there exists $r > 0$ such that $B(x;r) \cap S = \{x\}$. Show that the closure of a subset S of X is the disjoint union of the limit points of S and the isolated points of S.
12. Two metrics on X are *equivalent* if they determine the same open subsets. Show that two metrics d,ρ on X are *equivalent* if and only if the convergent sequences in (X,d) are the same as the convergent sequences in (X,ρ).
13. Define ρ on $X \times X$ by

$$\rho(x,y) = \min(1,d(x,y)), \qquad x,y \in X.$$

Show that ρ is a metric that is equivalent to d. (Hence every metric is equivalent to a bounded metric.)

14. The *boundary* ∂E of a set E is defined to be the set of points adherent to both E and the complement of E,

$$\partial E = \bar{E} \cap (\overline{X \backslash E}).$$

Show that E is open if and only if $E \cap \partial E$ is empty. Show that E is closed if and only if $\partial E \subseteq E$.

2. COMPLETENESS

The definition of a convergent sequence involves not only the sequence itself but also the limit of the sequence. We wish to develop a notion of convergence that is intrinsic to the sequence, that is, one that does not require having at hand an object that can be called the limit of the sequence. For this, we make the following definition.

A sequence $\{x_n\}_{n=1}^{\infty}$ in a metric space X is a *Cauchy sequence* if

$$\lim_{m,n\to\infty} d(x_n,x_m) = 0,$$

that is, if for each $\varepsilon > 0$, there is an N such that $d(x_n,x_m) < \varepsilon$ for all $n,m \geq N$.

2.1 Lemma: A convergent sequence is a Cauchy sequence.

Proof: Suppose that $\{x_n\}$ converges to x. Then

$$d(x_n,x_m) \leq d(x_n,x) + d(x,x_m),$$

and the right-hand side tends to zero as $n,m \to \infty$. □

2.2 Lemma: If $\{x_n\}$ is a Cauchy sequence and if there is a subsequence $\{x_{n_k}\}_{k=1}^{\infty}$ of $\{x_n\}$ that converges to x, then $\{x_n\}$ converges to x.

Proof: Let $\varepsilon > 0$. Choose $k > 0$ so large that $d(x_n,x_m) < \varepsilon/2$ whenever $n,m \geq k$. Choose $l > 0$ so large that $d(x_{n_j},x) < \varepsilon/2$ whenever $j \geq l$. Set $N = \max(k,n_l)$. If $m \geq N$ and $n_j \geq N$, then

$$d(x_m,x) \leq d(x_m,x_{n_m}) + d(x_{n_j},x) < \varepsilon/2 + \varepsilon/2 = \varepsilon. \qquad \square$$

A metric space X is *complete* if every Cauchy sequence in X converges.

Not every metric space is complete. For instance, the set $X = \{x \in \mathbb{R} : 0 < x < 1\}$ with metric $d(x,y) = |x - y|$ is not complete since $\{1/n\}$ is a Cauchy sequence that converges to no point (of X). However, many important metric spaces are complete; in particular, $\mathbb{R}^n$ is complete for all $n = 1, 2, 3, \ldots$. This completeness will be established in the next section.

The space $\{x \in \mathbb{R} : 0 < x < 1\}$, though not complete, is a subspace of the complete metric space $\tilde{X} = \{x \in \mathbb{R} : 0 \leq x \leq 1\}$. (That $\tilde{X}$ is complete will follow from the completeness of $\mathbb{R}$ and Theorem 2.3.) This example illustrates the general situation: every metric space X may be regarded as a subspace of a complete metric space $\tilde{X}$ in such a way that $\overline{X} = \tilde{X}$ (Exercise 7). Thus any noncomplete metric space can be thought of as a complete metric space with certain points deleted.

2.3 Theorem: A closed subspace of a complete metric space is complete.

Proof: Let Y be a closed subspace of a complete space X and let $\{y_n\}$ be a Cauchy sequence in Y. Then $\{y_n\}$ is also a Cauchy sequence in X. Since X is complete, there exists $x \in X$ such that y_n converges to x in X. Since Y is closed, x must belong to Y. Consequently $\{y_n\}$ converges to x in Y. □

2.4 Theorem: A complete subspace Y of a metric space X is closed in X.

Proof: Suppose $x \in X$ is adherent to Y. By Theorem 1.11, there is a sequence $\{y_n\}$ in Y that converges to x. By Lemma 2.1, $\{y_n\}$ is a Cauchy sequence in X; hence it is also a Cauchy sequence in Y. Since Y is complete, $\{y_n\}$ converges to some point $y \in Y$. Since limits of sequences are unique, $y = x$ and x belongs to Y. Consequently Y is closed. □

Let $\{f_n\}_{n=1}^{\infty}$ be a sequence of functions from a set S to a metric space X and let f be a function from S to X. The sequence $\{f_n\}$ *converges uniformly* to f on S if for each $\varepsilon > 0$ there exists an integer N such that $d(f_n(s),f(s)) < \varepsilon$ for all integers $n \geq N$ and for all $s \in S$. A sequence $\{f_n\}$ of functions from S to X is a *Cauchy sequence of functions* if for each $\varepsilon > 0$ there exists an integer N such that

$$d(f_n(s),f_m(s)) < \varepsilon, \qquad \text{all } s \in S,\ n,m \geq N.$$

The idea of a Cauchy sequence of functions can be related to the idea of a Cauchy sequence in a metric space by defining an appropriate metric on the set of all functions from S to X (Exercise 6).

2.5 Theorem: Let S be a set, and let X be a complete metric space. If $\{f_n\}_{n=1}^{\infty}$ is a Cauchy sequence of functions from S to X, then there exists a function f from S to X such that $\{f_n\}$ converges uniformly to f.

Proof: For each fixed $s \in S$, $\{f_n(s)\}_{n=1}^{\infty}$ is a Cauchy sequence in X. Since X is complete, $\{f_n(s)\}$ converges to some point of X, which we define to be $f(s)$. Thus f is a function from S to X.

Let $\varepsilon > 0$. Choose an integer $N \geq 1$ such that $d(f_n(s),f_m(s)) < \varepsilon$ for all $s \in S$ and all $n,m \geq N$. Then

$$\begin{aligned} d(f_n(s),f(s)) &\leq d(f_n(s),f_m(s)) + d(f_m(s),f(s)) \\ &< \varepsilon + d(f_m(s),f(s)) \end{aligned}$$

whenever $n,m \geq N$. Letting m tend to ∞, we obtain

$$d(f_n(s),f(s)) \leq \varepsilon, \qquad n \geq N,$$

for all $s \in S$. Consequently $\{f_n\}$ converges uniformly to f on S. □

A subset T of a metric space X is *dense* in X if $\overline{T} = X$. The following theorem involving this concept has many important applications.

2.6 Theorem *(Baire Category Theorem):* Let $\{U_n\}_{n=1}^{\infty}$ be a sequence of dense open subsets of a complete metric space X. Then $\bigcap_{n=1}^{\infty} U_n$ is also dense in X.

Proof: Let $x \in X$ and let $\varepsilon > 0$. It suffices to find $y \in B(x;\varepsilon)$ that belongs to $\bigcap_{n=1}^{\infty} U_n$. Indeed, then every open ball in X meets $\bigcap U_n$, so that $\bigcap U_n$ is dense in X.

Since U_1 is dense in X, there exists $y_1 \in U_1$ such that $d(x,y_1) < \varepsilon$. Since U_1 is open, there exists $r_1 > 0$ such that $B(y_1;r_1) \subset U_1$. By shrinking r_1, we can arrange that $r_1 < 1$, and $\overline{B(y_1;r_1)} \subset U_1 \cap B(x,\varepsilon)$. The same argument, with $B(y_1;r_1)$ replacing $B(x;\varepsilon)$, produces $y_2 \in X$ and $0 < r_2 < 1/2$ such that $\overline{B(y_2;r_2)} \subset U_2 \cap B(y_1;r_1)$.

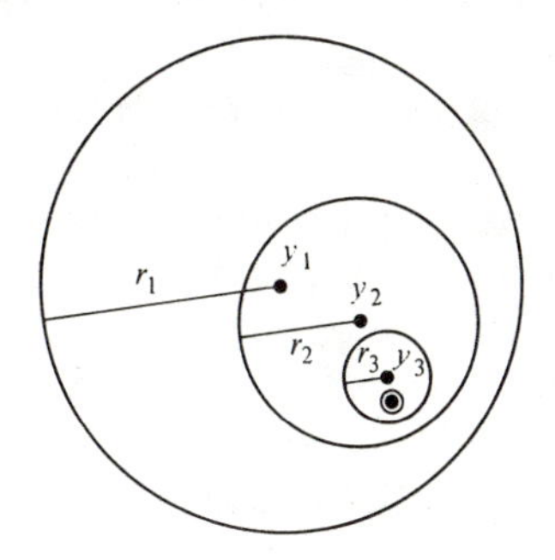

Continuing in this manner, we obtain a sequence $\{y_n\}_{n=1}^{\infty}$ in X and a sequence $\{r_n\}_{n=1}^{\infty}$ or radii such that $0 < r_n < 1/n$ and

$$\overline{B(y_n;r_n)} \subseteq U_n \cap B(y_{n-1};r_{n-1}). \tag{2.1}$$

It follows that

$$\overline{B(y_n;r_n)} \subset B(y_{n-1};r_{n-1}) \subset \ldots \subset B(y_1;r_1) \subset B(x;\varepsilon). \tag{2.2}$$

The nesting property (2.2) shows that $y_m \in B(y_n;r_n)$ if $m > n$, so that $d(y_m,y_n) < r_n \to 0$ as $m,n \to \infty$. Consequently $\{y_m\}$ is a Cauchy sequence. Since X is complete, there exists $y \in X$ such that $y_m \to y$. Since $y_m \in B(y_n;r_n)$ for $m > n$, we obtain $y \in \overline{B(y_n;r_n)}$. By (2.2) $y \in B(x;\varepsilon)$, and by (2.1) $y \in U_n$; this for all n, so that $y \in \bigcap_{n=1}^{\infty} U_n$. □

The hypothesis of completeness in the Baire Category Theorem is crucial. For example, consider the subspace of $\mathbb{R}$ consisting of the rational numbers $\mathbb{R}_0$. Arrange the rational numbers in a sequence $\{s_n\}_{n=1}^{\infty}$ and set $U_n = \mathbb{R}_0 \backslash \{s_n\}$. Then each U_n is a dense open subset of $\mathbb{R}_0$. However, $\bigcap_{n=1}^{\infty} U_n$ is empty.

A subset Y of X is *nowhere dense* if $\overline{Y}$ has no interior points, that is, if

$$\operatorname{int}(\overline{Y}) = \varnothing.$$

Evidently Y is nowhere dense if and only if $X \backslash \overline{Y}$ is a dense open subset of X. By taking complements of the sets in the Baire Category Theorem, we obtain the following equivalent version of the theorem.

2.7 Corollary: Let $\{E_n\}_{n=1}^{\infty}$ be a sequence of nowhere dense subsets of a complete metric space X. Then $\bigcup_{n=1}^{\infty} E_n$ has empty interior.

Proof: Apply the Baire Category Theorem to the dense open sets $U_n = X \backslash \overline{E}_n$. □

An explanation of the nomenclature is perhaps in order. A subset of a metric space X is of the *first category* (or *meager*) if it is the countable union of nowhere-dense subsets. A subset that is not of the first category is said to be of the *second category*. The Baire Category Theorem is then equivalent to the following statement. In a complete metric space X, the complement of a set of the first category is dense in X. The general idea of the terminology is that a nowhere dense set is a really small set and that a countable union of nowhere dense sets is still small in the sense that it has empty interior (by the Baire Category Theorem). Thus a set of the first category is a small set; a set of the second category is a set that is not small in this sense.

The Baire Category Theorem is a far-reaching generalization of the uncountability of the real numbers. Single-point subsets of the real numbers are nowhere dense, and the uncountability of the real numbers is just the fact that countable unions of such single-point sets cannot be the whole of the real numbers. Thus, the Baire Category Theorem has as a corollary the uncountability of the real numbers, once it is known that the real numbers are complete in their usual metric-space structure. This completeness is discussed in the next section.

The applications of the Baire Category Theorem go far beyond proving uncountability, however. Some additional applications to an entirely different situation are presented in Section 7. Some further applications to the real numbers are given in Exercises 3.5 to 3.7.

EXERCISES

1. A sequence $\{x_k\}_{k=1}^{\infty}$ in a metric space (X,d) is a *fast Cauchy sequence* if

$$\sum_{k=1}^{\infty} d(x_k,x_{k+1}) < \infty.$$

   Show that a fast Cauchy sequence is a Cauchy sequence.

2. Prove that every Cauchy sequence has a subsequence that is a fast Cauchy sequence.
3. Prove that the set of isolated points of a countable complete metric space X forms a dense subset of X.
4. Suppose that F is a subset of the first category in a metric space X and E is a subset of F. Prove that E is of the first category in X. Show by an example that E may not be of the first category in the metric space F.
5. Prove that any countable union of sets of the first category in X is again of the first category in X.
6. Let S be a nonempty set, let (X,d) be a metric space, and let $\mathcal{F}$ be the set of functions from S to X. For $f,g \in \mathcal{F}$, define

$$\rho(f,g) = \sup_{s \in S} \min(1,d(f(s),g(s))).$$

   Show that ρ is a metric in $\mathcal{F}$. Show that a sequence $\{f_n\}$ converges to f in the metric space $(\mathcal{F},\rho)$ if and only if $\{f_n\}$ converges uniformly to f on X. Show that $(\mathcal{F},\rho)$ is complete if and only if (X,d) is complete.

7. Let (X,d) be a metric space and let S be the set of Cauchy sequences in S. Define a relation "$\sim$" in X by declaring "$\{s_k\} \sim \{t_k\}$" to mean that $d(s_k,t_k) \to 0$ as $k \to \infty$.
   (a) Show that the relation "$\sim$" is an equivalence relation.
   (b) Let $\tilde{X}$ denote the set of equivalence classes of S and let $\tilde{s}$ denote the equivalence class of $s = \{s_k\}_{k=1}^{\infty}$. Show that the function

$$\rho(\tilde{s},\tilde{t}) = \lim_{k\to\infty} d(s_k,t_k), \qquad \tilde{s},\tilde{t} \in \tilde{X}$$

   defines a metric on $\tilde{X}$.
   (c) Show that $(\tilde{X},\rho)$ is complete.
   (d) For $x \in X$, define $\tilde{x}$ to be the equivalence class of the constant sequence $\{x,x,\ldots\}$. Show that the function $x \to \tilde{x}$ is an isometry of X onto a dense subset of $\tilde{X}$. (By an isometry, we mean that $d(x,y) = \rho(\tilde{x},\tilde{y})$, $x,y \in X$.)

   Note: If a complete metric space Y contains X as a dense subspace, we say that Y is a *completion* of X. The space $\tilde{X}$ of Exercise 7 can be regarded as a completion of X by identifying each $x \in X$ with the constant sequence $\{x,x,\ldots\}$. The next part of the exercise shows that the completion of X is unique, up to isometry.
   (e) Show that when Y is a completion of X, then the inclusion map $X \to Y$ extends to an isometry of $\tilde{X}$ onto Y.
8. The *diameter* of a nonempty subset E of a metric space (X,d) is defined to be

$$\operatorname{diam}(E) = \sup\{d(x,y) : x,y \in E\}.$$

   Show that if $\{E_k\}_{k=1}^{\infty}$ is a decreasing sequence of closed nonempty subsets of a complete metric space whose diameters tend to zero, then $\bigcap_{k=1}^{\infty} E_k$ consists of precisely one point. How much of the conclusion remains true if X is not complete? Can this property be used to characterize complete metric spaces? Justify your answer.

3. THE REAL LINE

The set of real numbers will always be denoted by $\mathbb{R}$. Unless stated otherwise, $\mathbb{R}$ will be regarded as a metric space, with the metric $d(s,t) = |s - t|$, $s,t \in \mathbb{R}$. The set of rational numbers is denoted by $\mathbb{R}_0$.

It is possible to construct the real number system $\mathbb{R}$ from the rational numbers; the construction is a special case of forming the completion of a metric space. The procedure for completing a metric space was outlined in Exercise 2.7. Familiarity with this construction will not be assumed. We shall assume familiarity only with certain properties of the real numbers, which can be taken as axioms.

A first batch of properties involves the algebraic structure of $\mathbb{R}$. These properties assert that the set $\mathbb{R}$, equipped with the operations of addition and multiplication, forms an algebraic entity called a *field*. These properties are just a guarantee that $\mathbb{R}$ follows the usual rules of arithmetic. These arithmetic properties will not be listed here.

The second batch of properties involves the ordering in $\mathbb{R}$. There is a relation "$<$" on $\mathbb{R}$ that obeys the following laws:

Trichotomy: If $r,s \in \mathbb{R}$, then exactly one of the following three possibilities must hold: $r < s$, $r = s$, or $s < r$.

Transitivity: If $r < s$ and $s < t$, then $r < t$.

Compatibility With Addition: If $r < s$ and $t \in \mathbb{R}$, then $r + t < s + t$.

Compatibility With Multiplication: If $r < s$ and $t > 0$, then $rt < st$.

A field with an ordering satisfying these axioms is a *totally ordered field*.

The properties we have mentioned so far do not characterize $\mathbb{R}$. Indeed, $\mathbb{R}_0$ is also a totally ordered field. That is, $\mathbb{R}_0$ satisfies the rules of arithmetic and has a relation "$<$" (the usual order of $\mathbb{R}_0$) that satisfies the rules listed. The final property, which characterizes $\mathbb{R}$ up to "isomorphism," is the Least Upper Bound Axiom. (Characterization up to isomorphism means that any other system having all these properties can be put into one-to-one correspondence with $\mathbb{R}$ in such a way that the correspondence preserves arithmetic operations and order. The fact that the properties given characterize $\mathbb{R}$ in this sense will not be further discussed here.) A subset S of $\mathbb{R}$ is *bounded above* if there exists $M \in \mathbb{R}$ such that $s < M$ for all $s \in S$; such an M is an *upper bound* for S. An upper bound M for S is a *least upper bound* for S if $M \leq M'$ for any other upper bound M' for S.

Least Upper Bound Axiom: If S is a nonempty subset of $\mathbb{R}$ that is bounded above, then S has a least upper bound.

Using the Least Upper Bound Axiom, we obtain the following property of $\mathbb{R}$, which is crucial for our purposes.

3.1 Theorem: The real line $\mathbb{R}$, with the usual metric $d(s,t) = |s - t|$, is a complete metric space.

Proof: Let $\{s_n\}_{n=1}^{\infty}$ be a Cauchy sequence in $\mathbb{R}$. Let S be the set of $y \in \mathbb{R}$ such that $s_n < y$ for only finitely many integers n. If $y \in S$ and $z < y$, then evidently $z \in S$. Consequently S includes the entire interval $(-\infty,y]$, just as soon as $y \in S$.

Let $\varepsilon > 0$. Since $\{s_n\}$ is Cauchy, there is an integer N such that $|s_n - s_m| < \varepsilon$ whenever $m,n \geq N$, so that all but finitely many terms of the sequence lie in the interval $(s_N - \varepsilon, s_N + \varepsilon)$. In particular, $s_N - \varepsilon \in S$, so that S is not empty. Also, no $t \geq s_N + \varepsilon$ belongs to S, so that $s_N + \varepsilon$ is an upper bound for S. By the Least Upper Bound Axiom, S has a least upper bound, call it b. Since $s_N + \varepsilon$ is an upper bound, $b \leq s_N + \varepsilon$. Since $s_N - \varepsilon \in S$, $s_N - \varepsilon \leq b$. Hence $|s_N - b| \leq \varepsilon$. If $m \geq N$, then

$$|s_m - b| \leq |s_m - s_N| + |s_N - b| \leq \varepsilon + \varepsilon = 2\varepsilon.$$

Since $\varepsilon > 0$ is arbitrary, $s_m \to b$ as $m \to \infty$. □

EXERCISES

1. Use the Least Upper Bound Axiom, together with the usual manipulations of algebraic identities and inequalities, to prove the following:
   (a) The set $\mathbb{Z}$ of integers is not bounded above.
   (b) For each $\varepsilon > 0$, there exists a rational number $r \in (0,\varepsilon)$.
   (c) If $a,b \in \mathbb{R}$ satisfy $a < b$, then there exists a rational number $s \in (a,b)$.
   (d) The set $\mathbb{R}_0$ of rational numbers is dense in $\mathbb{R}$.
2. Prove that the set of irrational numbers is dense in $\mathbb{R}$.
3. Regard the rational numbers $\mathbb{R}_0$ as a subspace of $\mathbb{R}$. Does the metric space $\mathbb{R}_0$ have any isolated points? Why does this not contradict Exercise 2.3?
4. Prove that every open subset of $\mathbb{R}$ is a union of disjoint open intervals (finite, semi-infinite, or infinite).
5. Prove that the set of irrational numbers cannot be expressed as the union of a sequence of closed subsets of $\mathbb{R}$.
6. Prove that the set of rational numbers cannot be expressed as the intersection of a sequence of open subsets of $\mathbb{R}$.
7. Let E_0 be the closed unit interval $[0,1]$. Let E_1 be the closed subset of E_0 obtained by removing the open, middle third of E_0, so that $E_1 = [0,1/3] \cup [2/3,1]$. Let E_2 be the closed subset of E_1 obtained by removing the open middle thirds of each of the two intervals in E_1, so that E_2 consists of four closed intervals, each of length $1/3^2$. Proceeding in this manner, we construct E_n to consist of 2^n closed intervals, and we obtain E_{n+1} by removing the open middle third from each interval of E_n. The *Cantor set* is defined to be the intersection of the E_n's. Prove the following:
   (a) The Cantor set is a closed subset of $\mathbb{R}$ with empty interior.
   (b) The Cantor set has no isolated points.
   (c) The Cantor set is uncountable. *Hint*: Use Exercise 2.3.

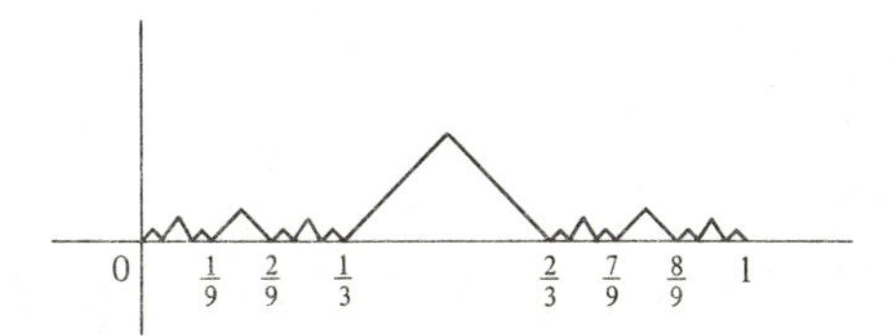

8. Define explicitly a continuous function f on the unit interval $[0,1]$ which is zero precisely on the Cantor set and which has a graph as suggested by the figure. Prove that f satisfies the following condition:

$$|f(s) - f(t)| \leq |s - t|, \qquad 0 \leq s, t \leq 1.$$

Remark: This type of condition is called a *Lipschitz condition* on f. Lipschitz conditions will arise again in Section 8.

9. Determine the interior, the closure, the limit points, and the isolated points of each of the following subsets of $\mathbb{R}$:
   (a) the interval $[0,1)$,
   (b) the set of rational numbers,
   (c) $\{m + n\pi : m$ and n positive integers$\}$,
   (d) $\left\{\frac{1}{m} + \frac{1}{n} : m \text{ and } n \text{ positive integers}\right\}$.
10. Let f be a real-valued function on $\mathbb{R}$. Show that there exist $M > 0$ and a nonempty open subset U of $\mathbb{R}$ such that for any $s \in U$, there is a sequence $\{s_n\}$ satisfying $s_n \to s$ and $|f(s_n)| \leq M$, $n \geq 1$.

4. PRODUCTS OF METRIC SPACES

Let $(X_1,d_1),\ldots,(X_n,d_n)$ be metric spaces. The product set $X = X_1 \times \cdots \times X_n$ consists of all n-tuples $(x_1,\ldots,x_n)$, where $x_k \in X_k$, $1 \leq k \leq n$. There are various ways of making X into a metric space. One possible choice of a metric for X is given by

$$d(x,y) = [d_1(x_1,y_1)^2 + \cdots + d_n(x_n,y_n)^2]^{1/2}, \tag{4.1}$$

where $x = (x_1,\ldots,x_n)$ and $y = (y_1,\ldots,y_n)$. This metric on the product stems from the formula for the metric on $\mathbb{R}^n = \mathbb{R} \times \cdots \times \mathbb{R}$ in terms of the metric on the component spaces $\mathbb{R}$. An open ball in $\mathbb{R}^2 = \mathbb{R} \times \mathbb{R}$ with the metric (4.1) is simply a disc.

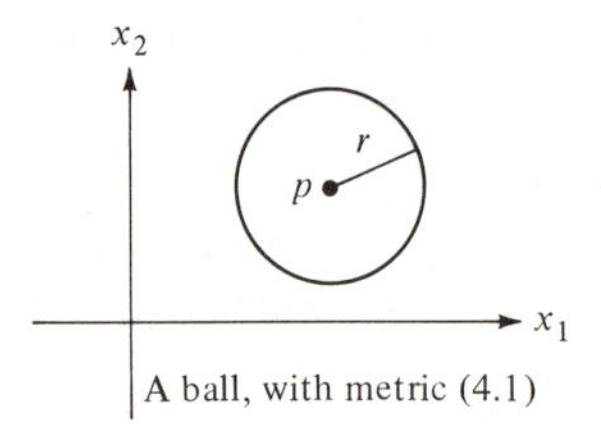

A ball, with metric (4.1)

There are other possible choices of a metric for X, which are convenient for certain problems. Two of them are given by

$$d(x,y) = \max(d_1(x_1,y_1),\ldots,d_n(x_n,y_n)), \tag{4.2}$$

$$d(x,y) = d_1(x_1,y_1) + \cdots + d_n(x_n,y_n). \tag{4.3}$$

Open balls in $\mathbb{R} \times \mathbb{R}$ with the metrics (4.2) and (4.3) are squares and diamonds, respectively. In dealing with many problems, it is not important which metric is

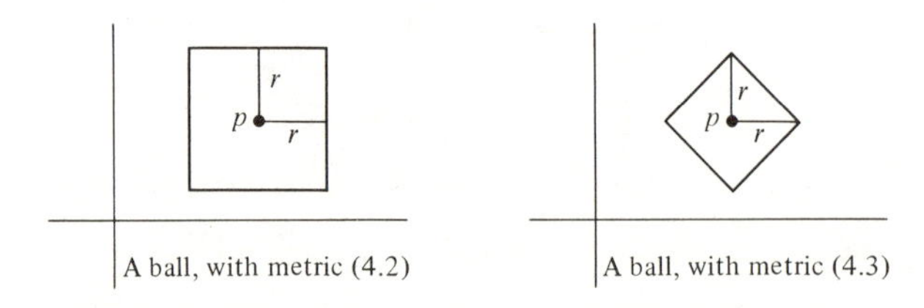

A ball, with metric (4.2)

A ball, with metric (4.3)

introduced in the product space, so long as the metric behaves reasonably. One property that each of the above metrics enjoys is the following.

(4.4) A sequence $\{x^{(j)} = (x_k^{(j)})\}_{j=1}^{\infty}$ converges to $x = (x_1,\ldots,x_n)$ in X if and only if for each k the sequence of component entries $\{x_k^{(j)}\}_{j=1}^{\infty}$ converges to x_k in X_k.

The following result shows that all metrics satisfying (4.4) determine the same family of open sets in X, and it describes the structure of these open sets explicitly.

4.1 Theorem: Suppose that d is a metric on $X = X_1 \times \cdots \times X_n$ that satisfies (4.4). Then the open sets in (X,d) are the unions of product sets of the form $U_1 \times \cdots \times U_n$, where U_j is an open subset of X_j, $1 \leq j \leq n$.

Proof: First note that if E_j is a closed subset of X_j, $1 \leq j \leq n$, then $E_1 \times \cdots \times E_n$ is closed in X. Indeed, if $\{x^{(i)}\}_{i=1}^{\infty}$ is a sequence in $E_1 \times \cdots \times E_n$ that converges to $x \in X$, then the coordinate sequences $\{x_k^{(i)}\}_{i=1}^{\infty}$ converge to x_k, $1 \leq k \leq n$. Since $x_k^{(i)} \in E_k$, also $x_k \in E_k$ and $x \in E_1 \times \cdots \times E_n$. It follows that $E_1 \times \cdots \times E_n$ is closed, as asserted.

Now suppose that U_k is an open subset of X_k, $1 \leq j \leq n$. By the preceding remarks, the set

$$X_1 \times \cdots \times X_{k-1} \times (X_k \backslash U_k) \times X_{k+1} \times \cdots \times X_n$$

is closed in X. Its complement

$$V^{(k)} = X_1 \times \cdots \times X_{k-1} \times U_k \times X_{k+1} \times \cdots \times X_n$$

is then open in X. Hence

$$U_1 \times \cdots \times U_n = V^{(1)} \cap \cdots \cap V^{(n)}$$

is open in X and any union of such products is open in X.

To complete the proof, it suffices to show that every open subset of X is a union of such product sets. For this, it suffices to show that if U is open in X and if $x \in U$, then there exist open sets U_k in X_k, $1 \leq k \leq n$, such that $x \in U_1 \times \cdots \times U_n \subseteq U$. We will argue by contradiction.

Let $x \in U$ and suppose that x is contained in no such product set. In particular, for each positive integer m, the product of open balls $B(x_1;1/m) \times \cdots \times B(x_n;1/m)$ is not contained in U. Hence there exists $x^{(m)} \in X \backslash U$ such that

$$d_k(x_k^{(m)},x_k) < 1/m, \qquad 1 \leq k \leq n.$$

Evidently, the sequence $\{x_k^{(m)}\}_{m=1}^{\infty}$ converges to x_k, $1 \leq k \leq n$. By (4.4), $x^{(m)}$ converges to x. Since $X \backslash U$ is closed, x belongs to $X \backslash U$. This contradiction establishes the theorem.
□

Another property of the metrics (4.1) to (4.3) is that

(4.5) $$d_k(x_k,y_k) \leq d(x,y), \qquad x,y \in X,\ 1 \leq k \leq n.$$

Condition (4.5) neither implies nor is implied by condition (4.4). It is clear though that, if a metric d on X satisfies (4.5) and if $\{x^{(m)}\}_{m=1}^{\infty}$ converges in (X,d) to x, then the component sequences $\{x_k^{(m)}\}_{m=1}^{\infty}$ converge to x_k, $1 \le k \le n$.

4.2 Theorem: Let $(X_1,d_1),\ldots,(X_n,d_n)$ be complete metric spaces. Let d be a metric on $X = X_1 \times \cdots \times X_n$ that satisfies (4.4) and (4.5). Then (X,d) is complete.

Proof: Let $\{x^{(j)}\}_{j=1}^{\infty}$ be a Cauchy sequence in X. The estimate (4.5) yields

$$d_k(x_k^{(i)},x_k^{(j)}) \le d(x^{(i)},x^{(j)}).$$

Since the right-hand side tends to 0 as $i,j \to \infty$, $\{x_k^{(i)}\}_{i=1}^{\infty}$ is a Cauchy sequence in X_k. Consequently there exists a point $x_k \in X_k$ such that $x_k^{(i)} \to x_k$ as $i \to \infty$. Set $x = (x_1,\ldots,x_n)$. By (4.4), $x^{(i)} \to x$ as $i \to \infty$. Hence X is complete. □

4.3 Corollary: The n-dimensional Euclidean space $\mathbb{R}^n$, with the usual metric

$$|x - y| = [(x_1 - y_1)^2 + \cdots + (x_n - y_n)^2]^{1/2}, \qquad x,y \in \mathbb{R}^n,$$

is complete.

In the sequel, we shall refer to the product metric space $X = X_1 \times \cdots \times X_n$ without specifying a metric for X. One can select one's own favorite metric for the product, as long as it satisfies (4.4) and (4.5).

EXERCISES

1. Prove that the functions on $X \times X$ given by (4.1), (4.2), and (4.3) are metrics.
2. For each of the metrics (4.1), (4.2), and (4.3) on the plane $\mathbb{R}^2$, sketch the open ball $B(0;1)$. What inclusion relations hold between these balls?
3. Prove that each of the metrics (4.1), (4.2), and (4.3) has property (4.4).
4. Determine the interior, the closure, the limit points, and the isolated points of each of the following subsets of $\mathbb{R}^2$:
   (a) $\{(x,y) : 0 < x^2 + y^2 < 1\}$,
   (b) $\{(x,0) : x \in \mathbb{R}\}$,
   (c) $\{(m,n) : m \text{ and } n \text{ positive integers}\}$,
   (d) $\left\{\left(\frac{1}{m},\frac{1}{n}\right) : m \text{ and } n \text{ positive integers}\right\}$.
5. Prove that if $\tilde{X}_j$ is the completion of X_j, $1 \le j \le n$, then $\tilde{X}_1 \times \cdots \times \tilde{X}_n$ is the completion of $X = X_1 \times \cdots \times X_n$.
6. Let (X_k,d_k), $1 \le k \le n$, be metric spaces. Let d be a metric on $X = X_1 \times \cdots \times X_n$ such that the open subsets of (X,d) are precisely the unions of sets of the form $U_1 \times \cdots \times U_n$, where U_k is an open subset of X_k, $1 \le$

$k \le n$. Show that d satisfies (4.4). *Note:* This establishes the converse to Theorem 4.1.

7. Let (X_k,d_k), $1 \le k < \infty$, be metric spaces and let $X = \Pi_{k=1}^{\infty} X_k$ be their Cartesian product, that is, let X be the set of sequences $(x_1,x_2,\ldots)$, where $x_j \in X_j$ for $1 \le j < \infty$.
   (a) Show that
   $$d(x,y) = \sum_{j=1}^{\infty} \frac{1}{2^j} \min(1,d_j(x_j,y_j))$$
   is a metric on X.
   (b) Show that a sequence $\{x^{(k)}\}_{k=1}^{\infty}$ converges in X if and only if $\{x_j^{(k)}\}_{k=1}^{\infty}$ converges in X_j for each $j \ge 1$.
   (c) Show that the open sets in X are precisely the unions of sets of the form $U_1 \times \cdots \times U_m \times X_{m+1} \times X_{m+2} \times \cdots$, where $m \ge 1$ and U_j is open in X_j, $1 \le j \le m$.

5. COMPACTNESS

In analysis it is often important to know when a property that is valid at each point of a space is in some sense uniformly valid over the space. For instance, we shall be concerned in Section 6 with the question of when a function that is continuous at each point of a space is in fact uniformly continuous on the space. The condition under which such uniformity tends to occur for sets in $\mathbb{R}^n$ is that the set be closed and bounded. However, this condition of being closed and bounded, for sets in an arbitrary metric space, does not guarantee that the set has the properties we desire. There is another condition, called *compactness,* that does guarantee that the set has these properties. There is a certain class of properties that finite sets have trivially, that are retained by compact sets. For instance, it will be seen that every continuous real-valued function on a compact space is bounded and assumes its maximum and minimum values.

Our aim in this section is to define the notion of compactness and to prove an important characterization (Theorem 5.1) of compact metric spaces. Most of the section will be devoted to the proof of Theorem 5.1, which is somewhat involved and conceptually difficult. During the course of the section, we shall return to subsets of $\mathbb{R}^n$ and show that the notion of compactness for a set in $\mathbb{R}^n$ is equivalent to the previously mentioned condition of being closed and bounded. This latter theorem is the celebrated Heine-Borel Theorem, and Theorem 5.1 may be regarded as a powerful extension of the Heine-Borel Theorem. We now formulate the appropriate ideas in a general metric space situation.

A family $\{U_\alpha\}_{\alpha \in A}$ of sets is said to *cover* a set S if S is contained in the union of the U_α's. An *open cover* of a metric space X is a family of open subsets of X that covers X. A metric space X is *compact* if every open cover has a finite subcover. In other words, X is compact if, whenever $\{U_\alpha\}_{\alpha \in A}$ is an open cover of X, there are finitely many of the U_α's, say $U_{\alpha_1},\ldots,U_{\alpha_m}$, such that

$$X \subset U_{\alpha_1} \cup \cdots \cup U_{\alpha_m}.$$

A metric space X is *totally bounded* if for each $\varepsilon > 0$, there exists a finite number of open balls of radius ε that cover X. Our main object in this section is to establish the following theorem. Both the theorem and the ideas entering into the proof are of the utmost importance.

5.1 Theorem: The following are equivalent for a metric space X:

(5.1) X is compact.

(5.2) Every sequence in X has a convergent subsequence.

(5.3) X is totally bounded and complete.

Proof: The proof of Theorem 5.1 will take most of this section. For now, we content ourselves with proving the straightforward implications (5.1) $\Rightarrow$ (5.2) $\Rightarrow$ (5.3).

Suppose that X is compact. Let $\{y_j\}_{j=1}^{\infty}$ be a sequence in X. Suppose that for each $x \in X$, there exists $\varepsilon = \varepsilon(x) > 0$ such that only finitely many terms of the sequence $\{y_j\}$ lie in $B(x;\varepsilon(x))$. The set of open balls $\{B(x;\varepsilon(x)) : x \in X\}$ forms an open cover of X. (Note the use of X as the index set for the cover!) Since X is compact, there is a finite subcover

$$X = B(x_1;\varepsilon(x_1)) \cup \cdots \cup B(x_m;\varepsilon(x_m)).$$

Since y_j belongs to $B(x_i;\varepsilon(x_i))$ for only finitely many indices j, we conclude that there is a grand total of only finitely many positive integers, an absurdity. This contradiction shows that there exists $x \in X$ such that each ball $B(x;\varepsilon)$ contains infinitely many terms of the sequence $\{y_j\}$.

Now we construct a convergent subsequence of $\{y_j\}$ as follows. Choose j_1 so that $y_{j_1} \in B(x;1)$. Since $y_j \in B(x;1/2)$ for infinitely many j, we can choose $j_2 > j_1$ such that $y_{j_2} \in B(x;1/2)$. Continuing in this manner, we find $j_1 < j_2 < \cdots$ such that $y_{j_n} \in B(x;1/n)$. Then $\{y_{j_n}\}_{n=1}^{\infty}$ is a subsequence of $\{y_j\}$ that converges to x. We have proved that (5.1) implies (5.2).

Next, suppose that (5.2) is valid. Let $\{x_j\}$ be a Cauchy sequence in X. By (5.2), $\{x_j\}$ has a convergent subsequence. By Lemma 2.2, the sequence $\{x_j\}$ itself converges. Hence X is complete.

Now assume still that (5.2) is valid and let $\varepsilon > 0$. We wish to cover X by a finite number of open balls of radius ε. Let $y_1 \in X$ be arbitrary. If $B(y_1;\varepsilon) \neq X$, let y_2 be any point in $X \backslash B(y_1;\varepsilon)$. Having chosen $y_1, \ldots, y_n$, we let y_{n+1} be any point in $X \backslash (\cup_{j=1}^{n} B(y_j;\varepsilon))$, provided the union does not coincide with X; otherwise we stop. The points $y_1, y_2, \ldots$ constructed in this manner satisfy

$$d(y_k,y_j) \geq \varepsilon, \qquad 1 \leq j < k.$$

If the procedure for selecting the y_j's does not terminate, then the sequence $\{y_j\}_{j=1}^{\infty}$ has no convergent subsequence. It follows that the choice of some y_{n+1} is impossible, so that X is covered by the n open balls of radius ε and centers $y_1, \ldots, y_n$. In particular, X is totally bounded, and (5.2) implies (5.3).

We shall return later to the proof of Theorem 5.1. First we discuss the notion of total boundedness in more detail and introduce various other notions.

A metric space X is *bounded* if there exists $b > 0$ such that $d(x,y) < b$ for all $x,y \in X$.

5.2 Lemma: A totally bounded metric space is bounded.

Proof: Let X be totally bounded. Choose $x_1, \ldots, x_m \in X$ such that the balls $B(x_j;1)$, $1 \le j \le m$, cover X. Let

$$b = 2 + \max\{d(x_j,x_k) : 1 \le j,k \le m\}.$$

An appropriate application of the triangle inequality shows that $d(x,y) < b$ for all $x,y \in X$, so that X is bounded. □

5.3 Lemma: Any subspace of a totally bounded metric space is totally bounded.

Proof: Let Y be a subspace of the totally bounded space X and let $\varepsilon > 0$. Since X is totally bounded, there are points $x_1, \ldots, x_n \in X$ such that the balls $B(x_j;\varepsilon/2)$ cover X. We can arrange the x_j's so that $B(x_j;\varepsilon/2) \cap Y \neq \emptyset$ for the indices $1 \le j \le m$, while $B(x_j;\varepsilon/2)$ does not meet Y for $m + 1 \le j \le n$. Then the balls $B(x_j;\varepsilon/2)$, $1 \le j \le m$, cover Y. Let y_j be any point in $B(x_j;\varepsilon/2) \cap Y$, $1 \le j \le m$. The triangle inequality shows that

$$B(x_j;\varepsilon/2) \subset B(y_j;\varepsilon).$$

Consequently the m open balls $B(y_j;\varepsilon) \cap Y$ in the subspace Y cover Y. □

5.4 Lemma: A subset E of $\mathbb{R}^n$ is totally bounded if and only if E is bounded.

Proof: The forward implication follows from Lemma 5.2. For the reverse implication, suppose that E is bounded. Then E is contained in a cube of the form $T = [-b,b] \times \cdots \times [-b,b]$ for some large $b > 0$. By Lemma 5.3, it suffices to show that T is totally bounded. For this, let $\varepsilon > 0$. Then the balls $B(\varepsilon\, j;\varepsilon)$ cover T, where $j = (j_1, \ldots, j_n)$ ranges over all integral lattice points of $\mathbb{R}^n$ which satisfy $\varepsilon|j_i| \le 2b$, $1 \le i \le n$. (An integral lattice point of $\mathbb{R}^n$ is by definition a point whose coordinates are integers.) Since there are only finitely many such lattice points, T is totally bounded. □

Now a subspace of $\mathbb{R}^n$ is complete if and only if it is closed, by Theorems 2.3 and 2.4. Consequently a subspace of $\mathbb{R}^n$ is totally bounded and complete if and only if it is closed and bounded. Theorem 5.1 then yields the following important theorem about subsets of $\mathbb{R}^n$.

5.5 Theorem *(Heine-Borel Theorem):* The following are equivalent for a subspace E of $\mathbb{R}^n$.

(5.1)′ E is compact.

(5.2)′ Every sequence in E has a convergent subsequence.

(5.3)′ E is closed and bounded.

Of course, the Heine-Borel Theorem is not established until the proof of Theorem 5.1 is complete.

The following theorem shows that (5.3) implies (5.2), so that the two are equivalent.

5.6 Theorem: Let X be a totally bounded metric space. Then every sequence in X has a Cauchy subsequence.

Proof: We shall use a diagonalization procedure to prove the theorem. It is important that one add this technique to one's bag of tricks.

Let $\{x_j\}_{j=1}^{\infty}$ be a sequence in X. For convenience, we rewrite the sequence in the form $\{x_{1j}\}_{j=1}^{\infty}$. By induction, we construct sequences $\{x_{kj}\}_{j=1}^{\infty}$, $k \geq 2$, with the following properties:

(i) $\{x_{kj}\}_{j=1}^{\infty}$ is a subsequence of $\{x_{k-1,j}\}_{j=1}^{\infty}$, $k \geq 2$.

(ii) $\{x_{kj}\}_{j=1}^{\infty}$ is contained in a ball of radius $1/k$, $k \geq 2$.

Indeed, suppose that $k \geq 2$ and that we already have the sequences $\{x_{ij}\}_{j=1}^{\infty}$ for $i < k$. Let $B_1, \ldots, B_n$ be a finite number of open balls of radius $1/k$ that cover X. Since there are infinitely many indices j and only finitely many balls, there must exist at least one ball, say B_m, such that $x_{k-1,j} \in B_m$ for infinitely many $j \geq 1$. Now let x_{k1} be the first of the $x_{k-1,j}$'s that belong to B_m, let x_{k2} be the second, etc. Then $\{x_{kj}\}_{j=1}^{\infty}$ has properties (i) and (ii).

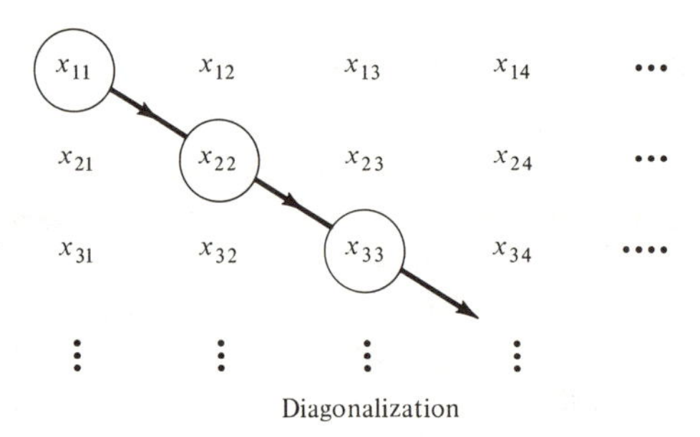

Diagonalization

Now set $y_n = x_{nn}$, so that $\{y_n\}_{n=1}^{\infty}$ is the *diagonal sequence*. It is a subsequence of the original sequence $\{x_j\}_{j=1}^{\infty}$. Furthermore, if we drop the first $k - 1$ terms, the resulting sequence $\{y_n\}_{n=k}^{\infty}$ is a subsequence of the sequence $\{x_{kj}\}_{j=1}^{\infty}$. In particular, (ii) shows that

$$d(y_n, y_m) < 2/k \qquad \text{if } n, m \geq k.$$

It follows that $\{y_n\}_{n=1}^{\infty}$ is a Cauchy sequence. □

A metric space X is *separable* if there is a dense subset of X that is countable. In other words, X is separable if and only if there is a sequence $\{x_j\}_{j=1}^{\infty}$ in X that is dense in X. (There may be repetitions among the x_j's.)

The real numbers $\mathbb{R}$ form a separable metric space. Indeed, the set $\mathbb{R}_0$ of rational numbers forms a countable dense subset of $\mathbb{R}$. Similarly, $\mathbb{R}^n$ is separable since the set of points of $\mathbb{R}^n$ with rational coordinates is countable.

5.7 Theorem: A subspace of a separable metric space is separable.

Proof: Suppose that X is a metric space with a dense subset $\{x_j\}_{j=1}^{\infty}$ and let Y be a subspace of X. Consider pairs of indices (j,n) such that

$$(*) \qquad B(x_j;1/n) \cap Y$$

is not empty. For such pairs (j,n), let y_{jn} be any point in the intersection (*). The $\{y_{jn}\}$ form a countable subset of Y. We claim that the y_{jn} are dense in Y.

Let $y \in Y$ and let $n \geq 1$. Then there exists j such that $d(x_j,y) < 1/n$. In particular, the intersection (*) is nonempty, so that y_{jn} is defined and $d(x_j,y_{jn}) < 1/n$. Then

$$d(y,y_{jn}) < d(y,x_j) + d(x_j,y_{jn}) < 2/n.$$

Letting $n \to \infty$, we find that y is adherent to the set $\{y_{jn}\}$, so that the closure of the double sequence is Y. □

5.8 Theorem: A totally bounded metric space is separable.

Proof: Let n be a positive integer. Then there exist $x_{n1},\ldots,x_{n,m_n}$ such that the open balls with centers at the x_{nj} and radii $1/n$ cover X. The family $\{x_{nj} : 1 \leq j \leq m_n, 1 \leq n < \infty\}$ is then a countable subset of X. For each $x \in X$ and each integer n, there is an x_{nj} such that $d(x_{nj},x) < 1/n$. Consequently the x_{nj} are dense in X. □

A *base of open sets* for a metric space X is a family $\mathcal{B}$ of open subsets of X such that every open subset of X is the union of sets in $\mathcal{B}$. For instance, the family of open balls forms a base of open sets for a metric space, as does the family of open balls with rational radii. Theorem 4.1 shows that the family of sets of the form $U_1 \times \cdots \times U_n$, where U_j is open in X_j, forms a base of open sets for a product metric space $X_1 \times \cdots \times X_n$.

5.9 Lemma: A family $\mathcal{B}$ of open subsets of a metric space X is a base of open sets if and only if for each $x \in X$ and each open neighborhood U of x, there exists $V \in \mathcal{B}$ such that $x \in V$ and $V \subset U$.

Proof: If $\mathcal{B}$ is a base, the set U above is the union of sets in $\mathcal{B}$, and so the condition of the lemma is evidently valid. Conversely, suppose that the condition given in the lemma holds. Let U be an open subset of X. For each $x \in U$, there is $V_x \in \mathcal{B}$ such that $x \in V_x$ and $V_x \subset U$. Then

$$U = \cup\{V_x : x \in U\}$$

represents U as a union of sets in $\mathcal{B}$, so that $\mathcal{B}$ is a base of open sets. □

A metric space satisfies the *second axiom of countability*, or is *second-countable,* if there is a base of open sets that is at most countable. In other words, X is second-countable if there is a sequence $\{U_n\}_{n=1}^{\infty}$ of open sets such that each open set U is the union of the U_n's it contains.

5.10 Theorem: A metric space is second-countable if and only if it is separable.

Proof: Suppose first that X is a separable metric space. Let $\{x_j\}_{j=1}^{\infty}$ be a dense sequence in X. Consider the family of open sets

$$\mathcal{B} = \{B(x_j;1/n) : j \geq 1, n \geq 1\}.$$

Let U be an open subset of X and let $x \in U$. For some $n \geq 1$, we have $B(x;2/n) \subset U$. Choose j so that $d(x_j,x) < 1/n$. Then $x \in B(x_j;1/n)$, and the triangle inequality shows that $B(x_j;1/n) \subset B(x;2/n) \subset U$. Consequently $\mathcal{B}$ is a base of open sets. Since $\mathcal{B}$ is countable, X is second-countable.

Conversely, suppose that X is second-countable. Let $\{U_n\}_{n=1}^{\infty}$ be a sequence of open sets in X that form a base. Let x_n be any point in U_n, $n \geq 1$. Then every nonempty open subset of X contains a point of the sequence $\{x_n\}$, so that the sequence is dense in X, and X is separable. □

The following theorem is the trickiest part of the circle of ideas covered in this section.

5.11 Theorem *(Lindelöf's Theorem):* Suppose the metric space X is second-countable. Then every open cover of X has a countable subcover.

Proof: Let $\{U_\alpha\}_{\alpha \in A}$ be an open cover of X, where A is some index set. Let $\mathcal{B}$ be a countable base of open sets. Let $\mathcal{C}$ be the subset of $\mathcal{B}$ consisting of those sets $V \in \mathcal{B}$ such that $V \subset U_\alpha$ for some α. We claim that $\mathcal{C}$ is a cover of X. Indeed, if $x \in X$, then there is some index α such that $x \in U_\alpha$. By Lemma 5.9, there is $V \in \mathcal{B}$ such that $x \in V$ and $V \subset U_\alpha$. In particular, $V \in \mathcal{C}$, so that $\mathcal{C}$ indeed covers X.

Now for each $V \in \mathcal{C}$, select one index $\alpha = \alpha(V)$ such that $V \subset U_{\alpha(V)}$. Then the sets $\{U_{\alpha(V)} : V \in \mathcal{C}\}$ cover X. Since $\mathcal{B}$ is countable, so is $\mathcal{C}$, so that the $U_{\alpha(V)}$'s form a countable subcover of X. □

Proof of Theorem 5.1 (concluded): As remarked earlier, Theorem 5.6 shows that (5.3) implies (5.2), so that the conditions (5.2) and (5.3) are equivalent.

Suppose that (5.2) and (5.3) are valid. It suffices to show that every open cover of X has a finite subcover.

By Theorem 5.8, X is separable, so that by Theorem 5.10, X is second-countable. By Theorem 5.11 every open cover of X has a countable subcover. It suffices to show then that if $\{U_n\}_{n=1}^{\infty}$ is a sequence of open subsets of X that cover X, then there is an integer m such that $X = U_1 \cup \cdots \cup U_m$.

Suppose that this is not the case, that is, that $U_1 \cup \cdots \cup U_m \neq X$ for all m. For each m, let x_m be any point in $X \backslash (U_1 \cup \cdots \cup U_m)$. By (5.3), the sequence $\{x_m\}$ has a subsequence that converges to some point $x \in X$. Since $x_j \in X \backslash (U_1 \cup \cdots \cup U_m)$ for all $j \geq m$ and since $X \backslash (U_1 \cup \cdots \cup U_m)$ is closed, $x \in X \backslash (U_1 \cup \cdots \cup U_m)$.

Since this is true for all m, $x \in X \backslash (\cup_{k=1}^{\infty} U_k)$ and the U_k do not cover X. This contradiction establishes the theorem. □

We state the following corollary to Theorems 5.1, 5.8, and 5.10 for emphasis.

5.12 Theorem: A compact metric space is separable and second-countable.

EXERCISES

1. Give an example of a totally bounded metric space which is not compact.
2. Give an example of a complete metric space which is not compact.
3. Show directly that a compact metric space is totally bounded.
4. Let $\{U_\alpha\}_{\alpha \in A}$ be a finite open cover of a compact metric space X.
   (a) Show that there exists $\varepsilon > 0$ such that for each $x \in X$, the open ball $B(x;\varepsilon)$ is contained in one of the U_α's.

Remark: Such an ε is called a *Lebesgue number* of the cover.

   (b) Show that if at least one of the U_α's is a proper subset of X, then there is a largest Lebesgue number for the cover.

5. Prove that the product of a finite number of compact metric spaces is compact.
6. A metric space is a *Polish space* if it is separable and if there is an equivalent metric for which it is complete. Prove that any closed subspace of a Polish space is Polish.

Remark: An important school of Polish mathematicians made deep contributions to the study of point-set topology during the period between the world wars. The terminology "Polish" recognizes the work of the Polish school on this important class of spaces. For more on Polish spaces, see Exercises 6.14 and 6.15.

7. Let $B(S)$ be the metric space of bounded real-valued functions on the set S, with the metric of uniform convergence on S (defined in Section 1). Show that $B(S)$ is separable if and only if S is finite.

   Hint: For any subset T of S, define $\chi_T \in B(S)$ by

$$\chi_T(s) = \begin{cases} 1, & s \in T, \\ 0, & s \notin T. \end{cases}$$

   Then the balls $B(\chi_T;1/2)$ are disjoint.

8. Let (X,d) be a bounded metric space and let $\mathscr{C}$ be the family of nonempty closed subsets of X. Show that

$$\rho(E,F) = \max\left(\sup_{x \in E} d(x,F), \sup_{y \in F} d(y,E) \right)$$

   defines a metric on $\mathscr{C}$. Show that $\mathscr{C}$ is compact whenever X is compact.

6. CONTINUOUS FUNCTIONS

Let (X,d) and (Y,ρ) be metric spaces. A function $f : X \to Y$ is *continuous at* $x \in X$ if whenever $\{x_n\}$ is a sequence in X such that $x_n \to x$, then $f(x_n) \to f(x)$. The function f is *continuous* if it is continuous at each $x \in X$.

6.1 Theorem: The function $f : X \to Y$ is continuous at the point $x \in X$ if and only if for each $\varepsilon > 0$, there exists $\delta > 0$ such that whenever $z \in X$ satisfies $d(x,z) < \delta$, then $\rho(f(x),f(z)) < \varepsilon$.

Proof: Suppose that the latter condition is violated. Then there exists $\varepsilon > 0$ such that for each $\delta > 0$, there exists $z \in X$ satisfying $d(x,z) < \delta$ and $\rho(f(x),f(z)) \geq \varepsilon$. Corresponding to $\delta = 1/k$, we choose such a point z_k, so that $d(x,z_k) < 1/k$ and $\rho(f(x),f(z_k)) \geq \varepsilon$. Then $z_k \to x$, but $f(z_k)$ does not converge to $f(x)$, so that f is not continuous at x.

Conversely, suppose that f is not continuous at x. Choose a sequence $\{x_k\}_{k=1}^{\infty}$ such that $x_k \to x$ but $f(x_k)$ does not converge to $f(x)$. Passing to a subsequence, we can assume that $\rho(f(x),f(x_k)) \geq \varepsilon$ for some $\varepsilon > 0$. Then there can exist no $\delta > 0$ as in the condition of the theorem, so that the condition is violated. □

6.2 Theorem: The following are equivalent for a function f from a metric space (X,d) to a metric space (Y,ρ):

(6.1) f is continuous.

(6.2) For each $x \in X$ and $\varepsilon > 0$, there exists $\delta > 0$ such that whenever $z \in X$ satisfies $d(x,z) < \delta$, then $\rho(f(x),f(z)) < \varepsilon$.

(6.3) $f^{-1}(V)$ is an open subset of X for every open subset V of Y.

Proof: The equivalence of (6.1) and (6.2) follows immediately from Theorem 6.1.

Suppose that (6.2) is valid. Let V be an open subset of Y and fix $x \in f^{-1}(V)$. Choose $\varepsilon > 0$ such that $B(f(x);\varepsilon) \subset V$. By (6.2), there exists $\delta > 0$ such that $f(B(x;\delta)) \subset B(f(x);\varepsilon) \subset V$. Consequently $B(x;\delta) \subseteq f^{-1}(V)$. Since $f^{-1}(V)$ contains an open ball about each of its points, $f^{-1}(V)$ is open and (6.3) is valid.

Conversely, suppose that (6.3) is valid. Let $x \in X$ and let $\varepsilon > 0$. Then $f^{-1}(B(f(x);\varepsilon))$ is an open neighborhood of x in X. Consequently there exists $\delta > 0$ such that $B(x;\delta) \subset f^{-1}(B(f(x);\varepsilon))$. Applying f, we obtain $f(B(x;\delta)) \subset B(f(x);\varepsilon)$, so that (6.2) is valid. □

A function $f : X \to Y$ is *uniformly continuous* if for each $\varepsilon > 0$, there exists $\delta > 0$ such that whenever $x,z \in X$ satisfy $d(x,z) < \delta$, then $\rho(f(x),f(z)) < \varepsilon$. The definition of uniform continuity is then the same as the characterization (6.2) of continuity except that the $\delta > 0$ must be chosen to work simultaneously for all $x \in X$. Evidently every uniformly continuous function is continuous.

6.3 Theorem: Let X and Y be metric spaces and suppose that X is compact. Then every continuous function f from X to Y is uniformly continuous.

Proof: Suppose that f is not uniformly continuous. Then there exist $\varepsilon > 0$ and (setting $\delta = 1/k$ in the definition) points $x_k,z_k \in X$ such that $d(x_k,z_k) < 1/k$ while $\rho(f(x_k),f(z_k)) \geq \varepsilon$. Passing to a subsequence, we can assume that $x_k \to x \in X$. Since $d(x_k,z_k) \to 0$, we also obtain $z_k \to x$. Consequently $f(x_k) \to f(x)$ and $f(z_k) \to f(x)$, so that $\rho(f(x_k),f(z_k)) \leq \rho(f(x_k),f(x)) + \rho(f(x),f(z_k)) \to 0$. This contradiction establishes the theorem. □

A function f from one metric space to another is a *homeomorphism* if f is continuous, one-to-one, and onto and if moreover the inverse function f^{-1} is continuous. We shall also say that such a function f is *bicontinuous*. A homeomorphism preserves all properties of a metric space that are definable in terms of open sets only. For example, if X is compact and there is a homeomorphism f from X to Y, then Y is also compact (Exercise 7). One property that need not be preserved under a homeomorphism is completeness (Exercise 3), but the properties that are definable in terms of open sets only and consequently preserved by homeomorphisms are numerous and important. The study of these properties in a generalized setting in which the metric as such is no longer considered will be taken up in Chapter II. The idea that open sets suffice to develop so much important structure is one of the most crucial observations for the foundations of topology.

EXERCISES

1. Show that if (X,d) is a metric space, then d is a continuous function from $X \times X$ to $\mathbb{R}$. Furthermore, show that for each fixed $x_0 \in X$, the function $x \to d(x,x_0)$ is a uniformly continuous function from X to $\mathbb{R}$.
2. Prove that two metrics d and ρ for X are equivalent if and only if the identity map $(X,d) \to (X,\rho)$ is bicontinuous (that is, it and its inverse are continuous).
3. Let E be the subspace of $\mathbb{R}^2$ obtained from the circle centered at $(0,1/2)$ of radius $1/2$ by deleting the point $(0,1)$. Define a function h from $\mathbb{R}$ to E so that $h(s)$ is the point at which the line segment from $(s,0)$ to $(0,1)$ meets E.

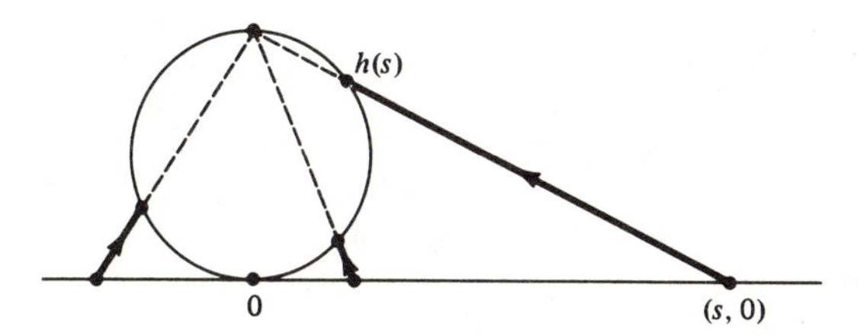

(a) Show that h is a bicontinuous function from $\mathbb{R}$ to E.

(b) Show that

$$\rho(s,t) = |h(s) - h(t)|, \qquad s,t \in \mathbb{R},$$

defines a metric ρ on $\mathbb{R}$ that is equivalent to the usual metric on $\mathbb{R}$.

(c) Show that $(\mathbb{R},\rho)$ is totally bounded but not complete.
(d) Identify the completion of $(\mathbb{R},\rho)$.

4. Let f be a continuous function from the metric space X to the metric space Y and let E be a subset of X. Show that the restriction of f to E is a continuous function from E to Y.

5. Let X_0, X_1, and X_2 be metric spaces, let f be a function from X_0 to X_1, and let g be a function from X_1 to X_2. Show that if f is continuous at $x_0 \in X_0$ and if g is continuous at $f(x_0)$, then the composition $g \circ f$ is continuous at x_0.

6. Let f be a continuous function from the closed unit interval $[0,1]$ to itself. Show that there exists $t \in [0,1]$ such that $f(t) = t$.

Remark: The point t is called a *fixed point* of f.

7. Show that if f is a continuous function from a compact metric space X to a metric space Y, then the image of f is a compact subset of Y. In particular, show that f is bounded, that is, that the range of f is a bounded subset of Y.

8. Prove that a continuous real-valued function on a compact metric space assumes its maximum value and its minimum value.

9. Prove that a metric space X is compact if and only if every continuous real-valued function on X is bounded.

10. Let (X,d) be a metric space and let S be a nonempty set. Let $B(S,X)$ denote the family of bounded functions from S to X.

(a) Show that

$$\rho(f,g) = \sup_{s \in S} d(f(s),g(s))$$

defines a metric ρ on $B(S,X)$. The metric ρ is called the *metric of uniform convergence on S*.

(b) Show that $B(S,X)$ is complete if and only if X is complete.
(c) Show that if S is itself a metric space, then the space $BC(S,X)$ of bounded continuous functions from S to X is a closed subspace of $B(S,X)$.

11. Show that there exists a continuous real-valued function h on $[0,1]$ such that

$$\limsup_{t \to 0+} \left| \frac{h(x+t) - h(x)}{t} \right| = \infty$$

whenever $0 \leq x < 1$.

Hint: Consider the space $C([0,2])$ of continuous real-valued functions on the interval $[0,2]$, with the metric of uniform convergence. Let E_m be the set of $f \in C([0,2])$ for which there exists $x \in [0,1]$ satisfying $|f(x+t) - f(x)|/|t| \leq m$ for $t > 0$, $x + t \leq 2$. Show that E_m is a closed nowhere-dense subset of $C([0,2])$.

12. Let X and Y be metric spaces such that X is complete. Show that if $\{f_\alpha\}_{\alpha\in A}$ is a family of continuous functions from X to Y such that $\{f_\alpha(x) : \alpha \in A\}$ is a bounded subset of Y for each $x \in X$, then there exists a nonempty open subset U of X such that $\{f_\alpha(x) : \alpha \in A,\ x \in U\}$ is a bounded subset of Y.

13. Let (X,d) and (Y,ρ) be metric spaces such that X is complete. Let $\{g_k\}_{k=1}^\infty$ be a sequence of continuous functions from X to Y such that $\{g_k(x)\}$ converges for each $x \in X$. Show that for each $\epsilon > 0$, there exist $N \geq 1$ and a nonempty open subset U of X such that $\rho(g_n(x),g_m(x)) < \epsilon$ for all $x \in U$ and all $n,m \geq N$.

14. (a) Prove that if X is a metric space and U is an open subset of X, then there is a positive continuous function f on U such that $f(x) \to +\infty$ as x tends to any point of ∂U.

 Hint: Set $f(x) = 1/\inf\{d(x,y) : y \in X\backslash U\}$.

 (b) Prove that an open subset of a Polish space is Polish.

Remark: Polish spaces were defined in Exercise 5.6. For the proof, consider a metric on U of the form $\rho(x,y) = d(x,y) + |f(x) - f(y)|$, where f is as in part (a).

 (c) Prove that if $\{U_j\}_{j=1}^\infty$ is a sequence of open subsets of a Polish space, then $\cap U_j$ is a Polish space.

 Hint: Extend your proof of (b) by considering

$$\rho(x,y) = d(x,y) + \sum_{j=1}^{\infty} \frac{1}{2^j} |f_j(x) - f_j(y)|/[1 + |f_j(x) - f_j(y)|],$$

 where f_j "blows up" at ∂U_j, as in (a).

 (d) Prove that the subset of $\mathbb{R}$ consisting of the irrational numbers is a Polish space.

Remark: It can be shown that a subspace of a Polish space is Polish if and only if it is the intersection of a sequence of open subsets.

15. Let N be the space of positive integers and let $N^\infty = N \times N \times \cdots$ be the countable product space, endowed with the metric introduced in Exercise 4.7.

 (a) Show that N^∞ is a Polish space.

 (b) Show that if X is any Polish space, there is a continuous function f from N^∞ into X such that f maps open sets onto open sets.

 Hint: Construct by induction open balls $B_{n_1n_2\cdots n_m}$ in X such that

 (i) $\overline{B_{n_1\cdots n_m}} \subseteq B_{n_1\cdots n_{m-1}}$,

 (ii) $\bigcup_{n_m=1}^\infty B_{n_1\cdots n_m} = B_{n_1\cdots n_{m-1}}$,

 (iii) the radius of $B_{n_1\cdots n_m}$ is less than $1/m$.

Remark: In Chapter II, Section 13, the notion of a quotient topological space will be defined and studied. Part (b) of this exercise shows that every Polish space is a quotient space of N^∞.

7. NORMED LINEAR SPACES

Many of the most useful metric spaces are vector spaces or subsets of vector spaces endowed with a metric that arises from a "norm." In this section, we define a norm on a vector space and discuss a circle of ideas related to the norm. We shall consider only real or complex vector spaces.

A *norm* on a vector space $\mathfrak{X}$ is a function $x \to \|x\|$ from $\mathfrak{X}$ to $\mathbb{R}$ that satisfies the following conditions:

$$\|x\| \geq 0, \quad \text{with equality if and only if} \quad x = 0, \tag{7.1}$$

$$\|cx\| = |c|\,\|x\|, \quad x \in \mathfrak{X},\ c \text{ scalar}, \tag{7.2}$$

$$\|x + y\| \leq \|x\| + \|y\|, \qquad x,y \in \mathfrak{X}. \tag{7.3}$$

Condition (7.3) is a version of the triangle inequality. It states that the norm is subadditive. Condition (7.2) is a homogeneity property of the norm with respect to scalar multiplication.

Examples of normed spaces abound. We give a few samples.

The Euclidean space $\mathbb{R}^n$ can be made into a normed space in various manners. The norm arising from the theorem of Pythagoras is given by

$$\|x\| = (x_1^2 + \cdots + x_n^2)^{1/2},$$

where $x = (x_1, \ldots, x_n)$. In this case, the verification of (7.1) and (7.2) is trivial, and (7.3) follows from Exercise 1.3. The reader should pause to verify that

$$\|x\|_1 = |x_1| + \cdots + |x_n|,$$

$$\|x\|_\infty = \max(|x_1|, \ldots, |x_n|)$$

also define norms on $\mathbb{R}^n$.

The norms $\|\cdot\|_1$ and $\|\cdot\|_\infty$ fit into a one-parameter family of norms $\|\cdot\|_p$, defined for $1 \leq p < \infty$ by

$$\|x\|_p = (|x_1|^p + \cdots + |x_n|^p)^{1/p}, \qquad x \in \mathbb{R}^n.$$

The verification of (7.1) and (7.2) is trivial, and the proof of (7.3) is laid out in Exercise 7. One may check (Exercise 2) that $\|x\|_p$ tends to $\|x\|_\infty$ as $p \to \infty$.

The norms above can be defined in the obvious manner on sequences $\{x_j\}_{j=1}^\infty$ provided that the expressions for the norms are finite. For $1 \leq p < \infty$, let l^p be the set of real or complex sequences $x = \{x_j\}_{j=1}^\infty$ such that

$$\|x\|_p = (\Sigma\, |x_j|^p)^{1/p} < \infty.$$

It turns out that $\|\cdot\|_p$ is a norm on l^p, so that l^p becomes a normed space. Similarly, the space of bounded sequences l^∞ becomes a normed space with norm

$$\|x\|_\infty = \sup_{1 \leq j < \infty} |x_j|.$$

Let S be a nonempty set and let $B(S)$ denote the space of bounded real-valued (or complex-valued) functions on S. Then

$$\|f\|_S = \sup_{s \in S} |f(s)|$$

defines a norm on $B(S)$ that is called the *sup norm* over S or the *norm of uniform convergence on S*. As a special case, we obtain the space l^∞ by taking S to be the set of positive integers.

Any vector subspace of $B(S)$ becomes a normed space, with the norm inherited from $B(S)$. For instance, let X be a metric space and let $BC(X)$ denote the space of bounded continuous real-valued (or complex-valued) functions on X. Endowed with the sup norm, $BC(X)$ becomes a normed space.

The space $C[0,1]$ of continuous functions on the unit interval $[0,1]$ can be made into a normed space in various ways. One norm is the sup norm. Other norms are given by

$$\|f\|_1 = \int_0^1 |f(s)|ds,$$

$$\|f\|_2 = \left(\int_0^1 |f(s)|^2 ds\right)^{1/2}.$$

The verification that $\|\cdot\|_1$ is a norm is straightforward (Exercise 10). The verification that $\|\cdot\|_2$ is a norm is outlined in Exercise 12. Again these norms can be fit into a one-parameter family of norms, defined for $1 \le p < \infty$ by

$$\|f\|_p = \left(\int_0^1 |f(s)|^p ds\right)^{1/p}.$$

That each of these is a norm is proved by developing a continuous version of Exercise 7 in which sums are replaced by integrals. The details are left to the (determined) reader.

7.1 Lemma: Let $\mathfrak{X}$ be a normed space. Then

$$d(x,y) = \|x - y\|, \qquad x,y \in \mathfrak{X}, \tag{7.4}$$

is a metric on $\mathfrak{X}$.

Proof: Setting $x = y - z$ in (7.1), we find that $d(y,z) \ge 0$, with equality if and only if $y = z$. Taking $c = -1$ and $x = y - z$ in (7.2), we find that $d(y,z) = d(z,y)$. Finally, substituting $x - y$ for x and $y - z$ for y in (7.3), we obtain

$$\|x - z\| = \|x - y + y - z\| \le \|x - y\| + \|y - z\|,$$

which is the triangle inequality for d. Thus d is a metric. □

The topology induced by the metric (7.4) is called the *norm topology* of $\mathfrak{X}$. A sequence $\{x_j\}_{j=1}^\infty$ in $\mathfrak{X}$ is said to *converge in norm* to $x \in \mathfrak{X}$ if it converges in the norm topology, that is, if

$$\lim_{j\to\infty} \|x_j - x\| = 0.$$

As an example, observe that the metric determined by the sup norm $\|\cdot\|_S$ is the metric of uniform convergence on S, treated in Exercises 6.10 and 5.7.

7.2 Lemma. The operations of addition and scalar multiplication on a normed space are continuous. The norm function $x \to \|x\|$ on a normed space is continuous.

Proof: The estimate

$$\|x + y - (x_n + y_n)\| \leq \|x - x_n\| + \|y - y_n\|$$

shows immediately that if $x_n \to x$ and $y_n \to y$, then $x_n + y_n \to x + y$. Hence addition is continuous. Similarly, the estimate

$$\|cx - c_n x_n\| \leq |c - c_n| \, \|x\| + |c_n| \, \|x - x_n\|$$

shows that scalar multiplication is continuous. Finally, from (7.3) we readily obtain the estimates

$$\|x_n\| - \|x - x_n\| \leq \|x\| \leq \|x - x_n\| + \|x_n\|.$$

These show that if $x_n \to x$, then $\|x_n\|$ converges to $\|x\|$. Hence the norm function $x \to \|x\|$ is continuous on $\mathfrak{X}$. □

Recall that a *linear transformation* from a vector space $\mathfrak{X}$ to a vector space $\mathfrak{Y}$ is a function T from $\mathfrak{X}$ to $\mathfrak{Y}$ that satisfies

$$T(x + y) = T(x) + T(y), \qquad x,y \in \mathfrak{X},$$

$$T(\lambda x) = \lambda T(x), \qquad x \in \mathfrak{X}, \ \lambda \text{ scalar}.$$

Linear transformations are also called *linear operators*. One particular choice of $\mathfrak{Y}$ is of special importance. If we take $\mathfrak{Y}$ to be the field of scalars ($\mathbb{R}$ or $\mathbb{C}$), then the linear operators from $\mathfrak{X}$ to $\mathfrak{Y}$ are called *linear functionals*.

The closed unit ball of $\mathfrak{X}$ is the set of $x \in \mathfrak{X}$ satisfying $\|x\| \leq 1$. A linear operator T from a normed space $\mathfrak{X}$ to a normed space $\mathfrak{Y}$ is a *bounded linear operator* if

$$\sup\{\|Tx\| : x \in \mathfrak{X}, \|x\| \leq 1\} < \infty.$$

In other words, T is a bounded linear operator if Tx has uniformly bounded norm as x varies over the unit ball in $\mathfrak{X}$. Note that a bounded linear operator is not bounded on all of $\mathfrak{X}$ unless it is identically zero. Indeed, if $Tx \neq 0$, then $\|T(cx)\| = |c| \, \|Tx\|$ becomes arbitrarily large with $|c|$. If T is a bounded linear operator, we define

$$\|T\| = \sup\{\|Tx\| : x \in \mathfrak{X}, \|x\| \leq 1\}. \tag{7.5}$$

7.3 Lemma: If T is a bounded linear operator from $\mathfrak{X}$ to $\mathfrak{Y}$, then

$$\|Tx\| \leq \|T\| \, \|x\|, \qquad x \in \mathfrak{X}.$$

Proof: If $x \in \mathfrak{X}$ is not zero, then $x/\|x\|$ has unit norm. Consequently $\|T(x/\|x\|)\| \leq \|T\|$, and the result follows by homogeneity. □

7.4 Lemma: The space of bounded linear operators from a normed space $\mathfrak{X}$ to a normed space $\mathfrak{Y}$ is itself a normed space, with norm defined by (7.5).

Proof: Let T be a bounded linear operator from $\mathfrak{X}$ to $\mathfrak{Y}$ and let c be a scalar. Then

$$\|(cT)(x)\| = \|cT(x)\| = |c|\ \|T(x)\|.$$

Taking the supremum over x in the unit ball of $\mathfrak{X}$, we find that cT is bounded and $\|cT\| = |c|\ \|T\|$.

If S and T are both bounded linear operators from $\mathfrak{X}$ to $\mathfrak{Y}$, then

$$\begin{aligned}\|(S + T)(x)\| &= \|S(x) + T(x)\| \le \|S(x)\| + \|T(x)\| \\ &\le \|S\|\ \|x\| + \|T\|\ \|x\|.\end{aligned}$$

Taking the supremum again over the unit ball of $\mathfrak{X}$, we find that $S + T$ is bounded and $\|S + T\| \le \|S\| + \|T\|$. In particular, the space of bounded linear operators from $\mathfrak{X}$ to $\mathfrak{Y}$ is closed under addition and scalar multiplication, so that it forms a vector space.

Evidently $\|T\| \ge 0$. Using Lemma 7.3, we see that $\|T\| = 0$ if and only if $T = 0$. This completes the verification that $\|\cdot\|$ is a norm. □

The space of bounded linear operators from $\mathfrak{X}$ to $\mathfrak{Y}$ will denoted by $\mathcal{B}(\mathfrak{X},\mathfrak{Y})$. The norm topology on $\mathcal{B}(\mathfrak{X},\mathfrak{Y})$ is referred to as the *uniform operator topology*. If we take $\mathfrak{X} = \mathfrak{Y}$, we obtain the space of bounded linear operators on $\mathfrak{X}$, which will be denoted simply by $\mathcal{B}(\mathfrak{X})$. Corresponding to the choice of $\mathfrak{Y}$ as the field of scalars, there is the space of bounded linear functionals on $\mathfrak{X}$, which is denoted by $\mathfrak{X}^*$ and called the *conjugate space* of $\mathfrak{X}$.

Next we show that boundedness is the same as continuity for linear operators and linear functionals.

7.5 Theorem: Let $\mathfrak{X}$ and $\mathfrak{Y}$ be normed spaces and let T be a linear operator from $\mathfrak{X}$ to $\mathfrak{Y}$. The following are equivalent:

(i) T is continuous,

(ii) T is continuous at 0,

(iii) T is bounded.

Proof: Evidently (i) implies (ii).

Suppose that (ii) is valid. From the definition of continuity at a point, there exists $\delta > 0$ such that $\|x\| \le \delta$ implies $\|T(x)\| \le 1$. If $x \in \mathfrak{X}$ satisfies $\|x\| \le 1$, then $\|\delta x\| \le \delta$, so that $\|T(\delta x)\| \le 1$ and $\|T(x)\| \le 1/\delta$. This establishes (iii).

Finally, suppose that (iii) is valid. Then the estimate

$$\|T(x) - T(x_0)\| \le \|T\|\ \|x - x_0\|$$

shows that T is continuous at any fixed $x_0 \in \mathfrak{X}$, so that (i) is valid. □

It turns out that every linear operator on a finite dimensional normed space is continuous (Exercise 5). On the other hand, every infinite dimensional normed space has a discontinuous linear functional. The construction of discontinuous functionals depends upon Zorn's Lemma (Section II.11 and Exercise II.11.5.).

A *Banach space* is a normed vector space that is complete in that norm's metric. For example, the space $\mathbb{R}^n$ is a real Banach space and $\mathbb{C}^n$ is a complex Banach space. It turns out that the spaces l^p, $1 \le p \le \infty$, are Banach spaces (Exercise 8). The space of bounded real-valued functions on a set S, with the sup norm $\|\cdot\|_S$, is a real Banach space (Exercise 6.10), and so is the space of bounded continuous real-valued functions on a metric space X. However, the space of continuous functions on $[0,1]$, with the norm $\|\cdot\|_p$, $1 \le p < \infty$, is not complete and hence not a Banach space (Exercise 11).

7.6 Theorem: Let $\{T_n\}_{n=1}^{\infty}$ be a sequence of continuous linear operators from a normed space $\mathfrak{X}$ to a Banach space $\mathfrak{Y}$. Suppose that

$$\sup_n \|T_n\| < \infty$$

and $\lim_{n\to\infty} T_n x$ exists for x belonging to a dense subset of $\mathfrak{X}$. Then there exists a continuous linear operator T from $\mathfrak{X}$ to $\mathfrak{Y}$ such that

$$Tx = \lim_{n\to\infty} T_n x, \qquad x \in \mathfrak{X}.$$

Moreover,

$$\|T\| \le \liminf_{n\to\infty} \|T_n\|.$$

Proof: Define $Tx = \lim_{n\to\infty} T_n x$ whenever the limit exists. Set $M = \sup \|T_n\|$. Let $z \in \mathfrak{X}$ and let $\epsilon > 0$. Choose $x \in \mathfrak{X}$ such that $\|x - z\| < \varepsilon$ and such that Tx exists. Choose N so large that $\|T_n x - T_m x\| < \varepsilon$ for all $n,m \ge N$. Then

$$\begin{aligned}\|T_n z - T_m z\| &\le \|T_n(z - x)\| + \|(T_n - T_m)x\| + \|T_m(x - z)\| \\ &\le M\varepsilon + \varepsilon + M\varepsilon\end{aligned}$$

for $n,m \ge N$. Hence $\{T_n z\}_{n=1}^{\infty}$ is a Cauchy sequence in $\mathfrak{Y}$. Since $\mathfrak{Y}$ is complete, $Tz = \lim T_n z$ exists and T is defined on all of $\mathfrak{X}$. Since the operations of addition and scalar multiplication are continuous on $\mathfrak{X}$, T is linear.

If $x \in \mathfrak{X}$ satisfies $\|x\| \le 1$, then

$$\|Tx\| = \lim \|T_n x\| \le \limsup \|T_n\|.$$

Consequently T is bounded and

$$\|T\| \le \limsup \|T_n\|.$$

Now the same estimate is valid if we let n tend to ∞ through any subsequence of positive integers. Choosing a subsequence $\{n_j\}$ for which $\lim \|T_{n_j}\| = \liminf \|T_n\|$, we obtain the desired estimate for $\|T\|$. □

7.7 Theorem: Let $\mathfrak{X}$ be a normed space and let $\mathfrak{Y}$ be a Banach space. Then the space $\mathcal{B}(\mathfrak{X},\mathfrak{Y})$ of bounded linear operators from $\mathfrak{X}$ to $\mathfrak{Y}$, with the norm (7.5), is a Banach space.

Proof: Let $\{T_n\}_{n=1}^{\infty}$ be a Cauchy sequence in $\mathcal{B}(\mathfrak{X},\mathfrak{Y})$. The estimate

$$\|T_n x - T_m x\| \leq \|T_n - T_m\| \, \|x\|$$

shows that $\{T_n x\}_{n=1}^{\infty}$ is a Cauchy sequence in $\mathfrak{Y}$. Since $\mathfrak{Y}$ is complete, $\{T_n x\}$ converges. Furthermore, $\sup \|T_n\|$ is finite, so that the hypotheses of Theorem 7.6 are met. By that theorem, there exists $T \in \mathcal{B}(\mathfrak{X},\mathfrak{Y})$ such that $T_n x$ converges to Tx for all $x \in \mathfrak{X}$. Furthermore, for fixed m, $\{(T_n - T_m)x\}_{n=1}^{\infty}$ converges to $(T - T_m)x$. By Theorem 7.6,

$$\|T - T_m\| \leq \liminf_{n\to\infty} \|T_n - T_m\|.$$

Since $\{T_m\}$ is Cauchy, this tends to zero as $m\to\infty$, and $\{T_m\}$ converges to T. □

The following theorem is one of the fundamental principles of functional analysis. The proof illustrates the power of the Baire Category Theorem.

7.8 Theorem *(Principle of Uniform Boundedness):* Let $\{T_\alpha\}_{\alpha\in A}$ be a family of continuous linear operators from a Banach space $\mathfrak{X}$ to a Banach space $\mathfrak{Y}$. If for each $x \in \mathfrak{X}$,

$$\sup_{\alpha\in A} \|T_\alpha x\| < \infty,$$

then

$$\sup_{\alpha\in A} \|T_\alpha\| < \infty.$$

Proof: For $n \geq 1$, let E_n be the set of $x \in \mathfrak{X}$ such that

$$\sup\{\|T_\alpha(x)\| : \alpha \in A\} \leq n.$$

The E_n's form an increasing sequence of closed subsets of $\mathfrak{X}$. The assumption is that $\cup E_n = \mathfrak{X}$.

By the Baire Category Theorem (Theorem 2.6), one of the E_n's contains an interior point. Thus there exist $x_0 \in \mathfrak{X}$, $\varepsilon > 0$, and an integer m such that E_m includes the open ball of radius ε centered at x_0.

Let $c = \sup\{\|T_\alpha(x_0)\| : \alpha \in A\}$. If $x \in \mathfrak{X}$ satisfies $\|x\| < 1$, then $\|x_0 + \varepsilon x - x_0\| < \varepsilon$, so that $x_0 + \varepsilon x \in E_m$. Hence

$$\varepsilon\|T_\alpha(x)\| = \|T_\alpha(x_0 + \varepsilon x) - T_\alpha(x_0)\| \leq m + c$$

and $\|T_\alpha\| \leq (m + c)/\varepsilon$ for all $\alpha \in A$. □

EXERCISES

1. Sketch the unit ball of $\mathbb{R}^2$ when it is endowed with each of the following norms: $\|\cdot\|_1$, $\|\cdot\|_2$, $\|\cdot\|_4$, and $\|\cdot\|_\infty$. Sketch the ball of radius 1 and center (2,1) in the metric determined by each of these norms.

2. Show the following for $x \in \mathbb{R}^n$:
   (a) $\|x\|_\infty \leq \|x\|_p, \qquad 1 \leq p < \infty,$
   (b) $\|x\|_p \leq n^{1/p}\|x\|_\infty, \qquad 1 \leq p < \infty,$
   (c) $\lim_{p\to\infty} \|x\|_p = \|x\|_\infty.$
3. Two norms $\|\cdot\|$ and $|||\cdot|||$ on a vector space $\mathfrak{X}$ are *equivalent* if there exists $c > 0$ such that

$$\frac{1}{c} \; |||x||| \leq \|x\| \leq c \; |||x|||, \qquad x \in \mathfrak{X}.$$

   (a) Show that the norms $\|\cdot\|$ and $|||\cdot|||$ are equivalent if and only if the identity map of $\mathfrak{X}$ is bicontinuous between the $\|\cdot\|$-topology of $\mathfrak{X}$ and the $|||\cdot|||$-topology of $\mathfrak{X}$.
   (b) Show that the norms $\|\cdot\|$ and $|||\cdot|||$ are equivalent if and only if $\sup\{\, |||x||| : \|x\| = 1\} < \infty$ and $\inf\{\, |||x||| : \|x\| = 1\} > 0$.
   (c) Prove that the notion of being equivalent norms is an equivalence relation on the set of norms on $\mathfrak{X}$.
4. Show that any two norms on $\mathbb{R}^n$ are equivalent.

   Hint: For $x = (x_1,\ldots,x_n)$, show that $\|x\| \leq c\|x\|_1$, where $\|\cdot\|$ is the given norm and $\|x\|_1 = \Sigma\, |x_j|$. Then observe that the unit sphere of $\mathbb{R}^n$ in the $\|\cdot\|_1$-norm is $\|\cdot\|$-compact and hence $\inf\{\|x\| : \|x\|_1 = 1\} > 0$.
5. Prove that any linear functional on a finite dimensional normed space is continuous. Use this result to prove that any linear operator from a finite dimensional normed space to another normed space is continuous.
6. A linear transformation T from a normed space $(\mathfrak{X},\|\cdot\|)$ to a normed space $(\mathfrak{Y}, |||\cdot|||)$ is an *isometric isomorphism* if T is one-to-one and onto and if

$$|||Tx||| = \|x\|, \qquad x \in \mathfrak{X}.$$

   (a) Show that the composition of isometric isomorphisms is an isometric isomorphism.
   (b) Show that the inverse of an isometric isomorphism is an isometric isomorphism.
   (c) Prove that a one-to-one linear transformation T from $\mathfrak{X}$ to $\mathfrak{Y}$ is an isometric isomorphism if and only if the image under T of the unit ball of $\mathfrak{X}$ coincides with the unit ball of $\mathfrak{Y}$.
7. Fix $1 < p < \infty$ and let $q = p/(p - 1)$ be the *conjugate index* of p, so that $(1/p) + (1/q) = 1$.
   (a) Show that if $s,t \geq 0$, then

$$st \leq \frac{s^p}{p} + \frac{t^q}{q}.$$

   (b) Show that if $x,y \in \mathbb{R}^n$, then

$$|\Sigma\, x_j y_j| \leq \|x\|_p \|y\|_q.$$

   Hint: First assume $\|x\|_p = 1 = \|y\|_q$.

(c) Show that $\|\cdot\|_p$ is a norm on $\mathbb{R}^n$.
Hint: Write
$$\Sigma |x_j + y_j|^p = \Sigma |x_j||x_j + y_j|^{p-1} + \Sigma |y_j||x_j + y_j|^{p-1}$$
and apply the result of (b) to each summand.

8. Show the following:
   (a) l^p is a Banach space, $1 \leq p \leq \infty$.
   (b) If $1 \leq p < r < \infty$, then l^p is a proper dense subset of l^r.
   (c) l^1 is not dense in l^∞. (Can you identify the closure of l^1 in l^∞?)

9. For each $s = (s_1, s_2, \ldots) \in l^\infty$, show that

$$L_s(x) = \sum_{j=1}^{\infty} s_j x_j, \qquad x \in l^1,$$

defines a continuous linear functional on l^1. Show that the correspondence $s \to L_s$ is an isometric isomorphism of l^∞ and the conjugate space $(l^1)^*$ of l^1. (This type of theorem is called a *representation theorem*, in that it represents the conjugate space of a Banach space as some more concrete Banach space.)

10. Show that $\|f\|_1 = \int_0^1 |f(s)|ds$ defines a norm on $f \in C[0,1]$.

11. Show that the space of continuous real-valued functions on the unit interval $[0,1]$ is not complete in the norm $\|\cdot\|_1$ defined earlier.

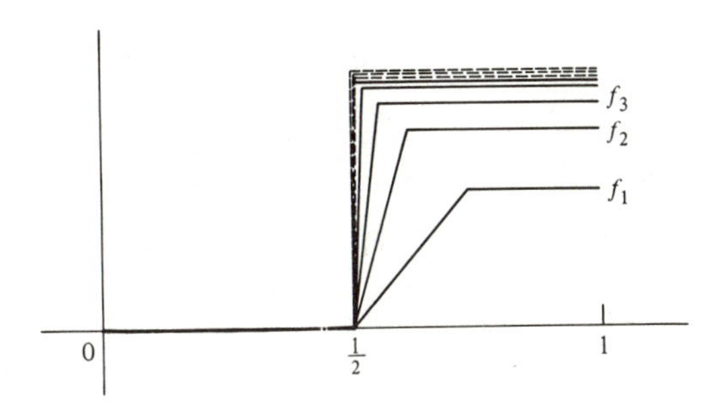

Hint: Consider a sequence of functions whose graphs are as indicated in the figure.

12. (a) Prove that if f and g are continuous real-valued functions on $[0,1]$, then

$$\int_0^1 f(s)g(s)ds \leq \left(\int_0^1 f(s)^2 ds\right)^{1/2} \left(\int_0^1 g(s)^2 ds\right)^{1/2}.$$

Remark: This is an integral version of the Cauchy-Schwarz inequality. For the proof, refer back to Exercise 1.3 and consider the integral

$$\int_0^1 [f(s) - \lambda g(s)]^2 ds,$$

which is nonnegative for all values of the real parameter λ.

(b) Using (a), show that the formula

$$\|f\|_2 = \left(\int_0^1 f(s)^2 ds\right)^{1/2}$$

defines a norm on the space of continuous real-valued functions on $[0,1]$.

(c) Prove that if f and g are continuous complex-valued functions on $[0,1]$, then

$$\int f(s)\overline{g(s)}ds \leq \left(\int |f(s)|^2 ds\right)^{1/2}\left(\int |g(s)|^2 ds\right)^{1/2}.$$

Remark: This is an integral form of the complex version of the Cauchy-Schwarz inequality.

(d) Show that the formula

$$\|f\|_2 = \left(\int_0^1 |f(s)|^2 ds\right)^{1/2}$$

defines a norm on the space of continuous complex-valued functions on $[0,1]$.

13. Show that there is a one-to-one correspondence between norms on a vector space $\mathfrak{X}$ and metrics d on $\mathfrak{X}$ that are translation-invariant and homogeneous in the sense that

$$d(x + w, y + w) = d(x,y), \qquad x,y,w \in \mathfrak{X},$$

and

$$d(cx,cy) = |c|d(x,y), \qquad x,y \in \mathfrak{X},\ c \text{ scalar}.$$

14. Let V be the vector space of all real-valued functions on a nonempty set S. Show that

$$d(f,g) = \sup_{s \in S} |f(s) - g(s)|/[1 + |f(s) - g(s)|]$$

is a metric on V that is translation-invariant. Does the metric arise from a norm?

15. Prove that a sequence in $\mathcal{B}(\mathfrak{X},\mathfrak{Y})$ converges to $T \in \mathcal{B}(\mathfrak{X},\mathfrak{Y})$ in the uniform operator topology if and only if it converges to T uniformly on each bounded subset of $\mathfrak{X}$.

16. Let $\mathfrak{X}$, $\mathfrak{Y}$, and $\mathfrak{Z}$ be normed spaces, let $S \in \mathcal{B}(\mathfrak{X},\mathfrak{Y})$, and let $T \in \mathcal{B}(\mathfrak{Y},\mathfrak{Z})$. Show that the composed operator TS from $\mathfrak{X}$ to $\mathfrak{Z}$ is bounded and satisfies

$$\|TS\| \leq \|T\|\,\|S\|.$$

17. Let $\mathfrak{X}$ and $\mathfrak{Y}$ be Banach spaces and let $\mathfrak{X} \oplus \mathfrak{Y}$ be their vector space direct sum. Show that

$$\|(x,y)\| = \max(\|x\|, \|y\|)$$

defines a norm on $\mathfrak{X} \oplus \mathfrak{Y}$ that makes $\mathfrak{X} \oplus \mathfrak{Y}$ into a Banach space.

18. Let $\mathfrak{X}$ and $\mathfrak{Y}$ be Banach spaces and let $\{T_n\}$ be a sequence in $\mathcal{B}(\mathfrak{X},\mathfrak{Y})$ such

that $\{T_n x\}$ converges for all $x \in \mathfrak{X}$. Show that there exists $T \in \mathcal{B}(\mathfrak{X},\mathfrak{Y})$ such that $\lim T_n x = Tx$ for all $x \in \mathfrak{X}$.

19. Consider the linear functional L on the real Banach space $C[0,1]$, endowed with the sup norm, defined by

$$L(f) = \sum_{j=1}^{m} a_j f(t_j),$$

where the t_j's are distinct points of $[0,1]$ and the a_j's are real numbers. Show that L is continuous, with norm

$$\|L\| = \sum_{j=1}^{m} |a_j|.$$

20. For each integer $n \geq 0$, let $\{t_j^{(n)}\}_{j=1}^{m_n}$ be points in $[0,1]$ and let $\{a_j^{(n)}\}_{j=1}^{m_n}$ be positive real numbers such that

$$\int_0^1 g(t)dt = \sum_{j=1}^{m_n} a_j^{(n)}\, g(t_j^{(n)})$$

for all polynomials g of degree at most n. Show that

$$\int_0^1 f(t)dt = \lim_{n\to\infty} \sum_{j=1}^{m_n} a_j^{(n)}\, f(t_j^{(n)})$$

for all $f \in C[0,1]$. *Hint:* Use Theorem 7.6

21. Let $h \in C[0,1]$, sup norm. Define a linear functional L on $C[0,1]$ by

$$L(f) = \int_0^1 f(t)h(t)dt, \qquad f \in C[0,1].$$

What is $\|L\|$? Prove your answer.

22. Let $k(s,t)$ be a continuous function on $[0,1] \times [0,1]$. Let T be the linear operator on $C[0,1]$ endowed with the sup norm, defined by

$$(Tf)(s) = \int_0^1 k(s,t)f(t)dt, \qquad 0 \leq s \leq 1.$$

What is $\|T\|$? Prove your answer.

Remark: The function $k(s,t)$ is called a *kernel function;* it is said to be the *kernel* of the integral operator T.

8. THE CONTRACTION PRINCIPLE

Let X be a metric space with metric d and let Φ be a function from X to X. A point $x \in X$ is a *fixed point* of Φ if $\Phi(x) = x$. The solutions of many classes of equations can be regarded as fixed points of appropriate maps, so that the study of fixed points is of some importance. In this section, we give conditions that guarantee the existence of fixed points of certain maps, and we give several applications to integral and differential equations. In the next section, we apply the principle to obtain an Implicit Function Theorem for Banach spaces.

The map Φ is a *contraction* if there exists a number c, $0 < c < 1$, such that

$$d(\Phi(x),\Phi(y)) \leq cd(x,y), \qquad x,y \in X. \tag{8.1}$$

A contraction can have at most one fixed point. Indeed, if x and y are both fixed points of Φ, then the estimate (8.1) becomes $d(x,y) \leq cd(x,y)$, from which we conclude that $d(x,y) = 0$ and $x = y$. For the existence of a fixed point, there is the following important theorem.

8.1 Theorem *(Contraction Mapping Theorem):* A contraction mapping of a complete metric space has exactly one fixed point.

Proof: Let Φ be a contraction on a complete space X. Define by induction maps Φ^m from X to X by

$$\Phi^0(x) = x,$$

$$\Phi^m(x) = \Phi^{m-1}(\Phi(x)), \qquad m \geq 1.$$

Applying (8.1) m times, we obtain

$$d(\Phi^m(x), \Phi^{m+1}(x)) \leq cd(\Phi^{m-1}(x),\Phi^m(x)) \leq \cdots \leq c^m d(x,\Phi(x)).$$

It follows that if $0 \leq m \leq m + k$, then

$$\begin{aligned} d(\Phi^m(x),\Phi^{m+k}(x)) &\leq d(\Phi^m(x),\Phi^{m+1}(x)) + \cdots + d(\Phi^{m+k-1}(x),\Phi^{m+k}(x)) \\ &\leq c^m d(x,\Phi(x)) + \cdots + c^{m+k}d(x,\Phi(x)). \end{aligned}$$

Since $1 + c + \cdots + c^k \leq 1/(1 - c)$, we obtain

$$d(\Phi^m(x),\Phi^{m+k}(x)) \leq \frac{c^m}{1 - c} d(x,\Phi(x)). \tag{8.2}$$

This estimate shows that the iterates $\{\Phi^m(x)\}$ form a Cauchy sequence in X. Since X is complete, $\{\Phi^m(x)\}$ converges to some point $x^* \in X$. Moreover,

$$\begin{aligned} d(x^*,\Phi(x^*)) &= \lim_{m\to\infty} d(\Phi^{m+1}(x),\Phi(x^*)) \\ &\leq \lim_{m\to\infty} cd(\Phi^m(x),x^*) \\ &= cd(x^*,x^*) = 0. \end{aligned}$$

Consequently $\Phi(x^*) = x^*$ and x^* is a fixed point for Φ. The uniqueness of a fixed point has already been established, and so the proof is complete. □

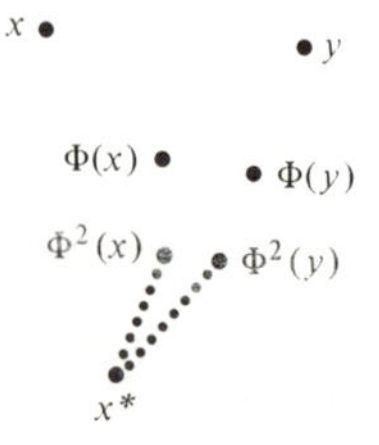

Thus one reaches the fixed point by following the iterates of any initial point $x \in X$. There is an important quantitative addendum to the Contraction Mapping Theorem, which we obtain immediately by sending k to ∞ in (8.2).

8.2 Theorem: Let Φ be a contraction mapping of a complete metric space X with contraction constant c and let x^* be the fixed point of Φ. Then for all $x \in X$,

$$d(\Phi^m(x),x^*) \leq \frac{c^m}{1-c} d(x,\Phi(x)), \qquad m \geq 0. \tag{8.3}$$

As an application, let T be a linear operator on a Banach space $\mathfrak{X}$ that satisfies $\|T\| < 1$. Such an operator is a contraction since

$$d(Tx,Ty) = \|Tx - Ty\| \leq \|T\| \, \|x - y\| = \|T\| \, d(x,y).$$

(What is the unique fixed point of T?) We wish to solve the equation

$$x = u + T(x) \tag{8.4}$$

where $u \in \mathfrak{X}$ is given and x is the unknown. Set

$$\Phi(x) = u + T(x), \qquad x \in \mathfrak{X}.$$

Then x^* is a solution of (8.4) if and only if x^* is a fixed point of Φ. Since $\|\Phi(x) - \Phi(y)\| \leq \|T\| \, \|x - y\|$, Φ is a contraction. Theorem 8.1 then yields a unique solution x^* of (8.4), and furthermore it gives an iterative procedure for obtaining the solution. If $x_0 = 0$ and x_n is defined so that

$$x_{n+1} = u + T(x_n), \qquad n \geq 0,$$

then by (8.3),

$$\|x_m - x^*\| \leq \|T\|^m \|u\| / [1 - \|T\|].$$

This method can be applied to what is called the Fredholm integral equation of the second kind,

$$v(s) = u(s) + \int_a^b K(s,t)v(t)dt, \qquad a \leq s \leq b. \tag{8.5}$$

All functions appearing in the equation will be complex-valued. The kernel function $K(s,t)$ is a given continuous function on the square $[a,b] \times [a,b]$, and u is a given continuous function on the interval $[a,b]$. The problem is to find a continuous function v on $[a,b]$ that satisfies (8.5). Without further hypotheses, the equation need not have a solution. The Fredholm equation does have a solution, however, if K is sufficiently small.

8.3 Theorem: Let the continuous complex-valued function $K(s,t)$ on $[a,b] \times [a,b]$ satisfy

$$|K(s,t)| < 1/(b - a), \qquad a \leq s,t \leq b.$$

Then for any continuous function u on $[a,b]$ there exists a unique continuous function v on $[a,b]$ that satisfies the Fredholm integral equation (8.5).

Proof: Let $C[a,b]$ denote the Banach space consisting of all continuous complex-valued functions on $[a,b]$, with the supremum norm

$$\|v\| = \sup\{|v(s)| : a \leq s \leq b\}.$$

The Fredholm equation (8.5) then has the form (8.4) for the linear operator T on $C[a,b]$ defined by

$$(Tv)(s) = \int_a^b K(s,t)v(t)dt, \qquad a \leq s \leq b.$$

In order to prove the theorem, it suffices, in view of our earlier work, to show that $\|T\| < 1$.

Define

$$M = \sup\{|K(s,t)| : a \leq s,t \leq b\}.$$

Let $v \in C[a,b]$. The obvious estimates yield

$$|(Tv)(s)| \leq M \int_a^b |v(t)|dt \leq M(b - a)\|v\|.$$

Taking the supremum over $a \leq s \leq b$, we obtain

$$\|Tv\| \leq M(b - a)\|v\|, \qquad v \in C[a,b].$$

Hence

$$\|T\| \leq M(b - a).$$

The hypothesis on K, and the compactness of the domain of K, show that $M(b - a) < 1$. Hence $\|T\| < 1$, as required. □

As another application of the Contraction Mapping Theorem, we prove an existence theorem for a first-order system of nonlinear ordinary differential equations. Let U be an open subset of $\mathbb{R}^n$ and let $[a,b]$ be an interval on $\mathbb{R}$. Let F be a continuous function from $U \times [a,b]$ to $\mathbb{R}^n$ and let $\xi \in \mathbb{R}^n$ be a fixed initial value. We consider the initial-value problem of finding a differentiable function $u(t)$ from $[a,b]$ to $\mathbb{R}^n$ that satisfies

$$(8.6) \qquad \begin{cases} \dfrac{du(t)}{dt} = F(u(t),t), & a \leq t \leq b, \\ u(a) = \xi. \end{cases}$$

If the components of u are $u_1,\ldots,u_n$ and those of F are $F_1,\ldots,F_n$, then the initial-value problem becomes

$$\begin{cases} \dfrac{du_j(t)}{dt} = F_j(u_1(t),\ldots,u_n(t),t), & a \leq t \leq b,\ 1 \leq j \leq n, \\ u_j(a) = \xi_j, & 1 \leq j \leq n. \end{cases}$$

Consider, for instance, the special case in which F is independent of t, that is, F is a function from U to $\mathbb{R}^n$. Such an F is called a *vector field* on U. It is visualized

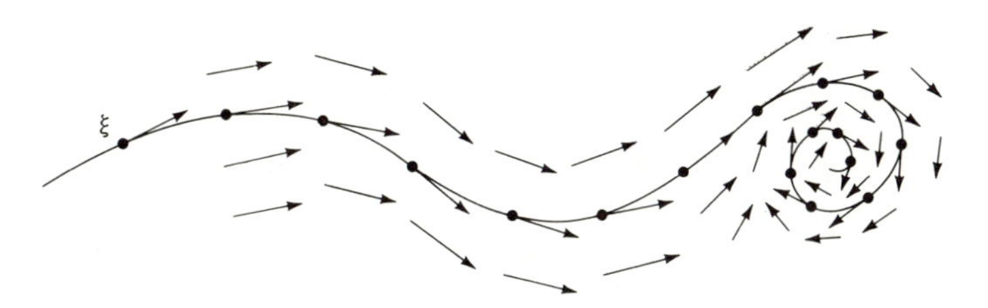

Integral curve of a vector field

by attaching to each $x \in U$ an arrow based at x with direction and length given by $F(x)$. A solution $u(t)$ of (8.6) can be regarded as a curve in U beginning at ξ, with the property that the tangent vector du/dt of the curve at the point $u(t)$ coincides with the value $F(u(t))$ of the vector field F at $u(t)$. The curve $u(t)$ is said to be an *integral curve* of the vector field F.

In order to obtain an existence theorem, we make some hypotheses on F. Choose $r > 0$ so that the closed ball $\{|x - \xi| \leq r\}$ centered at ξ with radius r is contained in U. We assume that F is continuous, so that in particular

$$M = \sup\{|F(x,t)| : |x - \xi| \leq r, a \leq t \leq b\} \tag{8.7}$$

is finite. We also assume that there is a constant $c > 0$ such that

$$|F(x,t) - F(y,t)| \leq c|x - y|, \qquad x,y \in B(\xi;r), a \leq t \leq b. \tag{8.8}$$

Such a condition is called a *Lipschitz condition*; we say that F satisfies a *Lipschitz condition in the first variable*. With these hypotheses, the theorem we obtain is the following.

8.4 Theorem *(Cauchy-Picard Existence Theorem):* Suppose that F and ξ are as above, so that (8.7) and (8.8) hold. Then there exists β, $a < \beta \leq b$, such that the initial-value problem (8.6) has a unique solution $u(t)$ defined on the interval $a \leq t \leq \beta$. Furthermore, β depends only on the parameters r, M, and c.

Proof: The function $u(t)$ satisfies the initial-value problem (8.6) on an interval $[a,\beta]$ if and only if it satisfies the integral equation

$$u(t) = \xi + \int_a^t F(u(s),s)ds, \qquad a \leq t \leq \beta. \tag{8.9}$$

In turn, u is a solution of (8.9) if and only if it is a fixed point of the integral operator Φ, defined by

$$(\Phi u)(t) = \xi + \int_a^t F(u(s),s)ds, \qquad a \leq t \leq \beta.$$

We must specify carefully the domain of the operator Φ.

Fix β, $a < \beta \leq b$, and let E be the set of continuous functions $u(t)$ from the interval $[a,\beta]$ to $\mathbb{R}^n$ that satisfy

$$|u(t) - \xi| \leq r, \qquad a \leq t \leq \beta.$$

Endowed with the metric of uniform convergence

$$d(u,v) = \sup\{|u(t) - v(t)| : a \le t \le \beta\}, \qquad u,v \in E,$$

E becomes a complete metric space. We aim to show that if β is chosen sufficiently near a, then Φ is a contraction mapping of E into E.

First we must arrange that $\Phi(E) \subseteq E$. Let $u \in E$. Using (8.7), we obtain

$$\begin{aligned} |(\Phi u)(t) - \xi| &= \left| \int_a^t F(u(s),s)ds \right| \\ &\le M(t - a) \\ &\le M(\beta - a). \end{aligned}$$

To place Φu in E, it suffices then to choose β so that

(8.10) $$\beta - a \le r/M.$$

Now let $u,v \in E$. Using (8.8), we obtain

$$\begin{aligned} d(\Phi u,\Phi v) &= \sup_{a\le t\le\beta} |(\Phi u)(t) - (\Phi v)(t)| \\ &= \sup_{a\le t\le\beta} \left| \int_a^t [F(u(s),s) - F(v(s),s)]ds \right| \\ &\le c \sup_{a\le t\le\beta} \int_a^t |u(s) - v(s)|ds \\ &\le c(\beta - a)d(u,v). \end{aligned}$$

In order that Φ be a contraction, it suffices then to choose β so that

(8.11) $$\beta - a < 1/c.$$

Choosing β to satisfy (8.10) and (8.11) and applying the Contraction Mapping Theorem (Theorem 8.1), we obtain the existence assertion for the Cauchy-Picard Theorem. To prove the uniqueness, observe that any solution of the initial-value problem (8.6) must lie in the space E defined above, provided β is chosen sufficiently near a. By the uniqueness assertion of the Contraction Principle, the solution must coincide with the solution produced above, at least in some small interval $[a,a + \varepsilon]$. Applying this local uniqueness assertion to each point of $[a,\beta]$, we deduce easily that the solution is unique. □

Actually the above proof contains information that goes substantially beyond the statement of the theorem. It provides us with a concrete iteration procedure for approximating the solution of (8.6), and the estimate (8.3) yields a specific bound on the error of the approximation. By keeping track of the error terms, one can prove, for instance, that if F and ξ depend continuously on some other parameters, then the solutions of (8.6) also depend continuously on those parameters. For instance, the solutions of (8.6) depend continuously on the initial condition ξ. We return briefly to the abstract situation, in order to give an indication of how one might approach the problem of dependence of solutions on parameters.

8.5 Theorem: Let X be a complete metric space with metric d, let S be a metric space, and let c, $0 < c < 1$, be fixed. Suppose that $(s,x) \to \Phi_s(x)$ is a continuous function from $S \times X$ to X such that

$$d(\Phi_s(x),\Phi_s(y)) \leq cd(x,y), \qquad x,y \in X,\ s \in S.$$

Then for each $s \in S$, there is a unique point $x_s^* \in X$ such that $\Phi_s(x_s^*) = x_s^*$. Furthermore, x_s^* depends continuously on s.

Proof: The first statement follows immediately from the Contraction Mapping Theorem. We must check that the fixed point x_s^* given by the proof of Theorem 8.1 depends continuously on s.

Let $s,t \in S$. We consider the estimate (8.3) in the case $m = 0$ for $\Phi = \Phi_s$, $x^* = x_s^*$, and $x = x_t^*$. Substituting these data into (8.3), we obtain

$$d(x_t^*,x_s^*) \leq \frac{1}{1-c}\, d(x_t^*,\Phi_s(x_t^*)). \tag{8.12}$$

As s tends to t, $\Phi_s(x_t^*)$ tends to $\Phi_t(x_t^*) = x_t^*$. Hence the right hand side of (8.12) tends to 0, so that x_s^* tends to x_t^* as s tends to t. □

EXERCISES

1. Show that if Φ is a function from a nonempty compact metric space X to itself such that

   $$d(\Phi(x),\Phi(y)) < d(x,y), \qquad x \neq y,$$

   then Φ has a unique fixed point.
2. Find a compact metric space X and a function $\Phi : X \to X$ such that

   $$d(\Phi(x),\Phi(y)) \leq d(x,y), \qquad x,y \in X, \tag{8.13}$$

   while Φ has no fixed point. Find also X and Φ satisfying (8.13) such that Φ has more than one fixed point.
3. Let X be a complete metric space and let Φ be a function from X to X such that Φ^m is a contraction for some $m \geq 1$. Show that Φ has a unique fixed point.
4. Let $\mathfrak{X}$ be a Banach space, let $m \geq 1$, and let T be a continuous linear operator on $\mathfrak{X}$ such that $\|T^m\| < 1$. Fix $u \in \mathfrak{X}$ and define

   $$\Phi(v) = u + T(v), \qquad v \in \mathfrak{X}.$$

   (a) Show that Φ^m is a contraction.
   (b) Show that the equation

   $$v = u + T(v)$$

   has a unique solution $v \in \mathfrak{X}$.

5. Let $\mathfrak{X}$ be a Banach space and let T be a continuous linear operator on $\mathfrak{X}$. Let $u \in \mathfrak{X}$, let $\lambda \in \mathbb{C}$, and define $\Phi(v) = (u/\lambda) + T(v/\lambda)$, $v \in \mathfrak{X}$.
   (a) By applying the Contraction Mapping Theorem to Φ, show that the equation

$$\lambda v = u + T(v) \tag{8.14}$$

   has a unique solution $v \in \mathfrak{X}$, provided $|\lambda| > \|T\|$.
   (b) Let $v_0 = 0$ and $v_m = \Phi(v_{m-1})$, $m \geq 1$. Show that

$$v_m = \sum_{k=0}^{m+1} \frac{T^k(u)}{\lambda^{k+1}}.$$

   (c) Show that if $|\lambda| > \|T\|$, then the series

$$\sum_{k=0}^{\infty} T^k/\lambda^{k+1} \tag{8.15}$$

   converges in norm to $(\lambda I - T)^{-1}$. How can this be used to solve (8.14)?
   (d) Show that if

$$|\lambda| > \limsup_{n\to\infty} \|T^n\|^{1/n},$$

   then the series (8.15) converges in norm to $(\lambda I - T)^{-1}$.
6. Consider the Volterra integral equation

$$v(s) = u(s) + \int_a^s K(s,t)v(t)dt, \qquad a \leq s \leq b, \tag{8.16}$$

   where K is continuous on $[a,b] \times [a,b]$. Show that (8.16) has a unique solution $v \in C[a,b]$ for each fixed $u \in C[a,b]$. *Hint:* Show that the integral operator T appearing in (8.16) satisfies

$$\|T^m\| \leq M^m(b - a)^m/m!, \qquad m \geq 0,$$

   where M is a bound for the kernel function K. Then apply Exercise 5 or Exercise 4.
7. Prove in detail the uniqueness assertion of the Cauchy-Picard Theorem.
8. Show that the initial-value problem

$$\begin{cases} \dfrac{du}{dt} = u^{1/2}, & t \geq 0, \\ u(0) = 0 \end{cases}$$

   has more than one solution. (*Hint:* Look for solutions of the form $u(t) = t^\alpha$.) Reconcile this fact with the Cauchy-Picard Theorem.
9. Formulate and prove a theorem to the effect that the solution of the initial-value problem (8.6) depends continuously on the initial data ξ.
10. Let $F : U \times [a,b] \to \mathbb{R}^n$ be continuous and suppose that F satisfies locally

a Lipschitz condition in the first variable, i.e., for each $\xi \in U$, there is a ball $B(\xi;r) \subset U$ on which F satisfies the Lipschitz condition (8.8), with c depending possibly on ξ and r. Show that either there is a continuously differentiable solution of (8.6) on $[a,b]$ or else there exists β, $a < \beta \leq b$, and a continuously differentiable solution u of (8.6) on $[a,\beta)$ such that $u(t) \to (\partial U) \cup \{\infty\}$ as t increases to β.

11. Let X, S, Φ_s, and $x_s^* = x^*(s)$ be as in Theorem 8.5. Fix $x_0 \in X$ and define $x_m(s) = \Phi_s^m(x_0)$ to be the mth iterate of x_0 under Φ_s. Show that if S is compact, then x_m converges uniformly on S to x^*.

9. THE FRECHET DERIVATIVE

Let U be an open subset of a Banach space $\mathfrak{X}$ and let G be a function from U to a Banach space $\mathfrak{Y}$. We say that G is *differentiable* at $x_0 \in U$ if there is a continuous linear transformation T from $\mathfrak{X}$ to $\mathfrak{Y}$ such that

$$\lim_{x \to x_0} \frac{G(x) - G(x_0) - T(x - x_0)}{\|x - x_0\|} = 0. \tag{9.1}$$

This means, roughly speaking, that G can be approximated near x_0 by an affine (linear plus constant) function, namely, $G(x_0) - T(x_0) + T$. The linear transformation T is the *Frechet derivative* of G at x_0, and it is denoted by $G'(x_0)$. The function G is *differentiable* if it is differentiable at each point of its domain U. It is *continuously differentiable* if it is differentiable and if the derivatives $G'(x)$ depend continuously on $x \in U$ in the uniform operator topology of $\mathcal{B}(\mathfrak{X},\mathfrak{Y})$.

In the case where $\mathfrak{X}$ and $\mathfrak{Y}$ are Euclidean spaces, any linear operator from $\mathfrak{X}$ to $\mathfrak{Y}$ can be represented by a matrix. In this case, the continuity of the operators $G'(x)$ in the uniform operator topology means simply that the matrix entries of $G'(x)$ are continuous functions of x.

Before treating this case in more detail, we derive some elementary properties of the Frechet derivative. In terms of $G'(x_0)$, condition (9.1) can be restated in the form

$$G(x) = G(x_0) + G'(x_0)(x - x_0) + R(x), \tag{9.2}$$

where the remainder term $R(x)$ satisfies

$$\lim_{x \to x_0} \frac{R(x)}{\|x - x_0\|} = 0. \tag{9.3}$$

Note that the derivative of G is uniquely defined. Indeed, if two continuous linear operators S and T satisfy (9.1), then

$$\lim_{x \to x_0} \frac{(T - S)(x - x_0)}{\|x - x_0\|} = 0.$$

Substituting $x - x_0 = tz$ and letting t tend to zero, we obtain $(S - T)(z) = 0$ for all $z \in \mathfrak{X}$, so that $S = T$.

We turn to three results that are immediate consequences of the definition of the Frechet derivative. The proofs are much the same as in the one-variable calculus.

9.1 Theorem: If G is differentiable at x_0, then G is continuous at x_0.

Proof: Condition (9.3) shows that $R(x)$ is continuous at x_0. Since the affine approximation $G(x_0) + G'(x_0)(x - x_0)$ is continuous at x_0, we see from (9.2) that $G(x)$ is continuous at x_0. □

9.2 Theorem: Differentiation in the sense of Frechet is a linear operation. In other words, if F and G are differentiable at x_0 and if c is a scalar, then $F + G$ and cG are differertiable at x_0 and

$$(cG)'(x_0) = cG'(x_0),$$

$$(F + G)'(x_0) = F'(x_0) + G'(x_0).$$

Proof: If (9.2) and (9.3) are valid, then also

$$cG(x) = cG(x_0) + cG'(x_0)(x - x_0) + cR(x),$$

where $cR(x)/\|x - x_0\|$ tend to 0 as x tends to x_0. Hence cG is differentiable at x_0, with derivative $cG'(x_0)$. The second formula is proved similarly. □

In the usual calculus, the derivative of a constant function is zero and the derivative of a linear function is constant. These rules are also valid for the Frechet derivative.

9.3 Theorem: Fix $y_0 \in \mathfrak{Y}$ and $T \in \mathcal{B}(\mathfrak{X},\mathfrak{Y})$ and define

$$G(x) = y_0 + Tx, \qquad x \in \mathfrak{X}.$$

Then G is differentiable at each $x_0 \in \mathfrak{X}$ and

$$G'(x_0) = T, \qquad x_0 \in \mathfrak{X}.$$

Proof: In this case,

$$G(x) - G(x_0) - T(x - x_0) = 0, \qquad x \in \mathfrak{X}.$$

Comparing this identity with (9.1), we see that $G'(x_0) = T$. □

Consider now the special case in which $\mathfrak{X} = \mathfrak{Y} = \mathbb{R}$. Then each linear transformation from $\mathfrak{X}$ to $\mathfrak{Y}$ is of the form $t \to ct$ for some $c \in \mathbb{R}$. The differentiability condition (9.1) means that there exists $c \in \mathbb{R}$ such that

$$\lim_{x \to x_0} \left| \frac{G(x) - G(x_0)}{x - x_0} - c \right| = 0.$$

This occurs if and only if G is differentiable at x_0, in the usual sense, and the number c is the usual value of the derivative associated with G. For the purposes of the Frechet derivative, we have changed our point of view, regarding $G'(x_0)$ not as a number c but as an operator "multiplication by c" on $\mathbb{R}$.

Suppose next that $\mathfrak{X} = \mathbb{R}^n$ and $\mathfrak{Y} = \mathbb{R}^m$. Each linear operator T from $\mathfrak{X}$ to $\mathfrak{Y}$ can then be represented by an $m \times n$ matrix (T_{jk}), defined so that if $x = (x_1,\ldots,x_n)$, then the jth component of Tx is given by

$$(Tx)_j = \sum_{k=1}^{n} T_{jk}x_k, \qquad 1 \le j \le m.$$

Let G be a function defined near $x^0 = (x_1^0,\ldots,x_n^0)$ in $\mathbb{R}^n$ with values in $\mathbb{R}^m$. Let $G_1,\ldots,G_m$ be the component functions of G, so that

$$G(x) = (G_1(x),\ldots,G_m(x)).$$

9.4 Theorem: If $G : \mathbb{R}^n \to \mathbb{R}^m$ is as above and if G is differentiable at x^0, then each of the partial derivatives $\partial G_j/\partial x_k$ exists at x^0 and the Frechet derivative of G at x^0 is represented by the Jacobian matrix

$$G'(x^0) = \left(\frac{\partial G_j}{\partial x_k}(x^0)\right).$$

Proof: The relation (9.1) becomes

$$\lim_{x\to x^0} \frac{G_j(x) - G_j(x^0) - \sum_{k=1}^{n} T_{jk}(x_k - x_k^0)}{|x - x^0|} = 0, \qquad 1 \le j \le m.$$

Substituting $x_k = x_k^0 + t$ for some fixed k, $x_i = x_i^0$ for $i \ne k$, we obtain

$$\lim_{t\to 0} \left| \frac{G_j(x_1^0,\ldots,x_{k-1}^0, x_k^0 + t, x_{k+1}^0,\ldots,x_n^0) - G_j(x_1^0,\ldots,x_n^0)}{t} - T_{jk} \right| = 0.$$

It follows that the partial derivative $\partial G_j/\partial x_k$ exists at x^0 and coincides with the matrix element T_{jk} of $G'(x^0)$. □

It turns out that the existence of the partial derivatives $\partial G_j/\partial x_k$ at x^0 is not sufficient to guarantee that G is differentiable at x^0, in the sense of Frechet (Exercise 1). However, it can be shown that G is continuously differentiable in U if and only if the partial derivatives $\partial G_j/\partial x_k$ exist and are continuous in U (Exercise 2).

A bounded linear operator A from a Banach space $\mathfrak{X}$ to a Banach space $\mathfrak{Y}$ is *invertible* if there is a bounded linear operator B from $\mathfrak{Y}$ to $\mathfrak{X}$ such that BA is the identity on $\mathfrak{X}$ and AB is the identity on $\mathfrak{Y}$. Evidently B is unique, and it is denoted by A^{-1}. For A to be invertible, it is necessary and sufficient that A be one-to-one and onto, and that

$$\|Ax\| \ge c\|x\|, \qquad x \in \mathfrak{X},$$

for some fixed $c > 0$. In this case $\|A^{-1}\| \le 1/c$, as can be seen by setting $Ax = y$ and $x = A^{-1}y$ above.

As a nontrivial infinite-dimensional example, we wish to differentiate the inverse function $A \to A^{-1}$. Note that the vector-valued function we are proposing to differentiate does not go from $\mathfrak{X}$ to $\mathfrak{Y}$ but rather from $\mathcal{B}(\mathfrak{X},\mathfrak{Y})$ to $\mathcal{B}(\mathfrak{Y},\mathfrak{X})$!

9.5 Theorem: Let $\mathfrak{X}$ and $\mathfrak{Y}$ be Banach spaces and let W be the set of the invertible operators in $\mathcal{B}(\mathfrak{X},\mathfrak{Y})$. Let G be the function from W to $\mathcal{B}(\mathfrak{Y},\mathfrak{X})$ defined by

$$G(A) = A^{-1}, \qquad A \in W.$$

Then W is an open subset of $\mathcal{B}(\mathfrak{X},\mathfrak{Y})$, the function G is differentiable on W, and the derivative of G at $A \in W$ is given by

$$G'(A)T = -A^{-1}TA^{-1}, \qquad T \in \mathcal{B}(\mathfrak{Y},\mathfrak{X}).$$

Proof: Note first that the proposed formula for $G'(A)$ is indeed a linear operator from $\mathcal{B}(\mathfrak{X},\mathfrak{Y})$ to $\mathcal{B}(\mathfrak{Y},\mathfrak{X})$, so that the theorem is not preposterous. The formula can be regarded as a generalization of the rule

$$\frac{d}{dt}\left(\frac{1}{t}\right) = -1/t^2.$$

Fix c such that $0 < c < 1$. Let $A \in W$. Let $T \in \mathcal{B}(\mathfrak{X},\mathfrak{Y})$ satisfy

$$\|T\| \le c/\|A^{-1}\|.$$

Then

$$\|TA^{-1}\| \le c$$

(see Exercise 7.16), so that

$$\|(TA^{-1})^n\| \le c^n, \qquad n \ge 0.$$

It follows that the partial sums of the series

$$S = \sum_{n=0}^{\infty} (-1)^n (TA^{-1})^n \tag{9.4}$$

form a Cauchy sequence in the space $\mathcal{B}(\mathfrak{Y})$ of continuous linear operators on $\mathfrak{Y}$. Since $\mathcal{B}(\mathfrak{Y})$ is complete (Theorem 7.7), the series converges to an element of $\mathcal{B}(\mathfrak{Y})$. Hence $A^{-1}S \in \mathcal{B}(\mathfrak{Y},\mathfrak{X})$. We claim that $A^{-1}S$ is an inverse for $A + T$. Indeed, multiplying the partial sums for $A^{-1}S$ by $A + T$, we obtain a telescoping series

$$(A + T)A^{-1} \sum_{n=0}^{N} (-1)^n (TA^{-1})^n = I + (-1)^N (TA^{-1})^{N+1},$$

where I is the identity operator on $\mathfrak{Y}$. Letting N tend to ∞, we see that $(A + T)A^{-1}S$ is the identity operator on $\mathfrak{Y}$. A similar calculation shows that $A^{-1}S(A + T)$ is the identity operator on $\mathfrak{X}$. We conclude that $A + T$ is invertible and that

$$(A + T)^{-1} = A^{-1} \sum_{n=0}^{\infty} (-1)^n (TA^{-1})^n, \qquad \|T\| < c/\|A^{-1}\|. \tag{9.5}$$

This shows that W is open since it includes a ball in $\mathcal{B}(\mathfrak{X},\mathfrak{Y})$ centered about any of its points.

To establish the assertion on differentiability, we must show that

$$H(T) = (A + T)^{-1} - A^{-1} + A^{-1}TA^{-1}$$

satisfies

$$\lim_{T\to 0} \frac{H(T)}{\|T\|} = 0.$$

From (9.5), we obtain the expansion

$$\begin{aligned} H(T) &= A^{-1} \sum_{n=2}^{\infty} (-1)^n (TA^{-1})^n \\ &= A^{-1}TA^{-1}TA^{-1} \sum_{n=0}^{\infty} (-1)^n (TA^{-1})^n. \end{aligned}$$

Consequently

$$\|H(T)\| \le \|T\|^2 \|A^{-1}\|^3 \sum_{n=0}^{\infty} c^n,$$

so that $H(T)/\|T\|$ does indeed tend to 0 with T. □

The use of the series in (9.4) might appear at first glance to be a trick. Its use is easy to motivate, though. We wished to construct an operator S to be the inverse of $I + TA^{-1}$. For this, we simply imposed a hypothesis to guarantee that $\|TA^{-1}\| < 1$, and we expanded $I/(I + TA^{-1})$ in a geometric series. This leads to (9.4).

Our next aim is to establish an Implicit Function Theorem for Banach spaces. It is one of the most important theorems in nonlinear functional analysis. The proof is quite difficult. It is not made simpler by replacing the Banach spaces by $\mathbb{R}^n$, though the reader may find the proof easier to follow conceptually in the finite-dimensional case. We prepare the way by proving several lemmas. These lemmas can be derived using calculus methods in the Euclidean case.

9.6 Lemma: Let h be a continuous real-valued function on the interval $[0,1]$ such that $h(0) = 0$ and

$$\limsup_{\delta\to 0+} \frac{h(s+\delta) - h(s)}{\delta} \le c, \qquad 0 \le s < 1.$$

Then

$$h(s) \le cs, \qquad 0 \le s \le 1. \tag{9.6}$$

Proof: Let $\varepsilon > 0$. Set $g(s) = h(s) - (c + \varepsilon)s$. Then

$$\limsup_{\delta\to 0+} \frac{g(s+\delta) - g(s)}{\delta} < 0, \qquad 0 \le s < 1. \tag{9.7}$$

Let s be a point of the interval $[0,t]$ at which g attains its minimum value over the interval $[0,t]$. If $s < t$, then by (9.7) there exists $\delta > 0$ such that $s + \delta < t$ and $g(s + \delta) - g(s) < 0$, contradicting the choice of s. It follows that the minimum of g over $[0,t]$ is attained at t. In particular, $g(t) \leq g(0) = 0$, and consequently $h(t) \leq (c + \varepsilon)t$. Since $\varepsilon > 0$ is arbitrary, we obtain (9.6). □

The following lemma contains a crucial estimate to be used in the proof of the Implicit Function Theorem.

9.7 Lemma: Let $\mathfrak{Y}$ and $\mathfrak{Z}$ be Banach spaces. Let U be an open ball in $\mathfrak{Y}$ and let G be a function from U to $\mathfrak{Z}$ that is differentiable at each point of U. Suppose the operator $T \in \mathcal{B}(\mathfrak{Y},\mathfrak{Z})$ satisfies

$$\|G'(y) - T\| \leq M$$

for all $y \in U$. Then

$$\|G(y_2) - G(y_1) - T(y_2 - y_1)\| \leq M\|y_2 - y_1\|$$

for all $y_1, y_2 \in U$.

Proof: Let $F = G - T$. Then $F'(y) = G'(y) - T$, so that $\|F'(y)\| \leq M$ for all $y \in U$. We must show that

$$\|F(y_2) - F(y_1)\| \leq M\|y_2 - y_1\|, \qquad y_1,y_2 \in U.$$

We shall use only the fact that F is differentiable at each point of the line segment joining y_1 to y_2 and that $\|F'\| \leq M$ on that segment.

Assume for convenience that $y_1 = 0$. Define

$$h(t) = \|F(ty_2) - F(0)\|, \qquad 0 \leq t \leq 1.$$

Using the triangle inequality, we obtain

$$h(t + \delta) - h(t) \leq \|F(t + \delta)y_2) - F(ty_2)\|$$

$$\leq \|F((t + \delta)y_2) - F(ty_2) - \delta F'(ty_2)y_2\| + \delta\|F'(ty_2)\| \, \|y_2\|.$$

Dividing by δ and letting δ tend to zero, we obtain

$$\limsup_{\delta \to 0+} \frac{h(t + \delta) - h(t)}{\delta} \leq \|F'(ty_2)\| \, \|y_2\| \leq M\|y_2\|.$$

From Lemma 9.2 we obtain the desired estimate:

$$\|F(y_2) - F(0)\| = h(1) \leq M\|y_2\|. \ \square$$

9.8 Theorem *(Implicit Function Theorem):* Let $\mathfrak{Y}$ and $\mathfrak{Z}$ be Banach spaces and let $\mathfrak{X}$ be a metric space with metric d. Let G be a continuous function defined in a neighborhood of $(x_0,y_0) \in \mathfrak{X} \times \mathfrak{Y}$, with values in $\mathfrak{Z}$, such that

$$G(x_0,y_0) = 0.$$

Suppose that for each fixed x near x_0, the slice function $y \to G(x,y)$ is differentiable, with derivative $G_2(x,y)$ depending continuously on x and y. Suppose, furthermore, that $G_2(x_0,y_0)$ is an invertible operator from $\mathfrak{Y}$ to $\mathfrak{Z}$.

(i) There exist $r > 0$ and $\rho > 0$ such that for each $x \in B(x_0;r)$, there is a unique $f(x) \in B(y_0;\rho)$ satisfying

$$G(x,f(x)) = 0.$$

(ii) The function f depends continuously on x.

(iii) Suppose in addition that $\mathfrak{X}$ is a Banach space and that the slice function $x \to G(x,y_0)$ is differentiable at x_0, with derivative $G_1(x_0,y_0)$. Then f is differentiable at x_0 and

$$f'(x_0) = -G_2(x_0,y_0)^{-1}G_1(x_0,y_0).$$

(iv) If, additionally, the partial derivative $G_1(x,y)$ exists and is continuous near (x_0,y_0), then f is continuously differentiable near x_0.

Proof: For convenience, set $T = G_2(x_0,y_0)$. Choose $r > 0$ and $\rho > 0$ so that $G(x,y)$ is defined when $\|y - y_0\| \leq \rho$ and $d(x,x_0) \leq r$ and such that the following two estimates are valid for such x and y:

$$\|G_2(x,y) - T\| < 1/(2\|T^{-1}\|), \tag{9.8}$$

$$\|G(x,y_0)\| < \rho/(2\|T^{-1}\|). \tag{9.9}$$

For each fixed x satisfying $d(x,x_0) \leq r$, define a map from the ball $\{\|y - y_0\| \leq \rho\}$ to $\mathfrak{Y}$ by

$$\Phi(x,y) = y - T^{-1}G(x,y).$$

Then $G(x,y) = 0$ if and only if y is a fixed point of Φ. We aim to show that for each fixed x, Φ is a contraction of the ball $\{\|y - y_0\| \leq \rho\}$. Assertion (i) of the theorem then follows from the Contraction Mapping Theorem (Theorem 8.1), and assertion (ii) follows from Theorem 8.5.

From (9.8) and Lemma 9.7, we obtain

$$\|G(x,y_2) - G(x,y_1) - T(y_2 - y_1)\| \leq \|y_2 - y_1\|/(2\|T^{-1}\|).$$

Consequently

$$\Phi(x,y_2) - \Phi(x,y_1) = -T^{-1}[G(x,y_2) - G(x,y_1) - T(y_2 - y_1)]$$

is estimated by

$$\|\Phi(x,y_2) - \Phi(x,y_1)\| \leq \|y_2 - y_1\|/2.$$

Hence Φ is a contraction with contraction constant $c = 1/2$.

It remains to show that $\Phi(x,\cdot)$ maps the ball $\{\|y - y_0\| \leq \rho\}$ into itself. This follows from the estimate

$$\begin{aligned}\|\Phi(x,y) - y_0\| &\leq \|\Phi(x,y) - \Phi(x,y_0)\| + \|\Phi(x,y_0) - y_0\| \\ &\leq \|y - y_0\|/2 + \|T^{-1}G(x,y_0)\| \\ &\leq \rho/2 + \rho/2 = \rho,\end{aligned}$$

where we have used (9.9) to estimate the second summand on the right.

The proof of (i) and (ii) is now complete. For future reference, we observe that

$$\|f(x) - y_0\| \le 2\|T^{-1}\| \, \|G(x,y_0)\| \tag{9.10}$$

whenever $d(x,x_0) \le r$. This follows immediately from (8.12), with $c = 1/2$.

Next we turn to part (iii), so that $\mathfrak{X}$ is now a Banach space. Let $\varepsilon > 0$. We restrict ourselves to (x,y) sufficiently near (x_0,y_0) so that $\|G_2(x,y) - G_2(x_0,y_0)\| < \varepsilon$. From (9.8) and Lemma 9.7, we then have

$$\|G(x,y) - G(x,y_0) - T(y - y_0)\| \le \varepsilon\|y - y_0\|.$$

If x is near x_0, we may substitute $y = f(x)$ in this estimate to obtain

$$\|G(x,y_0) + T(f(x) - y_0)\| \le \varepsilon\|f(x) - y_0\|.$$

With the aid of (9.10), we obtain

$$\begin{aligned}\|f(x) - y_0 + T^{-1}G(x,y_0)\| &\le \|T^{-1}\| \, \|G(x,y_0) + T(f(x) - y_0)\| \\ &\le 2\varepsilon\|T^{-1}\|^2\|G(x,y_0)\|.\end{aligned}$$

Substituting

$$G(x,y_0) = G_1(x_0,y_0)(x - x_0) + R(x),$$

where $\|R(x)\|/\|x - x_0\|$ tends to 0 as x tends to x_0, we find that

$$\|f(x) - y_0 + T^{-1}G_1(x_0,y_0)(x - x_0)\|$$

is bounded by

$$2\varepsilon\|T^{-1}\|^2\|G_1(x_0,y_0)\| \, \|x - x_0\| + 2\varepsilon\|T^{-1}\|^2\|R(x)\| + \|T^{-1}\| \, \|R(x)\|.$$

Hence

$$\limsup_{x\to x_0} \frac{\|f(x) - y_0 + T^{-1}G_1(x_0,y_0)(x - x_0)\|}{\|x - x_0\|} \le 2\varepsilon\|T^{-1}\|^2\|G_1(x_0,y_0)\|.$$

Since $\varepsilon > 0$ is arbitrary, the limes superior must be zero. Hence f is differentiable at x_0 and its derivative is $-T^{-1}G_1(x_0,y_0)$. This establishes (iii), and (iv) follows immediately from (iii). □

We mention in passing one application of the Implicit Function Theorem, to the initial-value problem treated in Section 8:

$$\begin{cases} \dfrac{du}{dt} = F(u,t), & a \le t \le b, \\[2ex] u(a) = \xi. \end{cases} \tag{9.11}$$

Here the initial data ξ is a fixed vector in $\mathbb{R}^n$, F is smooth, and u belongs to the space, call it C, of continuous functions from $[a,b]$ to $\mathbb{R}^n$. Consider the function Φ from $C \times \mathbb{R}^n$ to C defined by

$$\Phi(\xi,u)(t) = u(t) - \xi - \int_a^t F(u(s),s)ds, \qquad a \le t \le b.$$

Evidently u satisfies (9.11) if and only if $\Phi(\xi,u) = 0$. It turns out that Φ is differentiable with respect to the u variable, and in fact the hypotheses of the Implicit Function Theorem are met. The Implicit Function Theorem then shows that it is possible to solve for the solution of (9.11) in terms of the initial data ξ. The main ideas are given later, in Exercises 6 and 7.

In fact, one can also regard the function F appearing in (9.11) as a parameter, varying over an appropriate Banach space of smooth functions, and one can apply the Implicit Function Theorem to solve for the solution of (9.11) as a function of both ξ and F. (This is often a mind-boggling possibility for a student just becoming accustomed to thinking of functions as points in metric spaces.)

An important companion of the Implicit Function Theorem is the Inverse Function Theorem.

9.9 Theorem *(Inverse Function Theorem):* Let $\mathfrak{X}$ and $\mathfrak{Y}$ be Banach spaces and let $x_0 \in \mathfrak{X}$. Let F be a continuously differentiable function from some neighborhood of x_0 to $\mathfrak{Y}$ such that $F'(x_0)$ is invertible. Then there exists an open neighborhood U of x_0 such that $F(U)$ is open in $\mathfrak{Y}$ and F maps U one-to-one onto $F(U)$. Furthermore, the inverse function of $F|_U$, mapping $F(U)$ onto U, is continuously differentiable.

Proof: Set $y_0 = F(x_0)$. Let V be a small open neighborhood of x_0 such that F is defined and continuously differentiable on V. Since the set of invertible operators is open, by Theorem 9.5, we may assume that $F'(x)$ is invertible for each $x \in V$. Define a function G from $V \times \mathfrak{Y}$ to $\mathfrak{Y}$ by

$$G(x,y) = F(x) - y, \qquad x \in V,\ y \in \mathfrak{Y}.$$

The partial derivatives of G exist and are given by

$$G_1(x,y) = F'(x),$$

$$G_2(x,y) = -I,$$

where I is the identity operator on $\mathfrak{Y}$. Now we apply the Implicit Function Theorem (Theorem 9.8) with the variables x and y interchanged. By Theorem 9.8(i) there exist $r > 0$ and $\rho > 0$ such that for each $y \in B(y_0;r)$, there exists a unique point $x = f(y) \in B(x_0;\rho)$ satisfying $G(x,y) = 0$, that is, $F(x) = y$. Thus if

$$U = F^{-1}(B(y_0;r)) \cap B(x_0;\rho),$$

then F is one-to-one on U and F maps U onto the open subset $B(y_0;r)$ of $\mathfrak{Y}$. Moreover, f is the inverse of $F|_U$, so that the continuous differentiability of the inverse of $F|_U$ follows from Theorem 9.8(iv). □

EXERCISES

1. Define $G : \mathbb{R}^2 \to \mathbb{R}^1$ by $G(0,0) = 0$ and $G(x,y) = xy/(x^2 + y^2)^{1/2}$ if $(x,y) \neq (0,0)$. Show that G is continuous and the partial derivatives of G

with respect to x and y exist everywhere, but G is not Frechet differentiable at $(0,0)$.

2. Prove that if $G = (G_1, \ldots, G_m)$ maps $\mathbb{R}^n$ to $\mathbb{R}^m$ and if the partial derivatives $\partial G_j/\partial x_k$, $1 \le j \le m$, $1 \le k \le n$, exist everywhere and are continuous, then G is continuously differentiable in the sense of Frechet.
3. Prove that if $F : \mathfrak{X} \to \mathfrak{Y}$ is differentiable at $x_0 \in \mathfrak{X}$ and if $G : \mathfrak{Y} \to \mathfrak{Z}$ is differentiable at $y_0 = F(x_0) \in \mathfrak{Y}$, then $G \circ F$ is differentiable at x_0 and

$$(G \circ F)'(x_0) = G'(F(x_0))F'(x_0).$$

4. Prove or disprove that $\|A^{-1}\| = 1/\|A\|$.
5. Prove that the Frechet derivative of the operator function $A \to A^{-1}$, defined on the invertible operators from $\mathfrak{X}$ to $\mathfrak{Y}$, is continuous.
6. Fix an interval $[a,b]$ and let F be a continuously differentiable function on $\mathbb{R} \times [a,b]$. Define a nonlinear integral operator Ψ on the Banach space $C[a,b]$ of continuous real-valued functions on $[a,b]$ by

$$(\Psi u)(t) = \int_a^t F(u(s),s)ds, \qquad a \le t \le b.$$

Prove that Ψ is differentiable in the sense of Frechet and that the derivative $\Psi'(u_0)$ of Ψ at $u_0 \in C[a,b]$ is the Volterra integral operator T defined by

$$(Tv)(t) = \int_a^t F_1(u_0(s),s)v(s)ds, \qquad a \le t \le b,$$

where F_1 is the partial derivative of F with respect to the first variable.
7. Let F be a continuously differentiable real-valued function on $\mathbb{R} \times [a,b]$ and consider the initial-value problem

$$(9.12) \qquad \begin{cases} \dfrac{du}{dt} = F(u,t), & a \le t \le b, \\ u(a) = \xi, \end{cases}$$

treated in Section 8. The existence of a unique solution $u_\xi \in C[a,b]$ is guaranteed by the Cauchy-Picard Theorem (Theorem 8.4). Define a function Φ from $\mathbb{R} \times C[a,b]$ to $C[a,b]$ by

$$\Phi(\xi,u)(t) = u(t) - \xi - \int_a^t F(u(s),s)ds, \qquad a \le t \le b.$$

(a) Identify the partial derivatives Φ_1 and Φ_2 of Φ.
(b) Show that Φ satisfies the hypotheses of the Implicit Function Theorem. *Hint:* Use Exercise 6 and Exercise 8.6.
(c) Show that the solution u_ξ of (9.12) depends continuously on the initial value ξ.
8. Let U be an open subset of a Banach space $\mathfrak{X}$ and let F be a continuously differentiable function from U to a Banach space $\mathfrak{Y}$ such that $F'(x)$ is

invertible for all $x \in U$. Show that F is an open map, that is, that the image under F of any open subset of U is an open subset of $\mathcal{Y}$.

9. A function F from $\mathbb{R}^l$ to $\mathbb{R}^m$ is a *C^k-function* if each of the component functions of F has continuous partial derivatives of all orders $\leq k$. Suppose that F is a C^k-function ($k \geq 1$) from $\mathbb{R}^n \times \mathbb{R}^m$ to $\mathbb{R}^m$ such that $F(0,0) = 0$ and the Jacobian matrix

$$\left(\frac{\partial F_i}{\partial y_j}(0,0)\right)_{i,j=1}^{m}$$

is invertible. Prove that there is a C^k-function f, defined in a neighborhood U of 0 in $\mathbb{R}^n$, such that $F(x,f(x)) = 0$, $x \in U$.

10. Let $k \geq 2$ and let f be a real-valued C^k-function, defined on the interval $(-\delta,\delta)$ in $\mathbb{R}$, such that $f(0) = 0 = f'(0)$. Let $M = \{(t,f(t)) : |t| < \delta\}$ be the graph of f in $\mathbb{R}^2$. Let $h(x,y)$ denote the distance from a point $(x,y) \in \mathbb{R}^2$ to M, defined by

$$h(x,y)^2 = \inf\{(t - x)^2 + (f(t) - y)^2 : |t| < \delta\}.$$

(a) Prove that there is a neighborhood U of 0 such that each point (x,y) of U has a unique nearest point $(t(x,y),f(t(x,y)))$ in M.
(b) Prove that $t(x,y)$ is a C^{k-1}-function on U.
(c) Prove that h^2 is a C^k-function on U and h is a C^k-function in $U \backslash M$.
(d) Show that if $f(t) = |t|^{3/2}$, then points $(0,y)$, $y > 0$, do not have a unique nearest point in M and, moreover, h is not a C^1-function near the semiaxis $\{(0,y) : y > 0\}$.
(e) Extend parts (a), (b), and (c) to functions f from $\mathbb{R}^n$ to $\mathbb{R}^m$.

Hint: The distance-squared function $|(x,y) - (t,f(t))|^2$ is minimized where its t-derivative is zero. Thus $t(x,y)$ is obtained by solving implicitly the equation

$$F(x,y,t) = t - x + [f(t) - y]f'(t) = 0.$$

The Implicit Function Theorem yields a C^{k-1} solution of this equation (Exercise 9). To show that h^2 is a C^k-function requires an additional computation:

$$\frac{\partial}{\partial x} h^2 = 2(x - t(x,y)),$$

$$\frac{\partial}{\partial y} h^2 = 2(y - f(t(x,y))).$$

Topological Spaces

TWO

In Chapter I, an axiomatic framework for the purpose of generalizing the idea of distance in Euclidean space was developed. In this chapter, an even greater degree of generality will be obtained by introducing an axiomatic framework in which a notion of "nearness" is introduced without reference to a distance function or metric. The establishment of this framework is based on the observation that many of the concepts in Chapter I could be expressed solely in terms of open sets. Thus if the basic properties of open sets are taken as axioms, much of the material in Chapter I can still be retained. This process yields the concept of a "topological space," the fundamental concept for this chapter and for topology as a whole.

The consideration of topological spaces, rather than just metric spaces, it not an idle exercise in abstraction, though it is the case that metric spaces are the most common examples of topological spaces. As often happens in mathematics, however, the passage to a higher level of abstraction clarifies and simplifies the consideration of the more concrete situations. The reader has already seen how the concept of a metric space unifies many arguments in concrete analysis. The yet more abstract idea of a topological space offers additional unification and simplification. It often happens, for instance, that the clearest and shortest proof of a fact about a metric space is obtained by considering the metric space as a topological space. Another justification for consideration of topological spaces is the fact that many naturally occurring objects in mathematics have some natural topological space structure which does not arise from a metric structure. For instance, if one considers pointwise convergence of sequences of functions instead of uniform convergence, one finds that this idea of convergence is not expressible in terms of a metric on the space of functions. Thus one is quickly led beyond the notion of metric-space convergence to the consideration of more general notions of convergence and eventually to the idea of topological spaces in which the notions of nearness and convergence are expressed in terms of open sets. The great generality of the concept of topological space gives the subject of topology very wide applicability.

Precisely because the idea of a topological space is so general, the initial discussion in this chapter has a very formal character. Many definitions are introduced, and most of the results proved have an almost purely linguistic nature in the sense that the results amount to establishing formal relationships among the definitions. Readers can reduce the feeling of perhaps excessive formality by illustrating for themselves the meaning of the results in the concrete metric-space cases discussed in Chapter I. As noted, the generality of the discussion is justified by the consequent wealth of applications of the ideas. Later in this chapter, restrictions will be imposed on the topological spaces considered, restrictions which make it possible to establish results that are more concrete and more specific. What restrictions of this sort it is most fruitful to consider has become apparent only by the experience of mathematicians over some time; by now, certain specific sorts of topological spaces have proved most significant. However, the great generality offered by the unrestricted concept of a topological space remains very important.

1. TOPOLOGICAL SPACES

Let X be a set. A family $\mathcal{T}$ of subsets of X is a *topology* for X if $\mathcal{T}$ has the following three properties:

(1.1) Both X and the empty set belong to $\mathcal{T}$.
(1.2) Any union of sets in $\mathcal{T}$ belongs to $\mathcal{T}$.
(1.3) Any finite intersection of sets in $\mathcal{T}$ belongs to $\mathcal{T}$.

A *topological space* is a pair $(X,\mathcal{T})$, where X is a set and $\mathcal{T}$ is a *topology* for X. The sets in $\mathcal{T}$ are called *open sets*. In other words, a topology for X is a specification of certain subsets of X as open sets with the properties (1.1), (1.2), and (1.3) required. When no confusion can arise, we suppress the mention of $\mathcal{T}$ and refer to X as a topological space.

A set X usually has many topologies. Among them are two trivial ones, one of which is useful and the other of which is not. The useful topology is the *discrete topology*, consisting of all subsets of X. The useless topology is the *indiscrete topology*, which consists of only the empty set and the space X.

If X is a metric space, then the open subsets of X form a topology for X, by Theorems I.1.3 and I.1.5. This topology is called the *metric topology* for X. Of course, different metrics on X may determine different topologies. Furthermore, some topologies cannot be determined by any metric. For instance, the indiscrete topology for X cannot arise from a metric when X has more than one point. One way to see this is to note that the complement of a one-point set in a metric space is always open.

A topological space X is *metrizable* if the topology for X is the metric topology associated with some metric on X. For example, the discrete topology is always metrizable, and one metric from which it arises is given by setting $d(x,y) = 1$ if $x \neq y$. The indiscrete topology on a set of more than one element is not metrizable.

Now let $(X,\mathcal{T})$ be a fixed topological space. As noted, a subset U of X is by definition *open* if $U \in \mathcal{T}$. Recall from Theorem I.1.8 that a subset of a metric space is closed if and only if its complement is open. For topological spaces, we take this characterization as a definition: a subset S of X is defined to be *closed* if $X \backslash S$ is open. From (1.1), (1.2), (1.3), and deMorgan's Laws, one obtains the following properties of closed sets, as in Theorem I.1.9:

(1.1)′ Both $\varnothing$ and X are closed sets.

(1.2)′ Any intersection of closed sets is closed.
(1.3)′ A finite union of closed sets is closed.

A subset S of X is a *neighborhood* of a point x if there is an open set U such that $x \in U$ and $U \subset S$. A point $x \in X$ is an *interior point* of S if S is a neighborhood of x. The set of interior points of S is called the *interior* of S and is denoted by int(S). Evidently

$$\text{int}(S) \subset S.$$

1.1 Theorem: A subset S of a topological space X is open if and only if $S = \text{int}(S)$, that is, if and only if S is a neighborhood of each of its points.

Proof: If S is open, then S is an open neighborhood of each of its points, so that each point of S is an interior point of S and $S = \text{int}(S)$.

Conversely, suppose that $S = \text{int}(S)$. For each $x \in S$, there is then an open neighborhood U_x of x such that $U_x \subset S$. Then $\cup\{U_x : x \in S\}$ evidently coincides with S. Since the union of open sets is open, S is open. □

1.2 Theorem: If S is a subset of a topological space X, then int(S) is an open subset of X. In other words, $\text{int}(\text{int}(S)) = \text{int}(S)$.

Proof: Suppose $x \in \text{int}(S)$. Then there is an open neighborhood U of x such that $U \subset S$. Since U is an open neighborhood of each of its points, $U \subset \text{int}(S)$. Consequently x is an interior point of int(S), so that int(S) is open. □

A point $x \in X$ is *adherent* to a subset S of X if S meets every neighborhood of x. Any point of S is then adherent to S. The *closure of* S, denoted by $\overline{S}$, is the set of points in X which are adherent to S. Evidently $S \subseteq \overline{S}$.

1.3 Theorem: A subset S of a topological space X is closed if and only if $S = \overline{S}$.

Proof: Suppose that S is closed. Then $X \backslash S$ is open and $X \backslash S$ does not meet S, so that no point of $X \backslash S$ is adherent to S. Consequently $\overline{S} \subseteq S$ and $\overline{S} = S$.

Conversely, suppose that $\overline{S} = S$. Let $x \in X \backslash S$. Since $x \notin \overline{S}$, there is an open neighborhood U_x of x such that U_x is disjoint from S. Then $\cup\{U_x : x \in X \backslash S\}$ coincides with $X \backslash S$. Since the union of open sets is open, $X \backslash S$ is open and S is closed. □

1.4 Theorem: If S is a subset of a topological space X, then $\overline{S}$ is closed, that is, $\overline{\overline{S}} = \overline{S}$.

Proof: Suppose $x \in X \backslash \overline{S}$. Then there is an open neighborhood U_x of x that does not meet S. Since U_x is a neighborhood of each of its points, no point of U_x can be adherent to S, so that $U_x \subset X\backslash\bar{S}$. Then $\cup\{U_x : x \in X\backslash\bar{S}\}$ evidently coincides with $X\backslash\bar{S}$. Hence $X\backslash\bar{S}$ is open and $\bar{S}$ is closed. □

Theorems 1.1 through 1.4 show in effect that the formal properties of interior and closure are the same in the case of topological spaces as in the case of metric spaces. However, it is important to realize that many properties of metric spaces do not hold in general topological spaces even though these properties may be expressible in terms of open sets only (Exercise 6).

The following definition extends the idea of convergence of a sequence to the context of topological spaces: a sequence of points $\{x_i\}$ in a topological space X *converges to* $x \in X$ if, for every open neighborhood U of x, there is an integer N such that $x_i \in U$ for all $i > N$.

1.5 Theorem: If S is a subset of a topological space X and if a sequence $\{x_i\}_{i=1}^{\infty}$ in S converges to $x \in X$, then $x \in \bar{S}$.

Proof: By Theorem 1.4, $X\backslash\bar{S}$ is open. Since the sequence $\{x_i\}_{i=1}^{\infty}$ does not enter $X\backslash\bar{S}$, the sequence cannot converge to any point of $X\backslash\bar{S}$. □

Note that in general the converse of Theorem 1.5 is not true. There may be some point x in $\bar{S}$ such that no sequence $\{x_i\}$ in S converges to x. An example of this behavior is presented in Exercise 12. The converse does hold, though, in metrizable spaces (Theorem I.1.11). A more general condition under which the converse holds is given in Exercise 4.4 of this chapter.

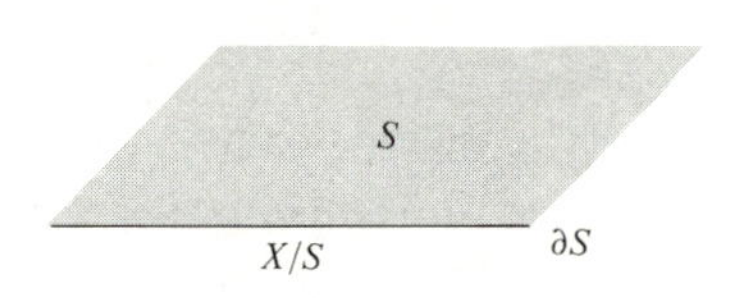

A point $x \in X$ is a *boundary point* of a subset S of X if x is adherent both to S and to $X\backslash S$. The *boundary of* S, denoted by ∂S, is the set of boundary points of S. Then

$$\partial S = \bar{S} \cap (\overline{X\backslash S}),$$

so that ∂S is closed. Furthermore,

$$\partial S = \partial(X\backslash S).$$

1.6 Theorem: $\bar{S}$ is the disjoint union of $\mathrm{int}(S)$ and ∂S.

Proof: Let $x \in \bar{S}$. Then every neighborhood U of x meets S. There are two mutually disjoint cases which can occur. Either there is a neighborhood U of x such that $U \subset S$, or else each neighborhood U of x meets $X\backslash S$. The former case occurs if and only if $x \in \mathrm{int}(S)$, whereas the latter case occurs if and only if $x \in \partial S$. □

EXERCISES

1. Show that the intersection of a family of topologies for X is a topology for X.
2. Let X be a set and let $\mathcal{T}$ be the family of subsets U of X such that $X\backslash U$ is finite, together with the empty set $\varnothing$. Show that $\mathcal{T}$ is a toplogy. ($\mathcal{T}$ is called the *cofinite topology* of X.)
3. Let $\mathcal{T}$ be the cofinite topology on the set $\mathbb{Z}$ of integers. Show that the sequence $\{1,2,3,\ldots\}$ converges in $(\mathbb{Z},\mathcal{T})$ to each point of $\mathbb{Z}$. Describe the convergent sequences in $(\mathbb{Z},\mathcal{T})$.
4. Let S be a subset of a set X. Describe the closure of S when (i) X has the discrete topology, (ii) X has the indiscrete topology, and (iii) X has the cofinite topology.
5. Let S be a subset of a topological space X in which sets consisting of one point are closed. A point $x \in X$ is a *limit point* of S if every neighborhood of x contains a point of S other than x itself. a point $s \in S$ is an *isolated point* of S if there is a neighborhood U of s such that $U \cap S = \{s\}$. Show that the set of limit points of S is closed. Show that $\bar{S}$ is the disjoint union of the set of limit points of S and the isolated points of S.
6. (a) Show that if X is a metrizable topological space and if p and q are distinct points of X, then there are open sets U and V such that $p \in U$, $q \in V$, and $U \cap V = \varnothing$.
   (b) Let X be an infinite set and let $\mathcal{T}$ be the cofinite topology on X (Exercise 2). Prove that the property described in part (a) does not hold for the open sets in X and hence that X with the cofinite topology is not metrizable.
7. Show that if S is a subset of a topological space X, then $\text{int}(S)$ is the union of all open sets contained in S.
8. Show that if S is a subset of a topological space X, then $\overline{S}$ is the intersection of all closed sets containing S.
9. Show that if S is a subset of a topological space X, then
   (a) $\overline{X\backslash S} = X\backslash\text{int}(S)$,
   (b) $\text{int}(X\backslash S) = X\backslash\overline{S}$.
10. If U is open, then prove that $U^{-\prime-\prime-} = U^{-}$, where the bar means *closure* and the prime means *complement*.
11. Prove the following result, originally noted by Kuratowski: If S is a subset of a topological space X, then there are at most 14 subsets of X that can be obtained from S by successively taking either complements or closures. Find a subset S of $\mathbb{R}$ such that exactly fourteen subsets of $\mathbb{R}$ can be obtained from S in this manner.

    Hint: For the proof, first observe that $S'' = S$ and $S^{--} = S^{-}$, so that we need consider only sets of the form $S'^{-\prime\cdots}$ and $S^{-\prime-\cdots}$. Now apply Exercise 10. For the example, use a subset of $\mathbb{R}$ with different pieces separated from each other: an open interval, a half-open interval, a sequence converging to a point, etc.

12. Let X be a set and let $\mathcal{T}$ be the family of subsets U of X such that $X \backslash U$ is at most countable, together with the empty set $\varnothing$.
    (a) Prove that $\mathcal{T}$ is a topology for X.
    (b) Describe the convergent sequences in X with respect to this topology.
    (c) Prove that if X is uncountable, then there is a subset S of X whose closure contains points that are not limits of convergent sequences in S. *Hint:* Take S to be any proper uncountable subset of X.

2. SUBSPACES

There is a natural way to make a subset of a metric space into a metric space: merely restrict the metric to the subset. The resulting metric space is called a *subspace,* and it was proved in Theorem I.1.6 that the open subsets of the subspace are simply the intersections of the subspace with the open subsets of the original space. This latter characterization of the open subsets of a subspace motivates us to define a subspace of a topological space as follows.

Let $(X,\mathcal{T})$ be a topological space and let S be a subset of X. Then the family

$$\mathcal{S} = \{U \cap S : U \in \mathcal{T}\}$$

of subsets of S is a topology for S called the *relative topology* inherited from $(X,\mathcal{T})$. The sets $V \in \mathcal{S}$ are *relatively open* subsets of S, and the sets $S \backslash V$, $V \in \mathcal{S}$, are *relatively closed* subsets of S. The topological space $(S,\mathcal{S})$ is a *subspace* of $(X,\mathcal{T})$. Again for convenience we refer to S as a subspace of X, it being understood that the topology for S is the relative topology inherited from S.

If X is a metric space and if Y is a subspace of X (in the metric sense), then the metric topology for Y coincides with the relative topology for Y inherited from the metric topology for X. This follows from Theorem I.1.6.

It is important to keep in mind that a relatively open subset of a subspace S of a topological space X need not be open in X. Indeed, S itself is always relatively open in S, but S need not be open in X. Similarly, S is relatively closed in S, but S need not be closed in X.

By definition, the open subsets of the subspace S of X are the intersections of S with the open subsets of X. An analogous statement holds for closed subsets.

2.1 Theorem: Let S be a subspace of a topological space X. A subset E of S is relatively closed in S if and only if E is the intersection of S and a closed subset of X.

Proof: If E is relatively closed in S, then $S \backslash E$ is relatively open in S, so that there is an open subset U of X such that $S \backslash E = U \cap S$. Then E is the intersection of S and the closed subset $X \backslash U$ of X.

Conversely, if $E = T \cap S$, where T is a closed subset of X, then $S \backslash E$ is the intersection of S and the open subset $X \backslash T$ of X, so that $S \backslash E$ is relatively open in S and E is relatively closed in S. □

2.2 Theorem: Let S be a subspace of a topological space X and let E be a subset of S. Then the relative closure of E in S is $\overline{E} \cap S$, where $\overline{E}$ is the closure of E in X.

Proof: Let $x \in S$. Then x lies in the relative closure of E if and only if every relatively open neighborhood of x in S meets E, that is, if and only if the intersection of S with every open neighborhood of x in X meets E. Since $E \subset S$, this occurs if and only if every open neighborhood of x in X meets E, and this occurs if and only if $x \in \overline{E} \cap S$. □

The relative topology is a trickier idea than it seems to be at first. Statements like Theorem 2.2 have such a plausible sound that one is inclined to accept them automatically, but similar plausible statements can turn out to be false. The exercises will help to clarify this point. In particular, Exercise 2 shows what happens if one tries to extend Theorem 2.2 to the situation where E is not a subset of S.

EXERCISES

1. Let X be a topological space, let S be a subspace of X, and let E be a subset of S. Show that the relative topology that E inherits from S coincides with the relative topology that E inherits from X.
2. Prove that if A and S are subsets of a topological space X, then the closure of $A \cap S$ in S in the relative topology for S is a subset of the intersection $\overline{A} \cap S$, where $\overline{A}$ is the closure of A in X. Give an example where the relative closure of $A \cap S$ is a proper subset of $\overline{A} \cap S$.
3. Let S be a subset of a topological space X. Show that a sequence $\{x_i\}$ in S converges to $x_0 \in S$ in the relative topology if and only if, considered as a sequence in X, the sequence $\{x_i\}$ converges to x_0.
4. Prove that a set S in X is open if and only if every relatively open subset of S is open in X. Is this statement true if "open" is replaced by "closed"?
5. Let X be a topological space and let S be a subset of X. Show that if A is a relatively open subset of S, then $A \cap T$ is a relatively open subset of $S \cap T$ for any subset T of X.
6. Let X be a topological space, let S and T be subsets of X, and let A be a subset of $S \cap T$ that is relatively open in S and in T. Is A relatively open in $S \cup T$? Justify your answer.

3. CONTINUOUS FUNCTIONS

Continuity of functions is one of the most important ideas in analysis, perhaps *the* most important idea. The fact that this idea can be expressed in a meaningful fashion in the metric-space setting is one of the main reasons that the concept of metric space is so useful in analysis. It will be seen in this section that the concept of continuity can be carried over successfully to functions from one arbitrary topological space to another. The possibility of so extending the concept of continuity already begins to show how appropriate the idea of a topological space is as a generalization of metric spaces.

The key to finding the appropriate idea of continuity of a function from one topological space to another is to recall that the metric-space idea of continuity could be expressed entirely in terms of open sets. According to Theorem I.6.2, a function f from a metric space X to a metric space Y is continuous if and only if $f^{-1}(V)$ is open in X whenever V is an open subset of Y. This characterization of continuous functions between metric spaces becomes the definition of continuous functions between topological spaces. Precisely, a function f from one topological space X to another Y is *continuous* if $f^{-1}(V)$ is open in X whenever V is an open subset of Y.

A continuous function from one topological space to another is often called a *map,* especially in the case where the topological spaces are regarded as geometric objects, such as subsets of $\mathbb{R}^n$. This terminology occurs often in Chapters III and IV, for instance.

Sometimes it is useful to have a notion of continuity, at a point of X, for a function f from X to Y. Intuitively, f is continuous at x if $f(w)$ is "near" $f(x)$ whenever w is "near" x. Recast in terms of open neighborhoods of x and $f(x)$, this definition becomes the following. A function $f : X \to Y$ is *continuous at a point* $x \in X$ if for every open set V in Y such that $f(x) \in V$, there exists an open set U in X such that $x \in U$ and $f(U) \subseteq V$. From Theorem I.6.1, it is a straightforward verification that this coincides with the corresponding concept for functions between metric spaces; this verification is left as Exercise 1.

The following theorem gives the relationship between continuity of a function $f : X \to Y$ and continuity of f at each point of X.

3.1 Theorem: Let X and Y be topological spaces. A function $f : X \to Y$ is continuous if and only if f is continuous at every point of X.

Proof: Suppose that f is continuous at every point of X and that V is open in Y. Let $x \in f^{-1}(V)$. Since f is continuous at x, there is an open set U_x containing x such that $U_x \subset f^{-1}(V)$, or $f(U_x) \subset V$. Set $U = \cup \{U_x : x \in f^{-1}(V)\}$. Then U is open and $f(U) \subset V$, so that $U \subset f^{-1}(V)$. Since U includes each point of $f^{-1}(V)$, U coincides with $f^{-1}(V)$ and $f^{-1}(V)$ is open. Thus f is continuous.

Conversely, suppose that f is continuous. Let $x \in X$ and let V be an open set containing $f(x)$. Then $U = f^{-1}(V)$ is an open set containing x and $f(U) \subset V$. Consequently f is continuous at x for all $x \in X$. □

The following important statement is an almost immediate consequence of the definitions.

3.2 Theorem: Let X_0, X_1, and X_2 be topological spaces and let $f : X_0 \to X_1$ and $g : X_1 \to X_2$ be continuous functions. Then $g \circ f : X_0 \to X_2$ is continuous.

Proof: Suppose that V is open in X_2. Since g is continuous, $g^{-1}(V)$ is open in X_1.

Since f is continuous, $f^{-1}(g^{-1}(V)) = (g \circ f)^{-1}(V)$ is open in X_0. It follows that $g \circ f$ is continuous. □

Note how much easier this proof was than the standard $\varepsilon - \delta$ proof of the continuity of the composition of continuous functions on metric spaces.

A *homeomorphism* from a topological space X to a topological space Y is a function $f : X \to Y$ that is one-to-one and onto, such that U is open in X if and only if $f(U)$ is open in Y. This latter condition can be restated as asserting that both f and f^{-1} are continuous.

The spaces X and Y are *homeomorphic* if there exists a homeomorphism of X and Y. Evidently the inverse of a homeomorphism is a homeomorphism, the composition of two homeomorphisms is a homeomorphism, and the identity map of a topological space to itself is a homeomorphism. Consequently the property of being homeomorphic is an equivalence relation on the family of topological spaces.

A property of a topological space is a *topological property* if it is preserved under homeomorphisms. An example of a topological property is discreteness: if X and Y are homeomorphic, then X is discrete if and only if Y is discrete. Another topological property is metrizability, though properties involving a specific metric are generally not topological. Roughly speaking, any property that can be expressed in terms of the open or closed subsets of a space is a topological property.

Homeomorphic spaces are identical from the viewpoint of topology. It is important in this context to realize that spaces which appear quite different—geometrically, for instance—may still be homeomorphic. For example, any open ball in $\mathbb{R}^n$ is homeomorphic to $\mathbb{R}^n$ itself. A homeomorphism of $B(x_0;r)$ and $\mathbb{R}^n$ is given by

$$g(x) = f(|x - x_0|)(x - x_0), \qquad x \in \mathbb{R}^n,$$

where f is the increasing homeomorphism of $[0,r)$ and $[0,\infty)$ given by

$$f(t) = t/(r - t), \qquad 0 \leq t < r.$$

Furthermore, any two open subintervals of $\mathbb{R}$, not necessarily bounded, are homeomorphic (Exercises 4 and 5). These examples indicate how a homeomorphism, which by definition preserves topological properties, might nonetheless drastically alter size and shape.

It is a nontrivial theorem (usually called the Invariance of Domain Theorem, for historical reasons which are rapidly becoming somewhat obscure) that a nonempty open subset of $\mathbb{R}^n$ cannot be homeomorphic to an open subset of $\mathbb{R}^m$ for $n \neq m$ (Exercise IV.4.14).

We conclude this section by discussing functions from a topological space to a metric space.

3.3 Theorem: Let $(X,\mathcal{T})$ be a topological space and let (Y,d) be a metric space. A function $f : X \to Y$ is continuous at $x \in X$ if and only if for each $\varepsilon > 0$, there exists a neighborhood U of x such that $d(f(y),f(x)) < \varepsilon$ whenever $y \in U$.

Proof: Suppose f is continuous at x. Let $\varepsilon > 0$. Then there exists a neighborhood U of x such that $f(U) \subset B(f(x);\varepsilon)$. In other words, $d(f(y),f(x)) < \varepsilon$ whenever $y \in U$.

Conversely, suppose that the criterion above is valid. let V be a neighborhood of $f(x)$. Then there exists $\varepsilon > 0$ such that $B(f(x);\varepsilon) \subset V$. By the criterion, there exists an open neighborhood U of x such that $f(U) \subset B(f(x);\varepsilon)$. In particular, $f(U) \subset V$, so that f is continuous at x. □

3.4 Theorem: Suppose the sequence $\{f_n\}_{n=1}^{\infty}$ of functions from a topological space X to a metric space Y converges uniformly to $f : X \rightarrow Y$. If each f_n is continuous at some point $x \in X$, then f is also continuous at x.

Proof: Let $\varepsilon > 0$. Fix an integer N such that $d(f_N(y),f(y)) < \varepsilon/3$ for all $y \in X$. Since f_N is continuous at x, there is an open neighborhood U of x such that $d(f_N(y),f_N(x)) < \varepsilon/3$ for all $y \in U$. Then

$$d(f(y),f(x)) \leq d(f(y),f_N(y)) + d(f_N(y),f_N(x)) + d(f_N(x),f(x))$$

$$< \varepsilon/3 + \varepsilon/3 + \varepsilon/3 = \varepsilon$$

for all $y \in U$. By Theorem 3.3, f is continuous at x. □

3.5 Corollary: The limit of a uniformly convergent sequence of continuous functions from a topological space X to a metric space Y is continuous.

EXERCISES

1. Let f be a function from a metric space X to a metric space Y and let $x \in X$. Show that the definition of continuity of f at x given in this section (for each open neighborhood V of $f(x)$, there exists an open neighborhood U of x such that $f(U) \subseteq V$) coincides with the definition given in Section I.6 (whenever $\{x_n\}$ is a sequence in X such that $x_n \rightarrow x$, then $f(x_n) \rightarrow f(x)$).
2. Show that a function $f : X \rightarrow Y$ is continuous if and only if $f^{-1}(E)$ is a closed subset of X for every closed subset E of Y.
3. Prove the following statements about continuous functions and discrete and indiscrete topological spaces.
   (a) If X is discrete, then every function f from X to a topological space Y is continuous.
   (b) If X is not discrete, then there is a topological space Y and a function $f : X \rightarrow Y$ that is not continuous.
   Hint: Let Y be the set X with the discrete topology.
   (c) If Y is an indiscrete topological space, then every function f from a topological space X to Y is continuous.
   (d) If Y is not indiscrete, then there is a topological space X and a function $f : X \rightarrow Y$ that is not continuous.
4. Prove that all open intervals in $\mathbb{R}$ (finite, semi-infinite, or infinite) are homeomorphic.

5. Prove that all semiopen intervals in $\mathbb{R}$ (finite or semi-infinite) are homeomorphic.
6. Show that the unit ball $\{(x,y) \in \mathbb{R}^2 : x^2 + y^2 < 1\}$ in $\mathbb{R}^2$ is homeomorphic to the open square $\{(x,y) : 0 < x < 1, 0 < y < 1\}$.
7. Show that the punctured plane $\mathbb{R}^2\backslash\{(0,0)\}$ is homeomorphic to the exterior of the closed unit ball $\mathbb{R}^2\backslash\{(x,y) : x^2 + y^2 \leq 1\}$.
8. Show that $\mathbb{R}^2\backslash\{(x,y) : x \text{ and } y \text{ integers}\}$ is homeomorphic to the space $\mathbb{R}^2\backslash\{(x,y) : \text{there are integers } n \text{ and } m \text{ such that } (x - n)^2 + (y - m)^2 < 1/10\}$.
9. Show that the open right half plane $\{(x,y) \in \mathbb{R}^2 : x > 0\}$ is homeomorphic to the unit ball $\{(x,y) \in \mathbb{R}^2 : x^2 + y^2 < 1\}$.
10. Develop analogues for $\mathbb{R}^n$, $n > 2$, of Exercises 6 through 9.
11. Prove that metrizability is a topological property.
12. Prove that if f and g are continuous functions from a topological space X to $\mathbb{R}$, then $f + g$ and fg are also continuous. Furthermore, if $g \neq 0$ on X, then f/g is continuous.
13. Prove that if $f : X \to Y$ is continuous and if S is a subspace of X, then the restriction $f|_S : S \to Y$ is continuous.
14. Let X be a topological space and let $BC(X)$ denote the set of bounded continuous real-valued functions on X, with the metric

$$d(f,g) = \sup\{|f(x) - g(x)| : x \in X\}.$$

Show that $BC(X)$ is a complete metric space.

4. BASE FOR A TOPOLOGY

In certain situations one often has some idea in advance that certain subsets of a space should be considered open, and then one tries to define a topology by declaring a set to be open if and only if it is a union of the particular sets that one has decided in advance should be open. For instance, in a metric space one certainly wishes the open balls to be open, and so one is led to declare a set to be open if and only if it is a union of open balls. The fact proved in Theorem I.1.5, which boils down to the statement that the intersection of any two open balls is a union of open balls, shows that the open sets thus defined satisfy the axioms for a topology. In general, this procedure of defining the open sets to be the unions of the "obvious" open sets does not yield a topology. It is easy to see, however, exactly when this procedure works. We now formalize this discussion and establish the circumstances under which this process does lead to a topology. In this case, the obvious open sets constitute what is called a *base for the topology*.

Let X be a topological space. A family $\mathcal{B}$ of open subsets of X is a *base for the topology* of X if every open subset of X is a union of sets in $\mathcal{B}$. This definition coincides with the definition given in Section I.5 for metric spaces. The analogue of Lemma I.5.9 is as follows.

4.1 Theorem: A family $\mathcal{B}$ of open subsets of a topological space X is a base for the topology of X if and only if for each $x \in X$ and each neighborhood U of x, there exists $V \in \mathcal{B}$ such that $x \in V$ and $V \subset U$.

Proof: If $\mathcal{B}$ is a base, then any open neighborhood U of x is a union of sets in $\mathcal{B}$, so that in particular there must exist $V \in \mathcal{B}$ satisfying $x \in V \subset U$.

Conversely, suppose that the condition is satisfied. Let U be an open subset of X. For each $x \in U$, there is then a set $V_x \in \mathcal{B}$ such that $x \in V_x$ and $V_x \subset U$. Evidently $U = \cup\{V_x : x \in U\}$, so that $\mathcal{B}$ is a base for the topology. □

As noted above, it is often convenient to define a topology on a set X by specifying a base for the topology. However, not every family of subsets of X can be a base for a topology. One must check that the proposed base satisfies the conditions of the following theorem.

4.2 Theorem: A family $\mathcal{B}$ of subsets of a set X is a base for a topology of X if and only if $\mathcal{B}$ has the following two properties:

(4.1) Each $x \in X$ lies in at least one set in $\mathcal{B}$.

(4.2) If $U,V \in \mathcal{B}$ and $x \in U \cap V$, then there exists $W \in \mathcal{B}$ such that $x \in W$ and $W \subset U \cap V$.

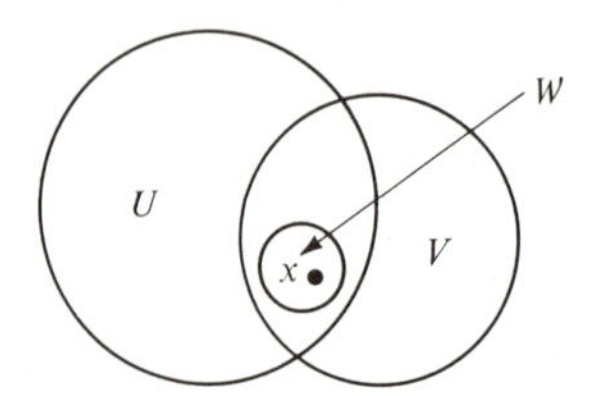

Proof: A base has property (4.1) because X is open, and property (4.2) because $U \cap V$ is open.

Conversely, suppose that the family $\mathcal{B}$ has properties (4.1) and (4.2). Let $\mathcal{T}$ be the family of all subsets of X that are unions of sets in $\mathcal{B}$. The empty set is then understood to be included in $\mathcal{T}$. It suffices to show that $\mathcal{T}$ is a topology, since then $\mathcal{B}$ is evidently a base for $\mathcal{T}$.

Condition (4.1) shows that X is the union of all sets in $\mathcal{B}$, so that $X \in \mathcal{T}$. Any union of sets in $\mathcal{T}$ evidently belongs to $\mathcal{T}$. It suffices then to show that a finite intersection of sets in $\mathcal{T}$ again belongs to $\mathcal{T}$. For this, it suffices to show that the intersection of any two sets in $\mathcal{T}$ again belongs to $\mathcal{T}$.

Suppose that $U,V \in \mathcal{T}$. Let $x \in U \cap V$. Since U and V are unions of sets in $\mathcal{B}$, there exist sets $U_0,V_0 \in \mathcal{B}$ such that $x \in U_0 \subset U$ and $x \in V_0 \subset V$. Then $x \in U_0 \cap V_0$. By (4.2), there exists $W_x \in \mathcal{B}$ such that $x \in W_x$ and $W_x \subset U_0 \cap V_0$. Then $W_x \subset U \cap V$ and evidently

$$U \cap V = \cup \{W_x : x \in U \cap V\},$$

so that $U \cap V \in \mathcal{T}$. □

A topological space X is *second-countable* if there is a countable family (sequence) of open sets that forms a base for the topology for X. This definition is identical to the definition of "second-countable" given for metric spaces in Section I.5. This type of condition, limiting the number (or cardinality) of the family of sets in a base, is a reasonable sort of restriction to impose in that it makes possible the transfer of certain properties of metric spaces to the setting of topological spaces, but still without assuming that the topological spaces are metrizable. We present two results on second-countable spaces now, the proofs of which are essentially contained in Section I.5. The results are important in their own right, and they also illustrate the concept of base in action.

First, we note that the definitions of open cover and subcover given in Section I.5 extend immediately from metric to topological spaces. An *open cover* of a topological space X is a family of open subsets of X whose union is X. A *subcover* of an open cover $\mathcal{C}$ of X is a subset of $\mathcal{C}$ whose union is X. The proof of Lindelöf's Theorem (Theorem I.5.11) also carries over verbatim and establishes the following.

4.3 Theorem *(Lindelöf's Theorem):* Let X be a second-countable topological space. Then every open cover of X has a countable subcover.

Further definitions that carry over to topological spaces are those of density and separability. A subset S of a topological space X is *dense* in X if $\overline{S} = X$. A topological space X is *separable* if there is a countable subset of X that is dense in X. Half of the proof of Theorem I.5.10 carries over verbatim to topological spaces and establishes the following result.

4.4 Theorem: If X is a second-countable topological space, then X is separable.

It turns out, though, that there are separable topological spaces that are not second-countable (Exercise 1). Of course, no such space can be metrizable.

EXERCISES

1. Let X be a topological space with the cofinite topology (Exercise 1.2). Show that X is separable. When is X second-countable?
2. Let X be a topological space with the discrete topology. Find a base $\mathcal{B}$ of open sets for X such that $\mathcal{B}$ is included in any other base of open sets for X.
3. Let X and Y be topological spaces and let $\mathcal{B}$ be a base of open sets for Y. Show that a function $f : X \to Y$ is continuous if and only if $f^{-1}(U)$ is an open subset of X for every $U \in \mathcal{B}$.
4. A topological space X satisfies the *first axiom of countability,* or is *first-countable,* if for each $x \in X$, there exists a sequence of open neighborhoods $\{U_n\}_{n=1}^{\infty}$ of x such that each neighborhood of x includes one of the U_n's.

Prove the following assertions:

(a) Any metric space is first-countable.

(b) Any second-countable space is first-countable.

(c) In a first-countable space, any point adherent to a set S is a limit of a sequence in S.

5. Let X be a set and let $\mathcal{S}$ be a family of subsets of X.

(a) Show that there exists a unique smallest topology $\mathcal{T}$ on X such that $\mathcal{S} \subset \mathcal{T}$. The topology $\mathcal{T}$ is called the *topology generated by* $\mathcal{S}$, and $\mathcal{S}$ is called a *subbase* for the topology $\mathcal{T}$.

(b) Let $\mathcal{B}$ be the family of subsets of X consisting of X, $\varnothing$, and all finite intersections of sets in $\mathcal{S}$. Show that $\mathcal{B}$ is a base of open sets for the topology generated by $\mathcal{S}$.

(c) Let X have the topology generated by $\mathcal{S}$, let Y be a topological space, and let $f : Y \to X$ be a function. Show that f is continuous if and only if $f^{-1}(S)$ is an open subset of Y for every set $S \in \mathcal{S}$.

6. Let $\mathcal{B}$ be the family of subsets of $\mathbb{R}$ of the form $[a,b)$, where

$$-\infty < a < b < \infty.$$

(a) Show that $\mathcal{B}$ is a base of open sets for a topology $\mathcal{T}$ of $\mathbb{R}$. The topology determined by $\mathcal{B}$ is the *half-open interval topology*.

(b) Show that every open subset of $\mathbb{R}$ (in the metric topology) is $\mathcal{T}$-open.

(c) Show that each interval $[a,b)$ is $\mathcal{T}$-closed.

(d) Show that a point $t \in R$ lies in the $\mathcal{T}$-closure of a subset S of $\mathbb{R}$ if and only if there is a sequence $\{t_n\}_{n=1}^{\infty}$ in S such that $t_n \geq t$ and $|t_n - t| \to 0$.

(e) Show that a function f from $(\mathbb{R},\mathcal{T})$ to $\mathbb{R}$ is continuous if and only if f is continuous from the right at each $t \in \mathbb{R}$, that is, if and only if

$$\lim_{\varepsilon>0,\varepsilon\to 0} f(t + \varepsilon) = f(t), \qquad t \in \mathbb{R}.$$

7. Prove that a subspace of a second-countable space is second-countable.

8. Regard $\mathbb{R}$ as a subset of $\mathbb{R}^2$ in the usual way. Let $\mathcal{T}$ be the family of all subsets U of $\mathbb{R}^2$ such that $U\backslash\mathbb{R}$ is an open subset of $\mathbb{R}^2$ (in the usual topology). Establish the following assertions:

(a) $\mathcal{T}$ is a topology for $\mathbb{R}^2$.

(b) $(\mathbb{R}^2,\mathcal{T})$ is separable.

(c) The subspace $\mathbb{R}$ of $(\mathbb{R}^2,\mathcal{T})$ is neither separable nor second-countable.

(d) $(\mathbb{R}^2,\mathcal{T})$ is not second-countable.

5. SEPARATION AXIOMS

We saw in the previous section how, by restricting our attention to topological spaces with a base of a certain type, we could obtain some interesting properties that do not hold for topological spaces in general. In this section, we shall consider another kind of restriction. The property of second-countability introduced in the previous section requires in effect that there not be too many open sets. The properties to be considered now require that there not be too few open sets. The reason for this latter type of requirement is that a topology with few open sets tends to reveal little information because it is too simple a structure. The extreme example is the indiscrete topology, which provides no structural information at all. Clearly, it is a good idea to direct attention away from this situation, where the topological space structure is essentially trivial.

In contrast to the previous section, in which overabundance of open sets was prevented by cardinality restrictions, the restrictions that guarantee abundance of open sets are not best expressed in terms of cardinality. Instead, they are expressed by what are called *separation properties*. These properties are some form of the requirement that pairs of disjoint subsets are always contained in disjoint open subsets. To require this in total generality is of course too strong, since such a requirement would require every set to be open. The precise forms that are appropriate have to do with the situation where the initially given disjoint sets are closed or are single points. The following separation properties turn out to be the most useful ones. As the reader probably feels by now is usual, they are generalizations of properties of metric spaces. Moreover, they rule out a number of unpleasant possibilities that we encountered earlier. For instance, one of them, the Hausdorff, or T_2, property, is designed to make limits of sequences unique, that is, to ensure that a given sequence has at most one limit.

A topological space X is a *T_1-space* if for each pair of distinct points $x,y \in X$, there exists an open set U containing y such that $x \notin U$. In this case, the complement of the singleton set $\{x\}$ can be expressed as the union of open sets, so that $X\backslash\{x\}$ is open and $\{x\}$ is a closed subset of X. Conversely, if $\{x\}$ is a closed subset of X, then $U = X\backslash\{x\}$ is an open set containing each $y \in X$, $y \neq x$. It follows that a topological space X is a T_1-space if and only if points are closed.

The space X is a *T_2-space*, or *Hausdorff space*, if for each pair of distinct points $x,y \in X$, there exist disjoint open sets U and V such that $x \in U$ and $y \in V$. Evidently every T_2-space is a T_1-space.

The space X is *regular* if for each closed subset E of X and each point $x \in X\backslash E$, there exist disjoint open sets U and V such that $E \subset U$ and $x \in V$. A *T_3-space* is a regular T_1-space. Evidently every T_3-space is a T_2-space.

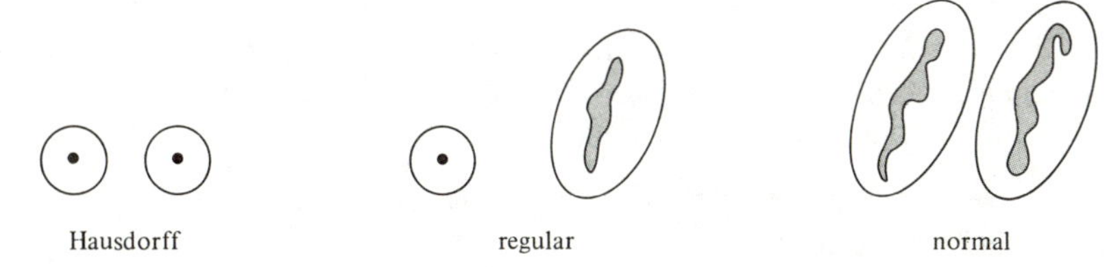

The space X is *normal* if for each pair E and F of disjoint closed subsets of X, there exist disjoint open sets U and V such that $E \subseteq U$ and $F \subseteq V$. A *T_4-space* is a normal T_1-space. In particular, every T_4-space is a T_3-space, so that

$$T_4 \Rightarrow T_3 \Rightarrow T_2 \Rightarrow T_1.$$

In the literature of topology, regular and normal topological spaces are sometimes defined to be T_3-spaces and T_4-spaces, respectively. This inconsistency does not lead to confusion, though, since the topological spaces of interest (in analysis) are Hausdorff spaces and for these the definitions coincide. The "T" notation stems from the German word "Trennungsaxiome" for separation axiom.

5.1 Theorem: Every metric space is a T_4-space.

Proof: Let (X,d) be a metric space. Since each singleton $\{x\}$ is the intersection of the closed sets $\overline{B(x;r)}$, $r > 0$, points in X are closed, and X is a T_1-space.

Suppose that E and F are closed disjoint subsets of X. For each $x \in E$, there is then $r(x) > 0$ such that $B(x;r(x)) \cap F = \emptyset$. Similarly, for each $y \in F$ there exists $r(y) > 0$ such that $B(y;r(y)) \cap E = \emptyset$. Define open sets U and V of X by

$$U = \cup\{B(x;r(x)/2) : x \in E\},$$

$$V = \cup\{B(y;r(y)/2) : y \in F\}.$$

Then U and V are open sets containing E and F, respectively. It suffices to show that U and V are disjoint.

Suppose that there exists $z \in U \cap V$. Then there exist $x \in E$ and $y \in F$ such that $d(x,z) < r(x)/2$ and $d(y,z) < r(y)/2$. Then

$$\begin{aligned} d(x,y) &\leq d(x,z) + d(y,z) \\ &< (r(x) + r(y))/2 \\ &\leq \max(r(x),r(y)). \end{aligned}$$

Consequently either $x \in B(y;r(y))$ or $y \in B(x;r(x))$, and each of these possibilities contradicts the choice of $r(x)$ and $r(y)$. □

5.2 Lemma: A topological space X is normal if and only if for each closed subset E of X and each open set W containing E, there exists an open set U containing E such that $\overline{U} \subset W$.

Proof: Suppose that X is normal and that E and W are as above. Then E and $X\backslash W$ are disjoint closed subsets of X. If U and V are disjoint open sets such that $E \subseteq U$ and $X\backslash W \subseteq V$, then $\overline{U} \subseteq X\backslash V \subseteq W$, so that U is the required open set containing E.

Conversely, suppose that the condition above is valid. Let E and F be disjoint closed subsets of X. Then $W = X\backslash F$ is an open set containing E. If U is an open set containing E such that $\overline{U} \subset W$, then U and $X\backslash\overline{U}$ are disjoint open sets satisfying $E \subset U$ and $F \subset X\backslash\overline{U}$. Hence X is normal. □

Now we arrive at the first theorem of this chapter that is not of a formal, definition-manipulative nature. In mathematical slang, formal, abstract theorems are referred to as "soft" and concrete theorems are called "hard." The words "hard" and "soft" are not to be taken in general as indications of difficulty or lack thereof; "hard" analysis is not necessarily more (or less) difficult than any other kind of mathematics. In any case, the next theorem is comparatively "hard."

The theorem we are about to state and prove is, as noted, rather more concrete than most of the previous results, and it has numerous concrete applications. First, it is used in the proof of results about extension of continuous functions to be defined on larger sets. This application will be considered in some detail after the theorem is proved. There is also a second kind of application, which we shall not consider in much detail here but which occurs in Chapter IV. In this kind of application, one "localizes" a continuous real-valued function by multiplying it by a function with values in [0,1] which is 1 near a point but 0 far from the point. Functions into [0,1] that are 1 everywhere near a given point and 0 far from the point occur so commonly that they have acquired an (informal) name; they are called *bump functions* because their graph looks like a bump! The following result is thus, in general terms, the assertion that bump functions exist around any closed set (instead of just around a point) in a normal space. This idea of bump functions and localization by multiplication by them has more ramifications than one might expect. For instance, the idea plays a vital role in the study of partial differential equations and also in differential geometry.

For the proof of the following theorem, recall that a dyadic rational number is a rational number of the form $p/2^n$, where p and n are integers. The dyadic rationals are dense in $\mathbb{R}$.

5.3 Theorem *(Urysohn's Lemma):* Let E and F be disjoint closed subsets of a normal topological space. Then there exists a continuous function f from X to the unit interval [0,1] such that $f = 0$ on E and $f = 1$ on F.

Proof: Set $V = X \backslash F$, an open set containing E. By Lemma 5.2, there exists an open set $U_{1/2}$ such that

$$E \subset U_{1/2} \subset \overline{U}_{1/2} \subset V.$$

Applying Lemma 5.2 to the open set $U_{1/2}$ containing E and the open set V containing $\overline{U}_{1/2}$, we obtain open sets $U_{1/4}$ and $U_{3/4}$ such that

$$E \subset U_{1/4} \subset \overline{U}_{1/4} \subset U_{1/2} \subset \overline{U}_{1/2} \subset U_{3/4} \subset \overline{U}_{3/4} \subset V.$$

Continuing in this manner, we construct, for each dyadic rational $r \in (0,1)$, an open set U_r, such that

$$\overline{U}_r \subset U_s, \qquad 0 < r < s < 1, \tag{5.1}$$

$$E \subset U_r, \qquad 0 < r < 1, \tag{5.2}$$

$$U_r \subset V, \qquad 0 < r < 1. \tag{5.3}$$

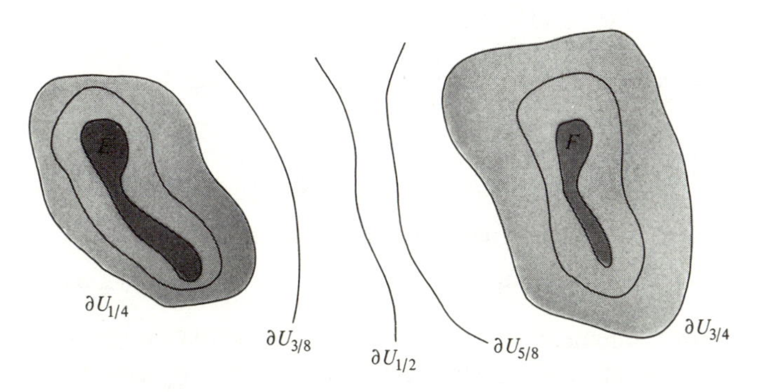

We define the function f now so that the sets ∂U_r are level sets of f on which f assumes the value r. This is done by defining $f(x) = 0$ if $x \in U_r$ for all $r > 0$ and

$$f(x) = \sup\{r : x \notin U_r\}$$

otherwise. Evidently $0 \leq f \leq 1$, $f = 0$ on E and $f = 1$ on F. It suffices to show that f is continuous.

Let $x \in X$. For convenience, we assume that $0 < f(x) < 1$; the cases $f(x) = 0$ and $f(x) = 1$ are easier. Let $\varepsilon > 0$. Choose dyadic rationals r and s such that $0 < r,s < 1$ and

$$f(x) - \varepsilon < r < f(x) < s < f(x) + \varepsilon.$$

Then $x \notin U_t$ for dyadic rationals between r and $f(x)$, so that by (5.1), $x \notin \overline{U}_r$. On the other hand, $x \in U_s$. Hence $W = U_s \backslash \overline{U}_r$ is an open neighborhood of x. If $y \in W$, then from the definition of f we see that $r \leq f(y) \leq s$. In particular, $|f(y) - f(x)| < \varepsilon$ for $y \in W$, so that f is continuous at x. □

A second important "hard" result is the following extension theorem.

5.4 Theorem *(Tietze Extension Theorem):* Let X be a normal topological space, let Y be a closed subset of X, and let f be a bounded continuous real-valued function on Y. Then there exists a bounded continuous real-valued function h on X such that $h = f$ on Y.

Proof: The function h will be constructed by an iteration scheme. For the first step, set

$$c_0 = \{\sup|f(y)| : y \in Y\},$$

$$E_0 = \{y \in Y : f(y) \leq -c_0/3\},$$

$$F_0 = \{y \in Y : f(y) \geq c_0/3\}.$$

Then E_0 and F_0 are closed disjoint subsets of X. Taking a linear combination of a constant function and the function appearing in Urysohn's Lemma, we find a continuous real-valued function g_0 on X such that $-c_0/3 \leq g_0 \leq c_0/3$, $g_0 = -c_0/3$ on E_0 and $g_0 = c_0/3$ on F_0. In particular,

$$|g_0| \leq c_0/3,$$

$$|f - g_0| \leq 2c_0/3 \qquad \text{on } Y.$$

Now we construct by induction functions $\{g_n\}_{n=0}^{\infty}$ such that

$$|g_n| \leq 2^n c_0/3^{n+1}, \tag{5.4}$$

$$|f - g_0 - g_1 - \cdots - g_n| \leq 2^{n+1}c_0/3^{n+1} \qquad \text{on } Y. \tag{5.5}$$

Indeed, suppose $g_0, \ldots, g_{n-1}$ have been constructed. Set

$$c_{n-1} = \sup\{|f(y) - g_0(y) - \cdots - g_{n-1}(y)| : y \in Y\}.$$

Repeat the above argument with c_{n-1} replacing c_0 and $f - g_0 - \cdots - g_{n-1}$ replacing f. This yields a continuous real-valued function g_n such that

$$|g_n| \le c_{n-1}/3,$$
$$|f - g_0 - \cdots - g_n| \le 2c_{n-1}/3 \qquad \text{on } Y.$$

Since $c_{n-1} \le 2^n c_0/3^n$, we obtain (5.4) and (5.5).

Now set

$$h_n = g_0 + \cdots + g_n, \qquad n \ge 1.$$

If $n \ge m$, then

$$|h_n - h_m| = |g_{m+1} + \cdots + g_n|$$
$$\le \left(\left(\frac{2}{3}\right)^{m+1} + \cdots + \left(\frac{2}{3}\right)^{n}\right)\frac{c_0}{3}$$
$$\le \left(\frac{2}{3}\right)^{m+1} c_0.$$

Consequently the sequence $\{h_n\}$ is Cauchy. By Theorem I.2.5, $\{h_n\}$ converges uniformly to a real-valued function h on X, and by Corollary 3.5, h is continuous. Furthermore,

$$|h| \le \sum_{n=0}^{\infty} |g_n| \le c_0,$$

so that h is bounded, and in fact the bound on h coincides with the bound on f. From (5.5) it is clear that $h = f$ on Y. □

The Tietze Extension Theorem just established is an especially useful result in the application of topological methods in concrete contexts, such as the study of the structure of sets in Euclidean spaces. The reason for this usefulness is that it is sometimes easier to study continuous functions defined on all of Euclidean-space $\mathbb{R}^n$ than it is to study a continuous function defined on some complicated closed subset of $\mathbb{R}^n$. The Tietze Extension Theorem guarantees that one need consider only functions defined on all of $\mathbb{R}^n$ and then consider the closed set and their restrictions to it. An instance in which these general ideas, and in particular the Tietze Extension Theorem itself, play a vital role will occur in the proof of the Jordan Curve Theorem in Chapter III. Whereas Urysohn's Lemma enables one to cut real-valued functions down by multiplying by a bump function, Tietze's Extension Theorem enables one to extend the domain to be the whole space. These two related, complementary techniques are both highly useful.

EXERCISES

1. Show that a sequence in a Hausdorff space cannot converge to more than one point.
2. Let X be a topological space and let X_0 be the topological space that is the set X with the cofinite topology. Show that the identity map of X to X_0 is continuous if and only if X is a T_1-space.

3. A property of a topological space is *hereditary* if, whenever a topological space X has that property, then every subspace of X has the property. Show that the properties of being a T_1-space, Hausdorff space, and regular space are hereditary.
4. Prove that a topological space X is normal if and only if the conclusion of Urysohn's Lemma is valid for X. Prove that this occurs if and only if the conclusion of the Tietze Extension Theorem is valid for X.
5. Let X be the subspace of $\mathbb{R}^2$ consisting of the closed upper half-plane minus the origin:

$$X = \{(x,y) : y \geq 0, (x,y) \neq (0,0)\}.$$

Show that $E = \{(x,0) : x < 0\}$ and $F = \{(x,0) : x > 0\}$ are disjoint closed subsets of X. Find a continuous function $f : X \to [0,1]$ such that $f = 0$ on E and $f = 1$ on F.
6. Let X be a normal topological space, let E be a closed subset of X, and let $f : E \to \mathbb{R}$ be continuous. Show that f can be extended to a continuous function from X to $\mathbb{R}$. *Hint*: Let $g : \mathbb{R} \to (-1,1)$ be a homeomorphism, extend $g \circ f$ to a map from X to $[-1,1]$, and decide what to do on the set where the extension assumes the values ± 1.
7. Let X have the cofinite topology. When is X a T_1-space? When is X a Hausdorff space?
8. Let $\mathcal{T}$ be the half-open interval topology for $\mathbb{R}$, defined in Exercise 4.6. Show that $(\mathbb{R},\mathcal{T})$ is a T_4-space.
9. Let $\mathcal{B}$ be the collection of subsets of $\mathbb{R}$ of the form (a,b), together with those of the form $(a,b) \cap \mathbb{R}_0$, where $-\infty < a < b < \infty$ and $\mathbb{R}_0$ is the set of rational numbers. Prove the following:
   (a) $\mathcal{B}$ is a base of open sets for a topology $\mathcal{T}$ for $\mathbb{R}$.
   (b) $(\mathbb{R},\mathcal{T})$ is a Hausdorff space.
   (c) $\mathbb{R}\backslash\mathbb{R}_0$ is $\mathcal{T}$-closed.
   (d) If $f : (\mathbb{R},\mathcal{T}) \to \mathbb{R}$ is a continuous function such that $f = 0$ on $\mathbb{R}\backslash\mathbb{R}_0$, then $f = 0$ everywhere.
   (e) $(\mathbb{R},\mathcal{T})$ is not regular.

Remark: Exercise 9 provides an example of a Hausdorff space that is not regular. For an example of a regular Hausdorff space that is not normal, see Exercises 10.2 and 10.5.

6. COMPACTNESS

It was seen in Chapter I that the property of compactness of metric spaces was of considerable interest and utility in applications of metric-space methods to analysis. Compactness of metric spaces can be expressed in various equivalent ways (Theorem I.5.1). Of these equivalent statements, the one singled out as the definition of compactness for metric spaces was precisely the property that involved only open sets and consequently could be carried over directly to topological spaces. Thus we declare a topological space X to be *compact* if every open cover of X has a finite subcover. In other words, X is compact if, whenever $\{U_\alpha\}_{\alpha\in A}$ is a family of open sets whose union is X, then there are finitely many of the U_α's whose union is X.

It is easy to see that compactness is a topological property. That this is so is really a statement about "metamathematics," that is, the language structure of mathematics. Specifically, the definition of compactness of a space is couched in terms of pure set-theory ideas (unions, set inclusions) and openness of sets. A homeomorphism, being a one-to-one onto function, preserves unions and inclusions, and it preserves openness by definition. Thus it preserves compactness. The fact that a homeomorphism preserves compactness can be deduced also as a consequence of the more general statement presented as Theorem 6 of this section. But the reader should be sure, however, to understand the "metamathematics" argument just given, in terms of just the language of the definition.

The compact Hausdorff spaces play an important role in analysis. Sometimes the Hausdorff property is built into the definition of compactness. This does not lead to confusion in analytic contexts because any topological space of real interest in analysis is Hausdorff. In applications to other fields, however, it is useful to develop the concept of compactness without the Hausdorff assumption.

A subset S of a topological space X is a *compact subset* of X if S is compact in the relative topology it inherits from X. This occurs if and only if S has the following property: if $\{U_\alpha\}_{\alpha\in A}$ is a family of open subsets of X that covers S, then there is a finite subfamily $U_{\alpha_1},\ldots,U_{\alpha_n}$ that covers S.

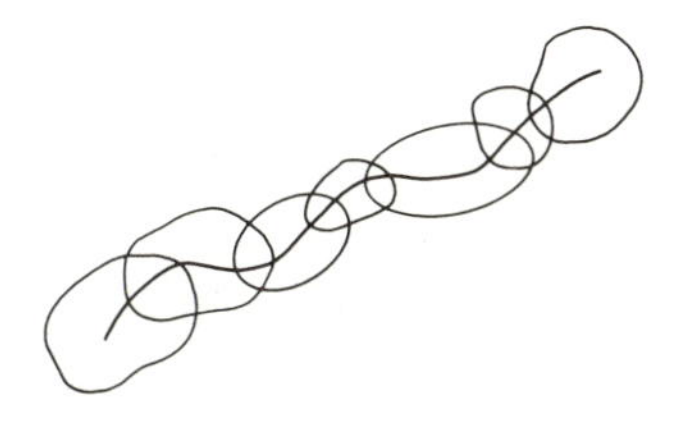

6.1 Theorem: Any finite union of compact subsets of a topological space is compact.

Proof: Let $S_1,\ldots,S_m$ be compact subsets of X and let $\{U_\alpha\}_{\alpha\in A}$ be a family of open subsets of X that covers $S_1 \cup \ldots \cup S_m$. For each j, there is a finite subcollection of the U_α's that covers S_j. Lumping these subcollections together, we obtain a finite subcover of $S_1 \cup \ldots \cup S_m$. □

6.2 Theorem: A closed subspace of a compact topological space is compact.

Proof: Let S be a closed subset of the compact space X and let $\{U_\alpha\}_{\alpha\in A}$ be a family of open subsets of X that covers S. Then $\{U_\alpha \cup (X\backslash S)\}_{\alpha\in A}$ is a family of open subsets of X that covers X. Since X is compact, there is a finite subfamily of this family that covers X. The corresponding subfamily of $\{U_\alpha\}$ then covers S. □

6.3 Lemma: Let S be a compact subset of a Hausdorff space X. For each $x \in X \backslash S$, there exist disjoint open neighborhoods U of x and V of S.

Proof: For each $y \in S$, there exist disjoint open neighborhoods U_y and V_y of x and y, respectively. The open sets $\{V_y\}_{y \in S}$ cover S. Since S is compact, there exist $y_1, \ldots, y_n \in S$ such that

$$S \subset V_{y_1} \cup \cdots \cup V_{y_n} = V.$$

Then

$$U_{y_1} \cap \cdots \cap U_{y_n} = U$$

is an open neighborhood of x that is disjoint from V. □

6.4 Corollary: A compact subset of a Hausdorff space is closed.

The next theorem is particularly important since it allows us to apply Urysohn's Lemma (Theorem 5.3) and the Tietze Extension Theorem (Theorem 5.4) to compact Hausdorff spaces. In particular, it shows that there is an abundance of continuous real-valued functions on a compact Hausdorff space.

6.5 Theorem: A compact Hausdorff space is normal.

Proof: Let S and T be disjoint closed subsets of a compact Hausdorff space X. By Theorem 6.2, S and T are compact. Applying Lemma 6.3, we find for each $x \in T$ disjoint open neighborhoods U_x of x and V_x of S. The open sets $\{U_x\}_{x \in T}$ cover T. Consequently there exist $x_1, \ldots, x_n \in T$ such that

$$T \subset V_{x_1} \cup \cdots \cup V_{x_n} = V.$$

Then $U = U_{x_1} \cap \cdots \cap U_{x_n}$ and V are disjoint open neighborhoods of S and T, respectively. □

6.6 Theorem: Let f be a continuous function from a compact space X to a topological space Y. Then $f(X)$ is a compact subset of Y.

Proof: Let $\{U_\alpha\}_{\alpha \in A}$ be a family of open subsets of Y that covers $f(X)$. Then $\{f^{-1}(U_\alpha)\}_{\alpha \in A}$ is a family of open subsets of X that covers X. Choose $U_{\alpha_1}, \ldots, U_{\alpha_n}$ so that

$$X = f^{-1}(U_{\alpha_1}) \cup \cdots \cup f^{-1}(U_{\alpha_n}).$$

Then

$$f(X) \subset U_{\alpha_1} \cup \cdots \cup U_{\alpha_n},$$

so that $f(X)$ is compact. □

6.7 Theorem: Let f be a continuous function from a compact space X to a Hausdorff space Y. If f is one-to-one, then f is a homeomorphism of X and $f(X)$.

Proof: It suffices to prove that $f(U)$ is relatively open in $f(X)$ whenever U is open in X. Therefore let U be open in X. By Theorem 6.2, $X \backslash U$ is compact. By Theorem 6.6, $f(X \backslash U) = f(X) \backslash f(U)$ is a compact subset of $f(X)$. By Corollary 6.4, $f(X) \backslash f(U)$ is closed in $f(X)$. Hence $f(U)$ is relatively open in $f(X)$. □

Despite its simple proof, Theorem 6.7 can be quite striking in specific instances. One such instance is in the consideration of continuous functions $f : [0,1] \rightarrow \mathbb{R}^2$, which one thinks of as curves. Curves, in this sense, can have startling properties. For instance, there are curves that "fill space" in the sense that the image $f([0,1])$ has nonempty interior (Exercise 11). This certainly defies the common, indeed well-nigh universal, intuition that a curve is one-dimensional. On the other hand, Theorem 6.7 guarantees that, if a curve $f : [0,1] \rightarrow \mathbb{R}^2$ is one-to-one, then it is a homeomorphism of the closed unit interval $[0,1]$ onto $f([0,1])$, the latter space being given the relative topology inherited from $\mathbb{R}^2$. With the aid of the notion of connectedness, it is quite easy to see that in this case the interior of $f([0,1])$ in $\mathbb{R}^2$ is empty, so that $f([0,1])$ is at least vaguely reminiscent of our intuitive idea of a one-dimensional subset of $\mathbb{R}^2$. The notion of connectedness will be introduced in Section 7; for the statement that $f([0,1])$ has empty interior, together with a hint for the proof, see Exercise 8.7.

EXERCISES

1. Prove in detail that compactness is a topological property.
2. Prove that a topological space X is compact if and only if it has the following property: if $\{E_\alpha\}_{\alpha \in A}$ is any family of closed subsets of X such that any finite intersection of the E_α's is nonempty, then $\bigcap_{\alpha \in A} E_\alpha$ is nonempty.
3. Show that any space with the cofinite topology is compact.
4. Show that a discrete topological space is compact if and only if it is finite.
5. Show that a continuous real-valued function on a compact space attains its maximum value and its minimum value. In particular, show that a continuous real-valued function on a compact space is bounded.
6. Prove that if X is a compact Hausdorff space and if $x,y \in X$ satisfy $x \neq y$, then there is a continuous real-valued function f on X such that $f(x) \neq f(y)$. (In other words, the continuous real-valued functions on X "separate the points" of X.)
7. Let X be a compact Hausdorff space and let $\{U_\alpha\}_{\alpha \in A}$ be an open cover of X. Show that there exist a finite number of continuous real-valued functions $h_1, \ldots, h_m$ on X with the following properties:
   (a) $0 \leq h_j \leq 1, \quad 1 \leq j \leq m,$
   (b) $\Sigma\, h_j = 1,$

(c) For each $1 \leq j \leq m$, there is an index α_j such that the closure of the set $\{x : h_j(x) > 0\}$ is contained in U_{α_j}.

Remark: The h_j's are called a *partition of unity* subordinate to the cover $\{U_\alpha\}$. They partition the function 1 into a number of bump functions, each of which "lives" inside one of the U_α's. For the proof, use the normality to construct, for each $x \in X$, a bump function g_x such that $g_x \geq 0$, $g_x(x) > 0$, and such that the open set $V_x = \{g_x > 0\}$ satisfies $\overline{V}_x \subseteq U_\alpha$ for some $\alpha = \alpha(x)$. Choose $x_1, \ldots, x_m$ such that the V_{x_i}'s cover X, and set $h_j = g_{x_j}/\sum g_{x_i}$.

8. A family F of real-valued functions on a topological space X is *equicontinuous* if for each $x \in X$ and $\varepsilon > 0$, there exists a neighborhood U of x such that $|f(y) - f(x)| < \varepsilon$ for all $y \in U$ and all $f \in F$. Let $\{f_k\}_{k=1}^{\infty}$ be a bounded sequence of real-valued functions on a compact space X such that the $\{f_k\}$ are equicontinuous. Show that there is a uniformly convergent subsequence of $\{f_k\}$. *Hint*: For each $n \geq 1$, find points $x_{n1}, \ldots, x_{nm_n}$ and open neighborhoods W_{nj} of the x_{nj} such that $X = W_{n1} \cup \cdots \cup W_{nm_n}$ and $|f_k(x_{nj}) - f_k(y)| < 1/n$ whenever $y \in W_{nj}$, $1 \leq j \leq m_n$, and $k \geq 1$. Use a diagonalization procedure to find a subsequence $\{f_{k_i}\}$ such that $\{f_{k_i}(x_{nj})\}_{i=1}^{\infty}$ converges for each fixed n and j. Then show that $\{f_{k_i}\}$ converges uniformly on X.
9. Let X be compact and let $C(X)$ denote the set of continuous real-valued functions on X. Show that

$$d(f,g) = \sup\{|f(x) - g(x)| : x \in X\}$$

defines a metric on $C(X)$, so that $C(X)$ becomes a complete metric space.
10. Let X be compact. Prove the Arzela-Ascoli Theorem, that a subset F of $C(X)$ is a compact subset of $C(X)$ if and only if F is closed, bounded, and equicontinuous. *Hint*: For the "if" half of the proof, use Exercise 8 and Theorem I.5.1.
11. Define functions f_1, f_2, f_3 from $[0,1]$ to the square $[0,1] \times [0,1]$ as suggested by the three diagrams, so that as t runs from 0 to 1 the images

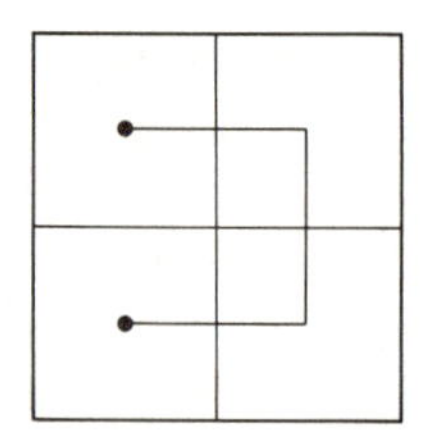
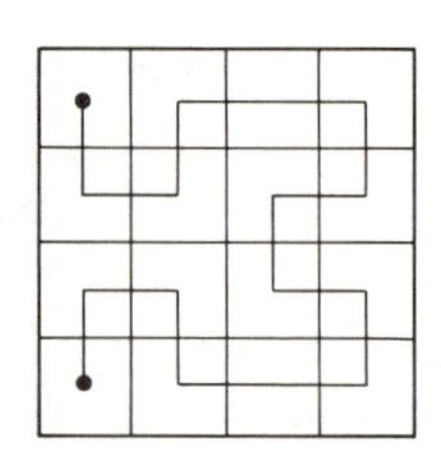

Approximations to a space-filling curve

$f_j(t)$ run at uniform speed along the polygonal curves. Continue in this fashion to obtain $f_j : [0,1] \to [0,1] \times [0,1]$ for all $j \geq 1$.

(a) Prove that the sequence $\{f_j\}$ converges uniformly on $[0,1]$ to some function f.

(b) Prove that the image of f is the square $[0,1] \times [0,1]$. (Such a space-filling curve is called a *Peano curve;* this particular procedure for constructing a space-filling curve is due to Hilbert.)

(c) The kth curve f_k can be traced by a sequence of pencil moves of equal length in one of the four directions up (U), down (D), left (L), or right (R). Find an algorithm for expressing f_k in terms of these simple moves (and in terms of the expression for f_{k-1}).

7. LOCALLY COMPACT SPACES

Although Euclidean space $\mathbb{R}^n$ is not itself compact, the closed balls in $\mathbb{R}^n$ are compact, so that a nearsighted observer at a point in $\mathbb{R}^n$ might think that $\mathbb{R}^n$ itself is in a compact space. This suggests that it might be worthwhile to formulate an idea of localized compactness, that is, compactness in small neighborhoods. In general, a topological space is said to have a property "locally" if the property holds, in an appropriate sense, in every sufficiently small neighborhood of each point, or more precisely, if for each point p and each open neighborhood U of p, there is an open neighborhood V of p such that $V \subseteq U$ and the property holds, in the appropriate sense, in V.

In the case of compactness, the appropriate localized version is defined as follows. A topological space X is *locally compact* if, for each $p \in X$, there is an open set W such that $p \in W$ and $\overline{W}$ is compact. For Hausdorff spaces, this definition is equivalent to one that strictly parallels the more general definition just given for localization of a property (Exercise 1). Since the notion of local compactness depends only on set-theory ideas and open sets, the metamathematical line of reasoning discussed in the preceding section shows that local compactness is a topological property (Exercise 2).

Any compact space is locally compact. Any nonempty open subset of $\mathbb{R}^n$ is locally compact, and any discrete space is locally compact.

Let Y be a compact Hausdorff space and let $y_0 \in Y$. Then the subspace $X = Y\backslash\{y_0\}$ is locally compact and Hausdorff. We shall show that every locally compact Hausdorff space arises in this manner, that is, by puncturing a compact Hausdorff space. This is advantageous because it is often easier to study compact spaces than ones that are just locally compact.

7.1 Theorem: Let X be a locally compact Hausdorff space and let Y be a set consisting of X and one other point. Then there exists a unique topology for Y such that Y becomes a compact Hausdorff space and the relative topology for X inherited from Y coincides with the original topology for X.

Proof: The point in $Y\backslash X$ will be labeled "∞," the point at infinity.

Consider the family $\mathcal{S}$ of those subsets U of Y such that either

(i) U is an open subset of X, or

(ii) $\infty \in U$ and $X\backslash U$ is a compact subset of X.

By Corollary 6.4, compact subsets of X are closed in X, so that any set U of type (ii) meets X in an open subset of X. It follows that the family of intersections of the sets in $\mathcal{S}$ with X coincides with the topology of X.

Next we show that $\mathcal{S}$ is a topology for Y. Evidently Y and the empty set $\varnothing$ belong to $\mathcal{S}$.

Let $\{U_\alpha\}_{\alpha \in A}$ be a family of sets in $\mathcal{S}$ and let $U = \cup U_\alpha$. If each U_α is contained in X, then U is open in X, so that $U \in \mathcal{S}$ by (i). If some U_{α_0} contains ∞, then $X \backslash U_{\alpha_0}$ is compact, so that

$$X \backslash U = (X \backslash U_{\alpha_0}) \cap (\bigcap_{\alpha \in A} (X \backslash U_\alpha))$$

is a closed subset of a compact set. By Theorem 6.2, $X \backslash U$ is compact, so that by (ii), $U \in \mathcal{S}$. In any event, the union of any subfamily of $\mathcal{S}$ again belongs to $\mathcal{S}$.

Suppose that $U_1, \ldots, U_n$ belong to $\mathcal{S}$ and set $U = U_1 \cap \cdots \cap U_n$. If $\infty \notin U_j$ for some j, then U is an open subset of X, so that $U \in \mathcal{S}$ by (i). If $\infty \in U_j$ for each j, then $\infty \in U$ and

$$X \backslash U = (X \backslash U_1) \cup \cdots \cup (X \backslash U_n)$$

is a finite union of compact subsets of X. By Theorem 7.1, $X \backslash U$ is compact, so that $U \in \mathcal{S}$ by (ii). That completes the proof that $\mathcal{S}$ is a topology for Y.

Let $\{V_\alpha\}_{\alpha \in A}$ be a subfamily of $\mathcal{S}$ that covers Y. Then for some index α_0, $\infty \in V_{\alpha_0}$. The open subsets $\{V_\alpha \cap X\}$ of X then form an open cover of the compact set $Y \backslash V_{\alpha_0}$, so that there are indices $\alpha_1, \ldots, \alpha_n$ such that $Y \backslash V_{\alpha_0} \subset V_{\alpha_1} \cup \cdots \cup V_{\alpha_n}$. The sets $V_{\alpha_0}, \ldots, V_{\alpha_n}$ then form a finite subcover of Y, so that Y is compact.

To show that Y is Hausdorff, it suffices to show that, if $x \in X$, then there are disjoint open neighborhoods U of x and V of ∞ in Y. It is at this point that we use the definition of local compactness. Choose U to be an open neighborhood of x such that $\overline{U}$ is compact and set $V = Y \backslash \overline{U}$. Then U and V are open subsets of Y which separate x and ∞. It follows that $(Y, \mathcal{S})$ is a compact Hausdorff space.

Let $\mathcal{T}$ be another topology for Y that has the properties of the theorem. Since $(Y, \mathcal{T})$ is a Hausdorff space, $\{\infty\}$ is $\mathcal{T}$-closed in Y and $X \in \mathcal{T}$. If U is an open subset of X, in the original topology of X, then there exists $V \in \mathcal{T}$ such that $V \cap X = U$. Since $X \in \mathcal{T}$, also $U \in \mathcal{T}$, and all sets in $\mathcal{S}$ of type (i) lie in $\mathcal{T}$. If $U \in \mathcal{S}$ is such that $\infty \in U$, then $Y \backslash U$ is compact in X. Hence $Y \backslash U$ is a compact subset of $(Y, \mathcal{T})$, so that $Y \backslash U$ is $\mathcal{T}$-closed and U is $\mathcal{T}$-open, that is, $U \in \mathcal{T}$. We have proved that $\mathcal{S} \subset \mathcal{T}$.

Now the identity map $f : (Y, \mathcal{T}) \to (Y, \mathcal{S})$ is continuous, one-to-one, and onto. By Theorem 6.7, f is a homeomorphism. It follows that $\mathcal{S} = \mathcal{T}$, so that the topology $\mathcal{S}$ is unique. □

If X is compact, then the space Y of the preceding theorem is obtained from X by adjoining a point $\{\infty\}$ that is both open and closed. In this case, consideration of the space Y is not more useful for obtaining information about X than consideration of X directly. On the other hand, if X is a locally compact Hausdorff space that is not compact, then consideration of the space $Y = X \cup \{\infty\}$ can be quite useful in understanding X. In this case, X is dense in Y (Exercise 6) and Y is called the *one-point compactification* of X.

More generally, a compact space Y is called a *compactification* of X if X is (homeomorphic to) a dense subspace of Y. There can be many compactifications of a space X. For instance, the closed unit interval $[0,1]$ is a compactification of the open unit interval $(0,1)$. This compactification is different from the one-point compactification of $(0,1)$, which is homeomorphic to a circle obtained from $[0,1]$ by gluing together the two endpoints (Exercises 7 and 8).

The construction of the one-point compactification perhaps seems a bit artificial, but it does have several good features. First, it works. Second, the space Y produced by it for reasonable X is often itself quite reasonable. Third, it is sometimes quite useful to know that compactification is possible even if it does turn out that the result is not too nice. Fourth, the existence of one compactification suggests the consideration of other ways of putting X inside a compact space. Sometimes one of these other compactifications is a nice space even if the one-point compactification is not itself nice.

EXERCISES

1. Prove that a Hausdorff topological space X is locally compact if and only if, for each $p \in X$ and each open neighborhood U of p, there is an open neighborhood V of p such that $V \subseteq U$ and $\overline{V}$ is compact.
2. Prove in detail that local compactness is a topological property.
3. Prove that a locally compact Hausdorff space is regular.
4. Prove that if every point of X has an open neighborhood U such that $\overline{U}$ is a compact Hausdorff space, then X is Hausdorff.
5. Let X be a topological space and define the family $\mathcal{S}$ of subsets of $Y = X \cup \{\infty\}$ as in the proof of Theorem 7.1. Which conclusions of Theorem 7.1 are valid without any further hypotheses on X? Which are valid whenever X is a Hausdorff space?
6. Let X be locally compact Hausdorff space that is not compact. Prove that X is dense in the one-point compactification $Y = X \cup \{\infty\}$ of X. (Thus the one-point compactification is indeed a compactification.)
7. Prove that the one-point compactification of $\mathbb{R}^n$ is homeomorphic to the n-sphere

   $$S^n = \{(y_1,\ldots,y_{n+1}) \in \mathbb{R}^{n+1} : y_1^2 + \cdots + y_{n+1}^2 = 1\}.$$

   Hint: It suffices to show that $S^n \backslash (0,\ldots,0,1)$ is homeomorphic to $\mathbb{R}^n$. See Exercise I.6.3.
8. Prove that the one-point compactification of any open ball in $\mathbb{R}^n$ is homeomorphic to the n-sphere S^n. *Hint:* See Exercise 7.

8. CONNECTEDNESS

If $(X_1,\mathcal{T}_1)$ and $(X_2,\mathcal{T}_2)$ are two topological spaces and if the sets X_1 and X_2 are disjoint, then one can make $X_1 \cup X_2$ into a topological space in a trivial way, namely, $U \subset X_1 \cup X_2$ is defined to be open if $U \cap X_1$ and $U \cap X_2$ are open. In this situation, information about $X_1 \cup X_2$ can be obtained by looking at X_1 and X_2 separately. Each of the spaces X_1 and X_2 exists independently of the other, and putting them together artificially contributes nothing to our understanding. If, on the other hand, we are presented with a topological space $(X,\mathcal{T})$ that is expressible as a disjoint union of two nonempty spaces $(X_1,\mathcal{T}_1)$ and $(X_2,\mathcal{T}_2)$, then clearly we should immediately decompose X in that way and study X by studying just the separate "pieces." In this section, we consider when such a decomposition is possible. If X is decomposable, one is also interested in subdividing the pieces as far as possible, into pieces that are irreducible in some sense and in particular that cannot be further subdivided. These considerations lead to the notions of a connected topological space and connected component of a topological space.

Specifically, a topological space X is defined to be *connected* if X cannot be expressed as a disjoint union of two nonempty subsets that are both open and closed. Thus X is *disconnected* if there are closed and open subsets U and V of X such that

$$U \cup V = X,$$

$$U \cap V = \emptyset,$$

$$U \neq \emptyset, \qquad V \neq \emptyset.$$

Another way of phrasing the definition is as follows. The topological space X is disconnected if there is a closed and open subset U of X such that $U \neq \emptyset$ and $U \neq X$. If there is such a U, then the complement $V = X\backslash U$ of U is also both closed and

Connected Disconnected

open and X is the disjoint union of the nonempty sets U and V. As with the notion of compactness, the usual metamathematical argument shows that connectedness is a topological property.

A subset of a topological space is a *connected subset* if it is connected in the relative topology.

8.1 Theorem: Let f be a continuous function from a connected topological space X to a topological space Y. Then $f(X)$ is connected.

Proof: Let E be a subset of $f(X)$ that is both closed and open. Then $f^{-1}(E)$ is a subset of X that is both closed and open. Since X is connected, either $f^{-1}(E) = \emptyset$ or $f^{-1}(E) = X$. In the first case $E = \emptyset$, and in the second case $E = f(X)$. □

In general, the union of two connected sets need not be connected. However, the following is true.

8.2 Theorem: Let $\{E_\alpha\}_{\alpha \in A}$ be a family of connected subsets of a topological space X such that $E_\alpha \cap E_\beta \neq \emptyset$ for each pair α,β of indices. Then $\cup\{E_\alpha : \alpha \in A\}$ is connected.

Proof: Let $E = \cup E_\alpha$ and let F be a subset of E that is both closed and open. Replacing F by $E \backslash F$, if necessary, we can assume that $F \neq \emptyset$. It suffices to show that $F = E$.

Let $x \in F$, say $x \in E_{\alpha_0}$. Since $F \cap E_{\alpha_0}$ is a nonempty closed and open subset of E_{α_0} and E_{α_0} is connected, $F \cap E_{\alpha_0} = E_{\alpha_0}$. Moreover, if β is any index, then $F \cap E_\beta$ is a closed and open subset of E_β that includes $E_{\alpha_0} \cap E_\beta$. Since $E_{\alpha_0} \cap E_\beta \neq \emptyset$, $F \cap E_\beta$ must be all of E_β, and $F \supset E_\beta$. Since this occurs for all β, $F = E$. □

Now let X be a topological space and let $x \in X$. The *connected component of* x *in* X, denoted by $C(x)$, is the union of all connected subsets of X that contain x. By Theorem 8.2, $C(x)$ is connected, and it is evidently the largest connected subset of X containing x.

Suppose E is a connected subset of X that meets $C(x)$. By Theorem 8.2, $E \cup C(x)$ is connected, so that $E \cup C(x)$ must be included in $C(x)$. Hence $C(x)$ includes each connected subset of X it meets. In particular, if $C(y)$ is the connected component of another point $y \in X$, then either $C(y)$ is disjoint from $C(x)$ or $C(y) \subset C(x)$. In the latter case, also $C(x) \subset C(y)$, so that $C(y) = C(x)$. We have proved the following.

8.3 Theorem: Two connected components of X either coincide or are disjoint. The connected components of X form a partition of X into maximal connected subsets.

For example, if X is a discrete topological space, then every subset of X is both open and closed, so that the connected components of X are precisely the sets consisting of one point. On the other hand, if X is connected, then X has only one connected component, X itself.

The most important example of a connected space is given by the following theorem.

8.4 Theorem: Any interval (closed, open, or semiopen; finite or infinite) in $\mathbb{R}$ is connected.

Proof: Any interval is homeomorphic to one of the three intervals $[0,1]$, $[0,1)$, or $(0,1)$, and so it suffices to prove that these are connected. We treat first the closed interval $[0,1]$.

Let E be a closed and open subset of $[0,1]$. Replacing E by its complement, if necessary, we can assume that $1 \notin E$. We must show that E is empty.

Suppose that E is not empty. Set

$$t = \sup\{s : s \in E\}. \tag{8.1}$$

Since E is closed, $t \in E$. Since $1 \notin E$, $t < 1$. Since E is open, there is then an interval $[t, t + \varepsilon)$ that is contained in E. This contradicts (8.1) and consequently $[0,1]$ is connected.

It follows that any compact interval is connected. Since $[0,1)$ and $(0,1)$ can be expressed as the union of an increasing sequence of compact intervals, $[0,1)$ and $(0,1)$ are also connected, by Theorem 8.2. □

Theorem 8.3 guarantees that every topological space can be decomposed into connected pieces in the sense that the underlying set is decomposed into disjoint subsets which are connected subsets. Note, however, that this decomposition does not necessarily mean that the space is decomposed in the sense that we considered at the beginning of this section. The components of the decomposition in Theorem 8.3 do not have to be open (Exercise 10). The question of when the components are open is treated in Exercise 11.

EXERCISES

1. Prove in detail that connectedness is a topological property.
2. Prove that every connected subset of $\mathbb{R}$ is an interval.
3. Consider the two tangent open discs $\{x^2 + y^2 < 1\}$ and $\{(x - 2)^2 + y^2 < 1\}$ in $\mathbb{R}^2$. Is the union of the discs a connected subset of $\mathbb{R}^2$? Is the union of their closures a connected subset of $\mathbb{R}^2$? Is the union of one disc and the closure of the other a connected subset of $\mathbb{R}^2$?
4. A point p of a topological space X is a *cut point* if $X\backslash\{p\}$ is disconnected. Show that the property of having a cut point is a topological property.
5. Show that no two of the intervals $[0,1]$, $(0,1)$, and $[0,1)$ are homeomorphic. *Hint:* Use Exercise 4.
6. Each of the topological spaces in the figure below is a union of four closed intervals. Are any of the spaces homeomorphic? Prove your answer. *Hint:* Extend the technique used in Exercise 5.

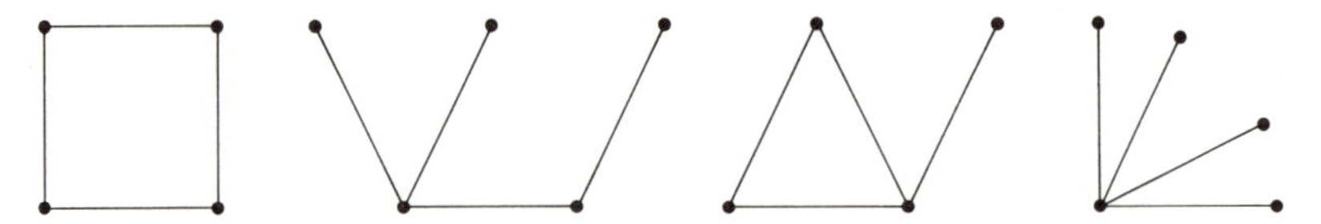

7. Let f be a continuous one-to-one function from the unit interval $[0,1]$ into $\mathbb{R}^n$ $(n \geq 2)$. Prove that the image of f has no interior. *Hint:* First note that f is a homeomorphism onto its range. Then prove and use the fact that a punctured ball in $\mathbb{R}^n$ is connected when $n \geq 2$.
8. A topological space is *totally disconnected* if the connected components are all singletons. Prove that any countable metric space is totally disconnected.
9. Prove that each connected component of a topological space is closed.
10. Show by counterexample that a connected component of a topological space is not necessarily open. *Hint:* See Exercise 8.
11. A topological space X is *locally connected* if, for each point $p \in X$ and each open set U containing p, there is a connected open set V with $p \in V$ and $V \subset U$. Show that each connected component of a locally connected topological space is open.
12. Let $\mathcal{T}$ be the half-open interval topology defined for $\mathbb{R}$ in Exercise 4.6. What are the connected components of $(\mathbb{R},\mathcal{T})$?

13. What are the connected components of $(\mathbb{R},\mathcal{T})$ where $\mathcal{T}$ is the topology introduced in Exercise 5.9, that is Hausdorff but not regular?
14. Let X be a locally compact Hausdorff space that is not compact and let $Y = X \cup \{\infty\}$ be its one-point compactification. If X is connected, is Y connected? If Y is connected, is X connected? Justify your answers.

9. PATH CONNECTEDNESS

There is another notion of the indecomposability of a topological space which is different from, though related to, the idea of connectedness. This new notion is essentially a formal version of the idea that a geometric figure is all one piece if it is possible to go from any one point to any other point by a curve that never leaves the figure. As might be expected, this idea is less prominent in analysis than is the concept of connectedness already discussed, but the new idea, which is called *path connectedness* or sometimes *arcwise connectedness,* is important in the geometric part of topology and its applications.

To make the concept of path connectedness precise, we introduce a preliminary definition. Let X be a topological space and let $x_0,x_1 \in X$. A *path* in X from x_0 to x_1 is a continuous function

$$\gamma : [0,1] \to X$$

such that $\gamma(0) = x_0$ and $\gamma(1) = x_1$. The space X is *path-connected* if, for every pair of points x_0 and x_1 in X, there is a path γ from x_0 to x_1.

We shall now develop the analogue of Theorem 8.3 for path connectedness. The essential result for this purpose is the following lemma.

9.1 Lemma: The relation "there is a path in X from x to y" is an equivalence relation on X.

Proof: For each $x \in X$, the constant map

$$\alpha(s) = x, \qquad 0 \leq s \leq 1,$$

is a path from x to x. Consequently the relation is reflexive.

Suppose that γ is a path from x to y. Then

$$\beta(s) = \gamma(1 - s), \qquad 0 \leq s \leq 1,$$

defines a path β from y to x that corresponds to running backwards along γ. Consequently the relation is symmetric.

Suppose that α is a path from x to y and β is a path from y to z. Define a path γ from x to z by running first along α at double speed, then along β at double speed:

$$\gamma(s) = \begin{cases} \alpha(2s), & 0 \leq s \leq 1/2, \\ \beta(2s - 1), & 1/2 \leq s \leq 1. \end{cases}$$

Since $\alpha(1) = y = \beta(0)$, the function γ is well defined. One checks that γ is continuous and that γ runs from x to z. The relation is then transitive. □

The equivalence classes corresponding to the above equivalence relation are called the *path components* of X. The space X is path-connected (in the sense already defined) if there is only one path component, that is, if any two points of X can be joined by a path in X. In general, the path components of a space are then the maximal path-connected subsets of the space. Note that the path components of a space need not be open (Exercise 8). Conditions under which they are open are considered in Exercise 5.

The path components of a space need not be the same as the connected components (Exercise 5). However, a fixed path component always lies entirely in a (necessarily unique) connected component; this is an immediate consequence of the following theorem.

9.2 Theorem: A path-connected space is connected.

Proof: Fix $x_0 \in X$. For each $x \in X$, let $\gamma_x : [0,1] \to X$ be a path from x_0 to x. By Theorems 8.1 and 8.4, each $\gamma_x([0,1])$ is a connected subset of X. Each $\gamma_x([0,1])$ contains x_0 and $X = \cup\{\gamma_x([0,1]) : x \in X\}$, so that Theorem 8.2 shows that X is connected. □

9.3 Corollary: Let X be a topological space. Then each connected component of X is a union of path components of X.

EXERCISES

1. Prove that any subinterval of $\mathbb{R}$ (closed, open, or semiopen) is path-connected.
2. Prove that path connectedness is a topological property.
3. Prove that if X is path-connected and $f : X \to Y$ is a map, then $f(X)$ is path-connected.
4. A space X is *locally path-connected* if, for each open subset V of X and each $x \in V$, there is a neighborhood U of x such that x can be joined to any point of U by a path in V. Prove that the path components of a locally path-connected space coincide with the connected components.
5. Suppose X is locally path-connected and locally connected (as defined in Exercise 8.11).
   (a) Show that each path component of X is open and closed.
   (b) Show that for each $x \in X$ and each neighborhood V of x, there is an open path-connected neighborhood U of x such that $U \subset V$.
6. An open subset of $\mathbb{R}^n$ is connected if and only if it is path-connected. *Hint:* Use Exercise 4.

7. Let X be the compact subset of $\mathbb{R}^2$ consisting of the vertical interval

$$E = \{(0,y) : -1 \leq y \leq 1\}$$

together with the portion of the graph of $\sin(1/x)$ given by

$$F = \{(x,\sin(1/x)) : 0 < x \leq 1\}.$$

(See the figure below.) Show that X is connected but not path-connected. What are the path components of X?

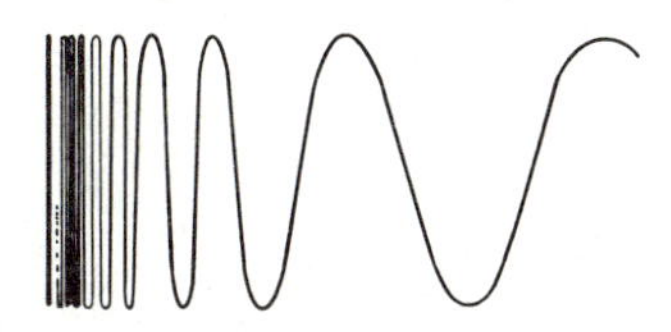

Connected but not path connected

8. Show that the path components of X are not necessarily closed subsets of X, nor are they necessarily open. *Hint:* Use the example in Exercise 7.

10. FINITE PRODUCT SPACES

It was seen in Chapter I that a (finite) product of metric spaces could be made into a metric space in a natural way. In fact, there were a number of easily obtainable metrics that could be defined on the product, and these were all equivalent in the sense of determining the same open sets in the product set. The purpose of this section is to present a construction that associates to a finite product of topological spaces a topology on the product of the spaces. This construction has the property that the topology it puts on the product of metric spaces is the same as the metric topology determined by any of the product metrics. The product-topology construction also has a number of other pleasant properties, which will also be presented in this section. The precise definition of the product topology is motivated by a simple geometric consideration, namely, that a set is open in $\mathbb{R}^2$ if and only if it is a (possibly infinite) union of products of open sets in $\mathbb{R}$. A similar statement holds for $\mathbb{R}^n$. Thus it becomes reasonable to define, in general, a set in a (finite) product of topological spaces to be open if it is a union of products of open sets. The definition is most easily given formally in terms of the concept of a base for a topology (Section 4) as follows.

Let $X_1,\ldots,X_n$ be topological spaces. The *product topology* on the Cartesian product $X = X_1 \times \cdots \times X_n$ is the topology for which a base of open sets is given by the "rectangles"

$$\{U_1 \times \cdots \times U_n : U_j \text{ open in } X_j,\ 1 \leq j \leq n\}. \tag{10.1}$$

That the family of such sets forms a base follows from Theorem 4.2 and from the observation that the intersection of two such sets is again a set of the same form:

$$(U_1 \times \cdots \times U_n) \cap (V_1 \times \cdots \times V_n) = (U_1 \cap V_1) \times \cdots \times (U_n \cap V_n).$$

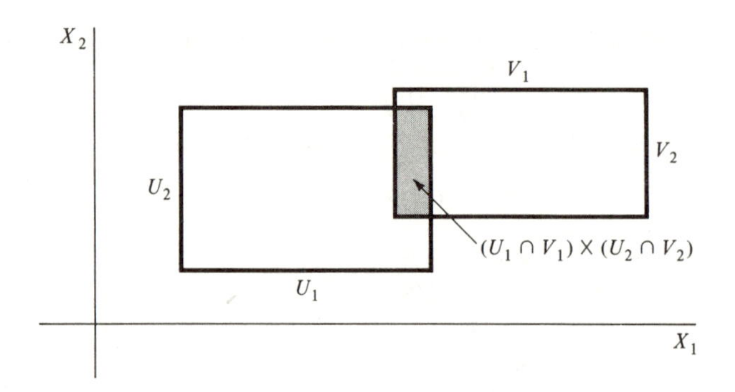

As remarked above, the product topology on a product of metric spaces is precisely the metric topology induced by any reasonable product metric, by Theorem I.4.1.

Unless specified otherwise, we shall always assume that the product of topological spaces is endowed with the product topology.

Let $\pi_j : X \to X_j$ be the projection of X onto the jth coordinate, defined by

$$\pi_j((x_1,\ldots,x_n)) = x_j, \qquad (x_1,\ldots,x_n) \in X.$$

If U_j is an open subset of X_j, then

$$\pi_j^{-1}(U_j) = X_1 \times \cdots \times X_{j-1} \times U_j \times X_{j+1} \times \cdots \times X_n$$

is a basic open set. It follows that each projection π_j is continuous.

10.1 Theorem: Let X be the product of the topological spaces $X_1,\ldots,X_n$ and let π_j be the projection of X onto X_j. The product topology for X is the smallest topology for which each of the projections π_j is continuous.

Proof: Let $\mathcal{T}$ be another topology for X such that the projections π_j are $\mathcal{T}$-continuous. Let $U_1,\ldots,U_n$ be open subsets of $X_1,\ldots,X_n$, respectively. Then each $\pi_j^{-1}(U_j)$ belongs to $\mathcal{T}$. Since

$$\pi_1^{-1}(U_1) \cap \cdots \cap \pi_n^{-1}(U_n) = U_1 \times \cdots \times U_n,$$

the basic set $U_1 \times \cdots \times U_n$ belongs to $\mathcal{T}$ and $\mathcal{T}$ includes the product topology. □

A function f from one topological space to another is *open* if the image of every open set is open.

10.2 Theorem: Let $X = X_1 \times \cdots \times X_n$ be the product of the topological spaces $X_1,\ldots,X_n$. Then each projection π_j is an open map of X onto X_j.

Proof: If $U = U_1 \times \cdots \times U_n$ is a nonempty basic open set, then $\pi_j(U) = U_j$ is open. Since maps preserve unions, the image of any open set is open. □

10.3 Theorem: Let E be a topological space and let f be a function from E to the product $X = X_1 \times \cdots \times X_n$. Then f is continuous if and only if each $\pi_j \circ f$ is continuous, $1 \leq j \leq n$.

Proof: If f is continuous, then $\pi_j \circ f$ is also continuous since the composition of continuous functions is continuous.

Conversely, suppose that each $\pi_j \circ f$ is continuous. Let $U = U_1 \times \cdots \times U_n$ be a basic open set in X. Then

$$f^{-1}(U) = (\pi_1 \circ f)^{-1}(U_1) \cap \cdots \cap (\pi_n \circ f)^{-1}(U_n)$$

is a finite intersection of open sets and hence is open. Since inverses of functions preserve unions, the inverse image of any open set is open. Hence f is continuous. □

It is important to know which properties of topological spaces are valid for a product $X = X_1 \times \cdots \times X_n$ whenever they are valid for the component spaces $X_1, \ldots, X_n$. One property that does not extend to products is normality. It turns out that a product of normal spaces need not be normal (Exercise 5). We give now several theorems showing that various properties of topological spaces are passed on to products of the spaces.

10.4 Theorem: If $X_1, \ldots, X_n$ are Hausdorff spaces, then $X = X_1 \times \cdots \times X_n$ is Hausdorff.

Proof: Let $x = (x_1, \ldots, x_n)$ and $y = (y_1, \ldots, y_n)$ be distinct points of X. Choose an index j such that $x_j \neq y_j$. Let U_j and V_j be disjoint open neighborhoods of x_j and y_j, respectively, in X_j. Then $\pi_j^{-1}(U_j)$ and $\pi_j^{-1}(V_j)$ are disjoint open neighborhoods of x and y, respectively, in X. □

10.5 Theorem: If X_j is a path-connected topological space, $1 \leq j \leq n$, then $X = X_1 \times \cdots \times X_n$ is path-connected.

Proof: Let $x = (x_1, \ldots, x_n)$ and $y = (y_1, \ldots, y_n)$ be points of X. For $1 \leq j \leq n$, let $\gamma_j : [0,1] \to X_j$ be a path from x_j to y_j. Then the path $\gamma : [0,1] \to X$, defined by

$$\gamma(t) = (\gamma_1(t), \ldots, \gamma_n(t)), \qquad 0 \leq t \leq 1,$$

is a path from x to y. Hence X is path-connected. □

Before going on to the next theorem, we make an observation concerning the coordinate slices of a product space. Let $X = X_1 \times \cdots \times X_n$ and fix points $x_2 \in X_2, \ldots, x_n \in X_n$. Then the embedding $h : x_1 \to (x_1, x_2, \ldots, x_n)$ maps X_1 homeomorphically onto the slice $X_1 \times \{x_2\} \times \cdots \times \{x_n\}$ of the product. Indeed, if $U = U_1 \times \cdots \times U_n$ is a nonempty basic open set in X, then $h^{-1}(U) = U_1$ is open, so that h is continuous. Since the inverse of h is the restriction of the projection π_1 to the slice,

h^{-1} is also continuous and h is a homeomorphism. Similarly, for each fixed j and fixed points $x_i \in X_i$, $i \neq j$, the natural embedding $X_j \to \{x_1\} \times \cdots \times \{x_{j-1}\} \times X_j \times \{x_{j+1}\} \times \cdots \times \{x_n\}$ is a homeomorphism of X_j and a slice of the product space.

10.6 Theorem: If X_j is a connected topological space, $1 \leq j \leq n$, then $X = X_1 \times \cdots \times X_n$ is also connected.

Proof: Let E be a closed and open subset of X. Replacing E by its complement, if necessary, we can assume that $E \neq \varnothing$. We must show that $E = X$.

Let $z = (z_1, \ldots, z_n) \in E$ and let $y = (y_1, \ldots, y_n) \in X$. The intersection $E \cap (X_1 \times \{z_2\} \times \cdots \times \{z_n\})$ is a relatively open and closed subset of the coordinate slice, which includes z. Since X_1 is connected, so is the coordinate slice. It follows that E must include the entire coordinate slice. In particular, $(y_1, z_2, \ldots, z_n) \in E$.

Now we repeat this argument in the second coordinate and continue until eventually we conclude that $y \in E$. Since $y \in X$ is arbitrary, we obtain $E = X$. □

Next we turn to the important fact that the product of compact spaces is compact. In Section 12, we shall define the product of an arbitrary family of topological spaces and prove that an arbitrary product of compact spaces is compact. This full result is known as Tychonoff's Theorem, and it is perhaps the most significant theorem in basic topology. The proof of Tychonoff's Theorem depends on certain axioms of set theory and in particular on a statement called "Zorn's Lemma," which is commonly taken as an axiom for set theory and which is equivalent to what is called the "Axiom of Choice." Zorn's Lemma is presented in the next section.

The proof that a finite product of compact spaces is again compact is reasonably straightforward. Thus we content ourselves for now with proving the version for finite products. We begin with the following lemma, whose proof is reminiscent of that of Lindelöf's Theorem (Theorem 4.3; cf. Theorem I.5.11).

10.7 Lemma: Let Y be a topological space and let $\mathcal{B}$ be a base for the topology of Y. If every open cover of Y by sets in $\mathcal{B}$ has a finite subcover, then Y is compact.

Proof: Let $\{U_\beta\}_{\beta \in B}$ be an open cover of Y. For each $y \in Y$, choose $V_y \in \mathcal{B}$ and an index β such that $y \in V_y \subset U_\beta$. The family $\{V_y : y \in Y\}$ forms an open cover of Y by sets in $\mathcal{B}$. By hypothesis, there exist a finite number of the V_y's that cover Y. Since each of these V_y's is contained in at least one of the U_β's, we obtain a finite number of U_β's that cover Y. Hence Y is compact. □

10.8 Theorem *(Tychonoff's Theorem, finite case):* If $X_1, \ldots, X_n$ are compact spaces, then $X_1 \times \cdots \times X_n$ is compact.

Proof: It suffices to prove the theorem in the case where $n = 2$, and so we consider only $X = X_1 \times X_2$.

Let $\mathcal{C}$ be a cover of $X_1 \times X_2$ by rectangular open sets of the form $U \times V$, where U is open in X_1 and V is open in X_2. By Lemma 10.7, it suffices to show that the cover $\mathscr{C}$ has a finite subcover.

Fix $z \in X_2$. The coordinate slice $X_1 \times \{z\}$ is compact. Hence there are a finite number of sets $U_1 \times V_1, \ldots, U_m \times V_m$ in $\mathscr{C}$ that cover $X_1 \times \{z\}$. We can assume that $z \in V_j$ for all j, $1 \leq j \leq m$. Then

$$V(z) = V_1 \cap \cdots \cap V_m$$

is an open neighborhood of z in X_2. Furthermore, $\pi_2^{-1}(V(z))$ is covered by a finite number of sets in $\mathscr{C}$, namely, the sets $U_j \times V_j$, $1 \leq j \leq m$.

The sets $\{V(z) : z \in X_2\}$ form an open cover of X_2. Since X_2 is compact, there exist $z_1, \ldots, z_k \in X_2$ such that $X_2 = V(z_1) \cup \cdots \cup V(z_k)$. Then $X = \pi_2^{-1}(V(z_1)) \cup \cdots \cup \pi_2^{-1}(V(z_k))$. Since each $\pi_2^{-1}(V(z_j))$ can be covered by a finite number of sets in $\mathscr{C}$ and since the aggregate of these sets covers X, $\mathscr{C}$ has a finite subcover. □

EXERCISES

1. Show that if E_j is a closed subset of X_j, $1 \leq j \leq n$, then $E_1 \times \cdots \times E_n$ is a closed subset of $X_1 \times \cdots \times X_n$.
2. Prove that the product of a finite number of regular spaces is regular.
3. Prove that the connected components of $X_1 \times \cdots \times X_n$ are the sets of the form $E_1 \times \cdots \times E_n$, where E_j is a connected component of X_j, $1 \leq j \leq n$. Show that a similar result holds for path components.
4. Suppose $X = X_1 \times \cdots \times X_n$, where each X_j is nonempty. Establish the following assertions:
   (a) If X is Hausdorff, then each X_j is Hausdorff.
   (b) If X is regular, then each X_j is regular.
   (c) If X is normal, then each X_j is normal.
   (d) If X is connected, then each X_j is connected.
   (e) If X is path-connected, then each X_j is path-connected.
   (f) If X is compact, then each X_j is compact.
5. Let $\mathscr{T}$ be the half-open interval topology on $\mathbb{R}$, introduced in Exercise 4.6. Let $\mathscr{S}$ be the product topology for $(\mathbb{R},\mathscr{T}) \times (\mathbb{R},\mathscr{T})$, so that a base for open sets of $(\mathbb{R}^2,\mathscr{S})$ is the family of sets of the form $[x,x+\varepsilon) \times [y,y+\varepsilon)$, where $x,y \in \mathbb{R}$ and $\varepsilon > 0$.
   (a) Prove that the line $L = \{(x,-x) : x \in \mathbb{R}\}$ is $\mathscr{S}$-closed.
   (b) Identify the relative topology on L inherited from $\mathscr{S}$.
   (c) Let S be a sequence in $\mathbb{R}$ that is dense in $\mathbb{R}$ (in the metric topology). Define

   $$E = \{(x,-x) : x \in S\},$$
   $$F = \{(x,-x) : x \in \mathbb{R} \backslash S\}.$$

   Prove that E and F are disjoint $\mathscr{S}$-closed subsets of $\mathbb{R}^2$.
   (d) Suppose $V \in \mathscr{S}$ contains F. Prove that there exist $\varepsilon > 0$, $b \in S$, and a sequence $\{x_n\}$ in $\mathbb{R}\backslash S$ such that $|x_n - b| \to 0$ and such that

$[x_n, x_n + \varepsilon) \times [-x_n, -x_n + \varepsilon) \subset V$ for all n. *Hint:* Apply the Baire Category Theorem to $\mathbb{R}$.

(e) Prove that there do not exist disjoint $\mathcal{S}$-open neighborhoods U of E and V of F. In particular, deduce that $(\mathbb{R}, \mathcal{T}) \times (\mathbb{R}, \mathcal{T})$ is not normal.

11. SET THEORY AND ZORN'S LEMMA

Most of the concepts of topology are based on concrete phenomena that are amply illustrated by sets in $\mathbb{R}^n$, that is, by what are in effect geometric objects in the intuitive sense. Even the discussion of infinite-dimensional vector spaces in Chapter I was primarily based on finite-dimensional intuitions, although infinite dimensionality as such certainly leads one far from geometric intuition. On occasion, however, it proves useful and important to consider situations that involve proof techniques not founded at all on the intuitions of geometry. These esoteric proof techniques tend to involve subtleties from mathematical logic, subtleties involving in particular the set-theoretic foundations of mathematics. The purpose of this section is to discuss briefly one aspect of these foundational set-theoretic questions, an aspect that occurs quite often in abstract mathematics and even occasionally in areas of concrete real analysis, such as measure theory, for example. In the next section, these set-theoretic matters will be applied to prove an important topological result, the compactness of (infinite) products of compact spaces.

In the development of mathematics as an axiomatic system, the starting point is usually chosen to be a set of axioms for set theory. From suitable axioms for set theory, the entire structure of mathematics can be built up. Naturally, this is a somewhat lengthy process, which we do not wish to enter into in detail. However, we do wish to focus attention on one particular axiom, called the Axiom of Choice, and specifically on an equivalent version of the Axiom of Choice, called Zorn's Lemma, from among the axioms often included as axioms for set theory.

The Axiom of Choice asserts that, given a family of nonempty sets, it is possible to select precisely one element from each member of the family. In other words, if $\{S_\alpha\}_{\alpha \in A}$ is a family of nonempty sets indexed by some set A, then there exists a function f on A satisfying $f(\alpha) \in S_\alpha$, $\alpha \in A$.

The Axiom of Choice is clearly an eminently reasonable assertion, and we shall accept it without qualms. It should be noted though that the Axiom of Choice, though highly plausible, leads logically to a number of somewhat peculiar and strong consequences. This has led mathematical logicians to consider the consequences, to various branches of mathematics, of omitting the Axiom of Choice from the set-theory axioms. The situation is somewhat analogous to that of the Parallel Postulate in Euclidean geometry; one can consider geometry without it, with it, or with its negation in various forms. These considerations lead, of course, to non-Euclidean geometry. Similarly, one can consider mathematics with the Axiom of Choice, without it, or without it but with some substitute axiom(s). Most of the work carried out in this book holds equally well no matter which (reasonable!) set-theory axioms are used; in particular, most of it does not depend on the Axiom of Choice. However, the Tychonoff Theorem (Theorem 12.4) for infinite products does depend on the Axiom of Choice; in fact, though it may come as a surprise, Tychonoff's Theorem is completely equivalent to the Axiom of Choice (Exercise 12.9).

Actually, the proof of Tychonoff's Theorem will be based not on the Axiom of Choice as stated, but on an equivalent though less intuitively clear version of the Axiom of Choice called Zorn's Lemma. By equivalent, it is meant that, in the presence of the usual elementary axioms of set theory, Zorn's Lemma implies and is implied by the Axiom of Choice. The proof of the equivalence requires some effort, though, and so we shall omit it. We confine ourselves to defining the prerequisite notions, stating Zorn's Lemma, and giving a simple illustration of the lemma's usefulness.

A *partially ordered set* is a nonempty set $\mathcal{S}$, together with a relation "$\leq$" defined between certain pairs of elements of $\mathcal{S}$, such that

(i) $x \leq x$ for all $x \in \mathcal{S}$,

(ii) if $x \leq y$ and $y \leq z$, then $x \leq z$,

(iii) if $x \leq y$ and $y \leq x$, then $x = y$.

Property (iii) is sometimes omitted in the definition of a partial ordering. If the relation satisfies only (i) and (ii), then one can pass to the quotient set obtained by identifying elements x and y such that $x \leq y$ and $y \leq x$ to arrive at a context for which (iii) is also valid.

The prototypal example of a partially ordered set is the family 2^S of subsets of a fixed set S, ordered by the set-inclusion relation. Another example of a partially ordered set is the set $\mathbb{R}$ of real numbers with the usual ordering.

Let $\mathcal{S}$ be a partially ordered set. An *upper bound* for a subset $\mathcal{E}$ of $\mathcal{S}$ is an element $x \in \mathcal{S}$ such that $y \leq x$ for all $y \in \mathcal{E}$. An element $x \in \mathcal{E}$ is a *maximal element* of $\mathcal{E}$ if whenever $y \in \mathcal{E}$ satisfies $x \leq y$, then $x = y$.

A subset E of $\mathcal{S}$ is *totally ordered* if, for each pair of points $x,y \in E$, either $x \leq y$ or $y \leq x$. The real numbers $\mathbb{R}$ form a totally ordered set. However, the family of subsets of a fixed set S is not totally ordered as soon as S has at least two elements.

The central result of this section is the following.

Zorn's Lemma: Let $\mathcal{S}$ be a partially ordered set. If each totally ordered subset of $\mathcal{S}$ has an upper bound, then $\mathcal{S}$ has a maximal element.

For example, each totally ordered subset $\mathcal{E}$ of 2^S has an upper bound, and the maximal element of 2^S is the set S itself. On the other hand, the real numbers $\mathbb{R}$, which are already totally ordered, have no maximal element.

In order to demonstrate the power of Zorn's Lemma and in order to practice applying it as preparation for the proof of Tychonoff's Theorem in the next section, we now use Zorn's Lemma to prove that every vector space has a basis.

Let V be a vector space over a field F (say V is a real or complex vector space, for instance). Recall that a subset A of V is *linearly independent* if whenever distinct elements $v_1,\ldots,v_m \in A$ (finite in number!) and $\alpha_1,\ldots,\alpha_m \in F$ satisfy

$$(*) \qquad \alpha_1 v_1 + \cdots + \alpha_m v_m = 0,$$

then

$$\alpha_1 = \cdots = \alpha_m = 0.$$

A *basis* for V is a linearly independent subset B for V such that every $v \in V$ can be written as a finite linear combination $v = \beta_1 v_1 + \cdots + \beta_n v_n$, where $v_1,\ldots,v_n \in B$. The theorem we wish to prove is the following.

Theorem: Every vector space has a basis.

Of course, if the vector space V is finite-dimensional, then this theorem is easy to establish. It is a standard fact from elementary linear algebra. The theorem is not at all obvious in the case where V is infinite-dimensional. Consider, for instance, the question of finding a basis for the vector space ℓ^2 of square-summable sequences, introduced in Chapter I (or in any of the infinite-dimensional Banach spaces introduced in Chapter I). The family of vectors $\{(1,0,0,0,\ldots),(0,1,0,0,\ldots),(0,0,1,0,\ldots),\ldots\}$ in ℓ^2 is a linearly independent set. However, it is not a vector-space basis for ℓ^2 since, for instance, the vector $(1,1/2,1/3,\ldots,1/n,\ldots) \in \ell^2$ is not a finite linear combination of elements in this linearly independent set. It is at first sight not at all clear that ℓ^2 has a basis.

Now we turn to the proof of the theorem. The proof is "nonconstructive" in the sense that we do not (in fact, cannot) exhibit a basis explicitly, but rather deduce its existence from Zorn's Lemma.

Proof: Let $\mathcal{S}$ be the family of linearly independent subsets of V. With the usual ordering of inclusion, $\mathcal{S}$ becomes a partially ordered set.

Let $\mathcal{T}$ be a totally ordered subset of $\mathcal{S}$. Define $B = \cup\{T : T \in \mathcal{T}\}$. We claim that B is a linearly independent subset of V. Indeed, let $v_1,\ldots,v_m \in B$ and let $\alpha_1,\ldots,\alpha_m \in F$ satisfy (*). For $1 \leq j \leq m$, choose $T_j \in \mathcal{T}$ so that $v_j \in T_j$. Since the T_j's are totally ordered and since there are only finitely many of the T_j's, one of the T_j's includes all of the others, say $T_j \subseteq T_\ell$ for $1 \leq j \leq m$. Then the v_j's all belong to T_ℓ. Since T_ℓ is a linearly independent set, the relation (*) implies that $\alpha_1 = \cdots = \alpha_m = 0$. Hence B is a linearly independent set.

Now $B \in \mathcal{S}$ and B is evidently an upper bound for $\mathcal{T}$, so that the hypothesis of Zorn's Lemma is met. According to Zorn's Lemma, there exists a maximal element A of $\mathcal{S}$.

We claim that A is a basis for V. Since A is a linearly independent set, it suffices to show that each $v \in V$ is a finite linear combination of elements in A. This is trivial if $v \in A$, so that we may assume that $v \notin A$. Since A is a maximal linearly independent set, $A \cup \{v\}$ is not linearly independent. Hence there exist distinct elements $v_1,\ldots,v_m$ of $A \cup \{v\}$ and scalars $\alpha_1,\ldots,\alpha_m$, not all zero, such that $\alpha_1 v_1 + \cdots + \alpha_m v_m = 0$. We may assume, in fact, that each α_j is nonzero. Since A is linearly independent, the vectors $v_1,\ldots,v_m$ cannot all belong to A. Hence one of them, say v_1, coincides with v. Then

$$v = (-\alpha_2/\alpha_1)v_2 + \cdots + (-\alpha_m/\alpha_1)v_m$$

expresses v as a linear combination of elements of A. □

EXERCISES

1. Let $\mathcal{S}$ be a family of subsets of a set S. Suppose that a subset E of S belongs to $\mathcal{S}$ if and only if every finite subset of E belongs to $\mathcal{S}$. Prove that $\mathcal{S}$ has a maximal element.

2. Let R be a commutative ring with an identity. Show that every proper ideal in R is contained in a maximal (proper) ideal in R.
   Note: The concepts of algebra occurring in this exercise are not needed elsewhere in this book.
3. Let A be a linearly independent subset of a vector space V. Show that there is a basis B for V such that $B \supseteq A$.
4. Let W be a subspace of a real vector space V. Show that every real-valued linear functional L on W is the restriction to W of a real-valued linear functional L on V. *Hint:* Take a basis for W and, by Exercise 3, extend it to a basis for V.
5. Let $\mathfrak{X}$ be a normed real vector space. Prove that if $\mathfrak{X}$ is infinite-dimensional, then there is a discontinuous linear functional on $\mathfrak{X}$ (cf. Exercise I.7.5).
6. Let $\mathfrak{X}$ be an infinite-dimensional Banach space. Show that any vector-space basis for $\mathfrak{X}$ is uncountable. *Hint:* Suppose $\{x_1,x_2,...\}$ is a basis for $\mathfrak{X}$. Define E_m to be the set of linear combinations of $\{x_1,...,x_m\}$ and apply the Baire Category Theorem.

12. INFINITE PRODUCT SPACES

The consideration of finite product spaces (as in Section 10) is motivated by geometric considerations, and the constructions introduced are motivated by the example of forming finite products of $\mathbb{R}$ with itself to yield Euclidean spaces $\mathbb{R}^n$. Infinite product spaces, which are the subject of this section, are in some ways just a natural extension of the finite product ideas to a situation of infinitely many factors. The extension to the case of infinitely many factors requires some care, however, even in the definition of the infinite product set. Moreover, the most useful definition of a topology on the infinite product of topological spaces is not the definition one might expect at first glance, that in which the open sets are unions of products of open sets in the individual coordinate spaces. The definition given for the topology of an arbitrary product space does, however, specialize to the definition given in Section 10 for products of finitely many topological spaces.

The first order of business is to make precise the idea of an infinite product of sets. The factor sets have to be listed in some way, or at least labeled, if there are uncountably many. We call the set of labels an *index set*. For instance, with $\mathbb{R}^n$ thought of as $\mathbb{R} \times \cdots \times \mathbb{R}$ (n times), the index set should contain n elements; the set $\{1,...,n\}$ would be appropriate. Formally, though, any set with n elements would do.

Now let A be an index set and suppose that to each $\alpha \in A$ there is associated a nonempty set X_α. The X_α need not be different sets for different choices of α. The *Cartesian product,* denoted by

$$\prod_{\alpha\in A} X_\alpha,$$

is by definition the set of all functions $x : A \to \cup_{\alpha\in A}X_\alpha$ with $x(\alpha) \in X_\alpha$ for all $\alpha \in A$. When the index set is understood, the product may also be denoted by ΠX_α. The elements $x(\alpha)$ are called the *components* of the point $x \in \Pi X_\alpha$. This rather odd-looking definition works out in practice to be what one would expect. For instance, in the case where $A = \{1,\ldots,n\}$ and $X_\alpha = \mathbb{R}$ for all $\alpha \in A$, ΠX_α is the set of all functions from $\{1,\ldots,n\}$ to $\mathbb{R}$. This set of functions is of course just the same as the set of ordered n-tuples of elements of $\mathbb{R}$ since a function from $\{1,\ldots,n\}$ can be thought of as an ordered n-tuple, the n-tuple of its values in order. Thus one recovers the usual concept of $\mathbb{R}^n$.

Now suppose that X_α, $\alpha \in A$, is an (indexed) family of topological spaces, that is, X_α is a topological space for each α in some index set A. We now define a topology on the Cartesian product ΠX_α by specifying a base for the topology. The base is to consist of all sets of the form

$$\{x \in X : x(\alpha_j) \in U_{\alpha_j},\ 1 \le j \le m\}, \tag{12.1}$$

where $\{\alpha_1,\ldots,\alpha_m\}$ is a finite subset of A and each U_{α_j} is an open subset of X_{α_j}, $1 \le j \le m$. In other words, the set U is defined by restricting a finite number m of its components to lie in open subsets of the corresponding spaces X_α while all other components are unrestricted. In this formulation, it is clear that the intersection of two sets of this form is again of this form. By Theorem 4.2, these sets form a base for a topology for the product set $X = \Pi X_\alpha$, which we call the *product topology* for X.

If A consists of the integers $\{1,\ldots,n\}$, then, as noted, the product set X as defined coincides with the usual product space $X_1 \times \cdots \times X_n$. The base of open sets described by (10.1) are in this case, since m can be taken to be n, precisely the sets that can be written in the form

$$U = U_1 \times \cdots \times U_n,$$

where U_j is an open subset of X_j, $1 \le j \le n$. Thus one sees that the product topology just defined for arbitrary product spaces is the same as the product topology on $X_1 \times \cdots \times X_n$ introduced in Section 10.

Note that in the case where the index set A is infinite, sets of the form $\Pi_{\alpha\in A} U_\alpha$, U_α open in X_α, need not be open in $\Pi_{\alpha\in A} X_\alpha$ in the product topology. The question arises of why one does *not* define the product topology in the infinitely-many-factors case by using as a base the set of sets of the form ΠU_α, U_α open in X_α. These sets do form a base for an interesting topology called the *box topology* (Exercise 11). However, the box topology contains so many open sets that it is simply not as useful as the product topology. Specifically, the Cartesian product of compact spaces is compact when endowed with the product topology (Tychonoff's Theorem), whereas it is usually not compact when it is endowed with the box topology (Exercise 11c). Furthermore, the box topology may be a totally disconnected topology even though each X_α is connected (Exercise 12).

Unless specified otherwise, we shall assume that the product of topological spaces is endowed with the product topology.

For each $\alpha \in A$, there is a natural projection

$$\pi_\alpha : X \to X_\alpha$$

defined by

$$\pi_\alpha(x) = x(\alpha), \qquad x \in X.$$

In terms of the projections π_α, the basic open set U given by (12.1) can also be written

$$U = \pi_{\alpha_1}^{-1}(U_{\alpha_1}) \cap \cdots \cap \pi_{\alpha_m}^{-1}(U_{\alpha_m}). \tag{12.2}$$

In particular, any set of the form $\pi_\alpha^{-1}(U_\alpha)$, where U_α is an open subset of X_α, is open in X. Hence each π_α is continuous. In fact, Theorem 10.1 remains valid for infinite products.

12.1 Theorem: The product topology for ΠX_α is the smallest topology for which each projection π_α, $\alpha \in A$, is continuous.

Proof: Let $\mathcal{T}$ be a topology for X for which the π_α's are continuous. If U_{α_j} is an open subset of X_{α_j}, $1 \leq j \leq m$, then $\pi_{\alpha_j}^{-1}(U_{\alpha_j})$ belongs to $\mathcal{T}$ and $\mathcal{T}$ includes all sets of the form (12.2). Since these sets form a base for the product topology, $\mathcal{T}$ includes the product topology. □

Each of the theorems stated in Section 10 remains valid for arbitrary product spaces. Most of these will be relegated to the exercises. The analogue of Theorem 10.3, however, is sufficiently important to merit a separate statement and proof.

12.2 Theorem: Let E be a topological space and let f be a function from E to ΠX_α. Then f is continuous if and only if $\pi_\alpha \circ f$ is continuous for all $\alpha \in A$.

Proof: If f is continuous, then each $\pi_\alpha \circ f$ is continuous, since the composition of continuous functions is continuous.

Conversely, suppose that each $\pi_\alpha \circ f$ is continuous. Let U be the basic open subset of X given by (12.2). Then

$$f^{-1}(U) = f^{-1}(\pi_{\alpha_1}^{-1}(U_{\alpha_1})) \cap \cdots \cap f^{-1}(\pi_{\alpha_m}^{-1}(U_{\alpha_m})).$$

Since each $\pi_{\alpha_j} \circ f$ is continuous, each $(\pi_{\alpha_j} \circ f)^{-1}(U_{\alpha_j}) = f^{-1}(\pi_{\alpha_j}^{-1}(U_{\alpha_j}))$ is open and $f^{-1}(U)$ is open. Since inverse images preserve unions, the inverse image of any open subset of X is open in E. Hence f is continuous. □

The remainder of this section will be devoted to Tychonoff's Theorem, asserting that an arbitrary product of compact spaces is compact. For this purpose, we introduce the notion of a subbase for a topology and we establish a theorem that will play a role analogous to that played by Lemma 10.7 in the case of finite products.

A family $\mathcal{S}$ of open subsets of a topological space Y is a *subbase* for the topology of Y if the family of finite intersections of sets in $\mathcal{S}$ forms a base for the topology of Y. For example, the family of open sets

$$\{\pi_\alpha^{-1}(U_\alpha) : \alpha \in A,\ U_\alpha \text{ an open subset of } X_\alpha\}$$

forms a subbase for the product topology of ΠX_α. The delicate part of the proof of Tychonoff's Theorem is included in the following result.

12.3 Theorem *(Alexander Subbase Theorem):* Let Y be a topological space and let $\mathcal{S}$ be a subbase for the topology of Y. If every open cover of Y by sets in $\mathcal{S}$ has a finite subcover, then Y is compact.

Proof: Let $\mathcal{C}$ be an open cover of Y such that no finite subfamily of $\mathcal{C}$ covers Y. It suffices to show that there exists a cover of Y by sets in $\mathcal{S}$ that has no finite subcover.

Consider families $\mathcal{D}$ of open subsets of Y such that $\mathcal{C} \subseteq \mathcal{D}$ and such that no finite collection of sets in $\mathcal{D}$ covers Y. The aggregate $\mathcal{P}$ of such families forms a partially ordered set, with the ordering of inclusion. Let $\{\mathcal{D}_\beta\}_{\beta \in B}$ be a totally ordered subset of $\mathcal{P}$. Let $\mathcal{D}$ be the family of subsets of Y that is the union of the $\mathcal{D}_\beta$'s. We claim that $\mathcal{D} \in \mathcal{P}$, that is, that no finite collection of sets in $\mathcal{D}$ covers Y. Indeed, any finite collection of sets in $\mathcal{D}$ is included in one of the $\mathcal{D}_\beta$'s, because of the total ordering, and no finite family of sets in any $\mathcal{D}_\beta$ covers Y. (Compare this argument with the corresponding point of the proof that every vector space has a basis.) Evidently $\mathcal{D}$ is an upper bound for $\{\mathcal{D}_\beta\}_{\beta \in B}$. Hence the hypotheses of Zorn's Lemma are met. By Zorn's Lemma, there is a maximal element $\mathcal{E}$ in $\mathcal{P}$.

It suffices now to show that the family $\mathcal{E} \cap \mathcal{S}$ covers Y. Indeed, by the defining property of $\mathcal{P}$, no finite subfamily of $\mathcal{E} \cap \mathcal{S}$ covers Y.

Let $x \in Y$. It suffices to show that there exists $V \in \mathcal{E} \cap \mathcal{S}$ such that $x \in V$.

Choose $U \in \mathcal{E}$ such that $x \in U$. Since $\mathcal{S}$ is a subbase, there exist $V_1, \ldots, V_m$ in $\mathcal{S}$ such that $x \in V_1 \cap \cdots \cap V_m \subseteq U$. Suppose that no V_j belongs to $\mathcal{E}$. Because of the maximality of $\mathcal{E}$, there exist for each j a finite number of open sets $W_{j1}, \ldots, W_{j,n_j}$ in $\mathcal{E}$ such that the W_{ji}'s, $1 \le i \le n_j$, together with V_j, cover Y. (Otherwise we could add V_j to $\mathcal{E}$, contradicting the maximality of $\mathcal{E}$.) Then $V_1 \cap \cdots \cap V_m$, together with all the W_{ji}'s, $1 \le i \le n_j$, $1 \le j \le m$, cover Y. Hence U together with the W_{ji}'s forms a finite cover of Y from $\mathcal{E}$, contradicting the fact that $\mathcal{E}$ belongs to $\mathcal{P}$. We conclude that some V_j belongs to $\mathcal{E}$, and we take this V_j to be our V. □

12.4 Theorem *(Tychonoff's Theorem):* Any product of compact spaces is compact.

Proof: Let $X = \Pi X_\alpha$, where each X_α is compact. Let $\mathcal{D}$ be a family of subbasic sets of the form $\pi_\alpha^{-1}(U_\alpha)$, where $\alpha \in A$ and U_α is an open subset of X_α. Suppose that no finite subfamily of $\mathcal{D}$ covers X. In view of the Alexander Subbase Theorem, it suffices to show that $\mathcal{D}$ does not cover X.

Fix an index β and consider the subbasic sets in $\mathcal{D}$ of the form $\pi_\beta^{-1}(V)$, where V is an open subset of X_β. The aggregate of such open sets V cannot cover X_β. (Otherwise we could extract a finite subcover $\{V_1, \ldots, V_m\}$ of X_β, and the corresponding sets $\pi_\beta^{-1}(V_j)$, $1 \le j \le m$, in $\mathcal{D}$ would cover X, contradicting our hypothesis on $\mathcal{D}$.) Choose a point $x(\beta) \in X_\beta$ such that $x(\beta)$ is not included in the union of the V's. The various choices $x(\beta)$ determine a point $x \in \Pi X_\alpha$, which is evidently not included in any of the sets in $\mathcal{D}$. □

Note that we have used the Axiom of Choice in the proof of Theorem 12.4 and that the proof of Theorem 12.3 depends on Zorn's Lemma. Can one give a proof of

Tychonoff's Theorem that does not involve some form of transfinite induction? The answer turns out to be negative. As mentioned in the preceding section, it turns out that the Axiom of Choice is equivalent to Tychonoff's Theorem. The proof that Tychonoff's Theorem implies the Axiom of Choice is outlined in Exercise 9.

EXERCISES

1. Prove that if E_α is a closed subset of X_α for all α, then ΠE_α is a closed subset of ΠX_α.
2. Prove that each projection π_β of ΠX_α onto a coordinate space X_β is an open map.
3. Prove that the restriction of the projection π_β to the coordinate slice $X_\beta \times \{y\}$ is a homeomorphism from $X_\beta \times \{y\}$ to X_β, $X_\beta \times \{y\}$ being given the relative topology as a subset of ΠX_α.
4. Prove that the product of Hausdorff spaces is Hausdorff.
5. Prove that the product of regular spaces is regular.
6. Prove that the product of connected spaces is connected. *Hint:* Show that a connected component of the product is dense and apply Exercise 8.9.
7. Identify the connected components of $X = \Pi X_\alpha$ in terms of those of the X_α's. Prove your answer.
8. Prove that the product of path-connected spaces is path-connected.
9. Let S_α be a nonempty set for each α belonging to an index set A. Using Tychonoff's Theorem and elementary reasoning, prove that ΠS_α is not empty. Deduce that Tychonoff's Theorem implies the Axiom of Choice. *Hint:* Let X_α be obtained from S_α be adjoining one point p_α. Endow X_α with the topology whose open sets are the cofinite subsets of X_α together with the empty set and the singleton $\{p_\alpha\}$. Show that the subsets $\pi_\alpha^{-1}(S_\alpha)$ of ΠX_α have the finite intersection property.
10. Let S be a set. For each $s \in S$, let X_s be the discrete topological space consisting of two points $\{0,1\}$ and let $X = \Pi\{X_s : s \in S\}$.
    (a) Show that X may be identified with the family of all subsets of S.
    (b) Prove that X is a compact Hausdorff space.
    (c) When is X second-countable?
    (d) When does X satisfy the first axiom of countability (defined in Exercise 4.4)?
11. Let X_α, $\alpha \in A$, be a collection of topological spaces. Let $\mathcal{B}$ be the family of subsets of ΠX_α of the form ΠU_α, where each U_α is an open subset of X_α.
    (a) Show that $\mathcal{B}$ forms a base for a topology for ΠX_α. (This topology is called the *box topology*.)
    (b) Show that if each X_α is discrete, then the box topology is the discrete topology for ΠX_α.

(c) Show that ΠX_α with the box topology, need not be compact, even though each X_α is compact. *Hint:* Let each X_α be the discrete space consisting of two points.

(d) Show that if each X_α is Hausdorff, then the box topology for ΠX_α is a Hausdorff topology. Is the same assertion true for regularity? for normality? Justify your answers.

12. Let A be an index set, and for each $\alpha \in A$, let $X_\alpha = \mathbb{R}$.

(a) Let w be any real-valued function on A such that $w(\alpha) > 0$ for all $\alpha \in A$. Fix $x \in \Pi X_\alpha$, and let E be the set of $y \in \Pi X_\alpha$ such that $|x(\alpha) - y(\alpha)|w(\alpha)$ is bounded on A. Prove that E is a closed and open subset of ΠX_α, with the box topology from Exercise 11.

(b) If the index set A is infinite, prove that ΠX_α, with the box topology, is disconnected.

13. QUOTIENT SPACES

The product construction introduced in Section 10 could be viewed as a generalization of the familiar geometric process of the construction of $\mathbb{R}^2$ and $\mathbb{R}^3$ from the real line $\mathbb{R}$. This section is devoted to the mathematical version of another familiar geometric process, that of gluing pieces of a geometric object together to form another geometric object. For instance, one can think of a triangle as being formed by the gluing together of three closed line segments at their endpoints. One can form a circle from a closed line segment by bending the segment around and gluing the ends together.

Set-theoretically, it is easy to formulate a precise version of the idea of gluing two points together. Let X be a set and let p and q be the two points of X that are to be attached. One defines an equivalence relation on X so that p and q are equivalent, while each remaining point of X is equivalent only to itself. The set Y of equivalence classes of this relation is then just what we want. There is an equivalence class for each point distinct from p and q, while p and q together form a single equivalence class. The set Y is said to be "obtained from X by identifying p and q."

There are more complicated ways of gluing pieces of spaces together. For instance, one can form a cylindrical pipe from a rectangular plate by bending the plate around and welding two edges together, as in the figure. By further bending the

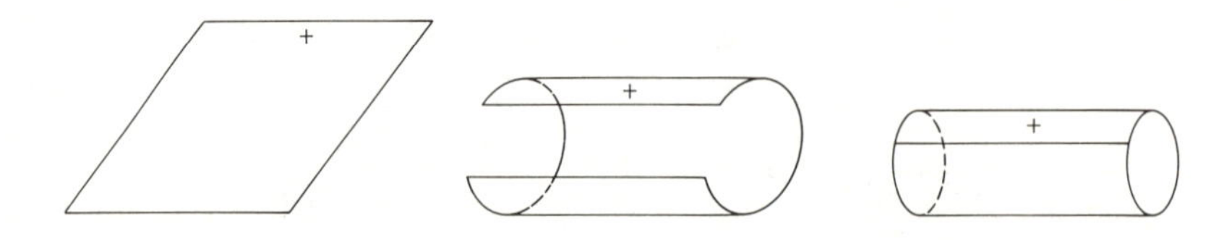

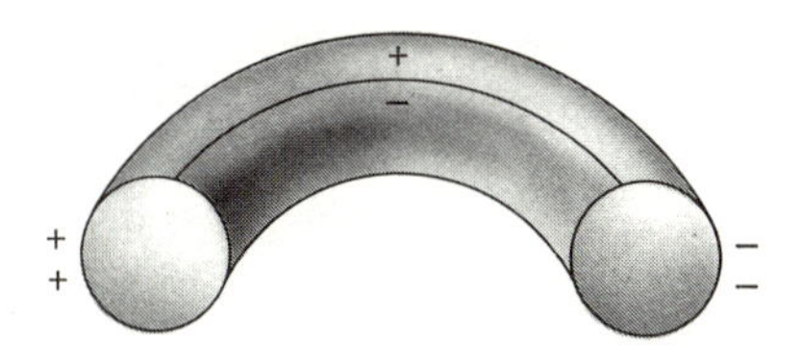

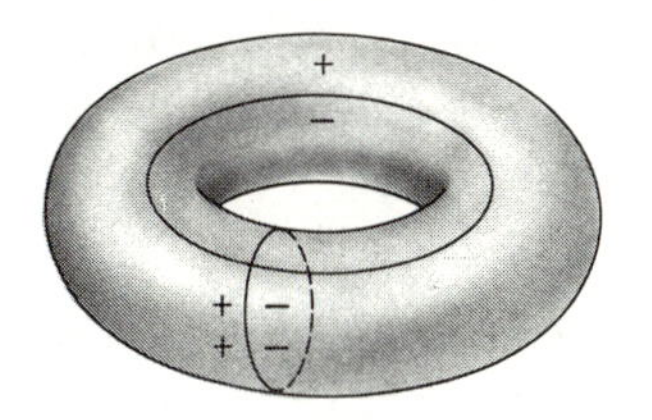

cylinder around and welding the two circular rims together, one obtains a metal doughnut, or torus.

Spaces obtained from a given space by welding or gluing subsets together are called *quotients* of the given space. Thus the torus is the quotient space of a rectangle, obtained by identifying pairs of points that lie opposite each other on parallel edges. Note that the quotient construction is not as such an inverse of the product construction. The terminology is not really appropriate, though it is by now standard. We turn now to the formal definition of a quotient space.

Let X be a set and let "$\sim$" be an equivalence relation on X. The equivalence classes corresponding to $\sim$ form a partition of X into pairwise disjoint subsets whose union is X. The set of equivalence classes will be denoted by $X/\sim$ and called the *quotient set* of X modulo the equivalence relation. There is a natural projection

$$\pi : X \to X/\sim,$$

such that $\pi(x)$ is the equivalence class of x.

Now suppose that X is a topological space. We wish to introduce a reasonable topology on the quotient set $X/\sim$. A minimum requirement is that the projection π be continuous. Of course, the indiscrete topology on $X/\sim$ will always satisfy this requirement, but the indiscrete topology gives us no information about $X/\sim$.

If π is to be continuous and if U is an open subset of $X/\sim$, then $\pi^{-1}(U)$ must be open in X. This leads us to define the *quotient topology* for $X/\sim$ to consist of all sets $U \subset X/\sim$ such that $\pi^{-1}(U)$ is open in X. Since inverse functions preserve unions and intersections, the quotient topology is indeed a topology. From the discussion above, we obtain the following characterization of the quotient topology, which is analogous to Theorems 10.1 and 12.1.

13.1 Theorem: Let X be a topological space and let $\sim$ be an equivalence relation on X. Then the quotient topology for $X/\sim$ is the largest topology for which the projection $\pi : X \to X/\sim$ is continuous.

Theorem 12.2 also has an analogue.

13.2 Theorem: Let X be a topological space, let $\sim$ be an equivalence relation on X, and let $\pi : X \to X/\sim$ be the projection map. A function f from $X/\sim$ to a topological space Y is continuous if and only if $f \circ \pi$ is continuous.

Proof: Since the composition of continuous functions is continuous, the continuity of f implies the continuity of $f \circ \pi$. Conversely, suppose that $f \circ \pi$ is continuous. Let

V be an open subset of Y. Then $(f \circ \pi)^{-1}(V) = \pi^{-1}(f^{-1}(V))$ is an open subset of X. By the definition of quotient topology, $f^{-1}(V)$ is an open subset of $X/\sim$. Consequently f is continuous. □

13.3 Theorem: Let f be a continuous function from a topological space X to a topological space Y. Let $\sim$ be an equivalence relation on X such that f is constant on each equivalence class. Then there exists a continuous function $g : X/\sim \to Y$ such that $f = g \circ \pi$.

Proof: The picture which accompanies Theorem 13.3 is the following:

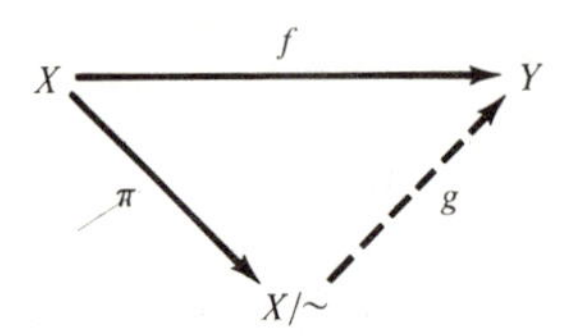

Define the value of g on an equivalence class to be the value of f at any element of the equivalence class. Then g is well defined and $g \circ \pi = f$. The continuity of g follows from Theorem 13.2. □

13.4 Theorem: Let X and Y be compact Hausdorff spaces and let f be a continuous function from X onto Y. Define an equivalence relation $\sim$ in X by declaring $x_0 \sim x_1$ if and only if $f(x_0) = f(x_1)$. Then $X/\sim$ is homeomorphic to Y.

Proof: Since $X/\sim$ is the image of X under the continuous function π, $X/\sim$ is compact. Let g be the continuous function from $X/\sim$ to Y which satisfies $g \circ \pi = f$. Then g is continuous, one-to-one, and onto. By Theorem 6.7, g is a homeomorphism. □

As a simple application of Theorem 13.4, consider the topological space Y obtained from the closed unit interval $[0,1]$ by identifying 0 and 1. In other words, $Y = [0,1]/\sim$, where the equivalence classes of $\sim$ are the singletons $\{s\}$, $0 < s < 1$, and the set $\{0,1\}$. Consider the map $f : [0,1] \to S^1$ defined by

$$f(t) = e^{2\pi i t}, \qquad 0 \leq t \leq 1.$$

The equivalence classes are precisely the level sets of f. By Theorem 13.4, f induces a homeomorphism of Y and S^1.

As another example, let $X = [0,1] \times [0,1]$ and let f be the continuous function of X onto the torus $S^1 \times S^1$ defined by

$$f(s,t) = (e^{2\pi i s}, e^{2\pi i t}), \qquad 0 \leq s, t \leq 1.$$

The level sets of f are then the singletons "inside" the square X, the pairs of opposite points on the interiors of the boundary intervals of X, and the set of four corners of X. Consequently the quotient space obtained from X by making these identifications is the torus, as can be seen from the figure discussed earlier in this section.

Some properties, such as compactness, connectedness, and path-connectedness, are passed on from any topological space having the property to all of its quotient spaces (Exercise 1). However, most properties are not passed on to quotient spaces (Exercises 2 and 3), and in fact quotient spaces are simply not very tractable. It is more important to understand the several basic examples of quotient spaces rather than to concentrate on a general theory of quotient spaces.

EXERCISES

1. Let $X/\sim$ be the quotient space determined by an equivalence relation $\sim$ on a topological space X. Prove the following assertions:
   (a) If X is compact, then $X/\sim$ is compact.
   (b) If X is connected, then $X/\sim$ is connected.
   (c) If X is path-connected, then $X/\sim$ is path-connected.
2. Let $\sim$ be an equivalence relation on a topological space X. Prove that $X/\sim$ is a T_1-space if and only if each equivalence class is closed. Give an example of a T_1-space X and an equivalence relation $\sim$ such that $X/\sim$ is not a T_1-space.
3. Define an equivalence relation in $X = [0,1] \times [0,1]$ by declaring $(s_0,t_0) \sim (s_1,t_1)$ if and only if $t_0 = t_1 > 0$. Describe the quotient space $X/\sim$ and show that it is not a Hausdorff space.
4. Let f be an open continuous map of a topological space X onto a topological space Y. Show that Y is homeomorphic to the quotient space of X obtained by identifying each level set of f to a point.
5. Let E be a closed subset of a compact Hausdorff space X. Prove that the quotient space obtained from X by identifying E to a point is homeomorphic to the one-point compactification of $X \backslash E$.
6. Let B^n be the closed unit ball in $\mathbb{R}^n$. Prove that the quotient space obtained from B^n by identifying its boundary S^{n-1} to a point is homeomorphic to the n-sphere S^n.
7. Let $X = X_1 \times \cdots \times X_n$ be a product of topological spaces. Define an equivalence relation $\sim$ on X by declaring $(x_1,\ldots,x_n) \sim (y_1,\ldots,y_n)$ if and only if $x_1 = y_1$. Show that $X/\sim$ is homeomorphic to X_1. Prove an analogous result for an infinite product space $X = \Pi_{\alpha\epsilon A} X_\alpha$.
8. For $n \geq 1$, define $P^n = S^n/\sim$, where the equivalence relation is defined by declaring $x \sim y$ if and only if $x = y$ or $x = -y$. In other words, P^n is obtained from S^n by identifying pairs of antipodal points. The space P^n is called *real projective space* of dimension n, and it can be regarded as the set of lines in $\mathbb{R}^{n+1}$ which pass through the origin. Establish the following assertions:
   (a) P^n is a compact Hausdorff space.
   (b) The projection $\pi : S^n \to P^n$ is a local homeomorphism, that is, each $x \in S^n$ has an open neighborhood that is mapped homeomorphically by π onto an open neighborhood of $\pi(x)$.

(c) P^1 is homeomorphic to the circle S^1.

(d) P^n is homeomorphic to the quotient space obtained from the closed unit ball B^n in $\mathbb{R}^n$ by identifying antipodal points of its boundary S^{n-1}.

9. Let $n \geq 1$ and let $X = \mathbb{C}^{n+1}\backslash\{0\}$. Let $\sim$ be the equivalence relation in X obtained by declaring $x \sim y$ if $x = \lambda y$ for some complex number λ. Then $CP^n = X/\sim$ is called *complex projective space* of dimension n. It can be regarded as the set of one-dimensional subspaces of the complex vector space $\mathbb{C}^{n+1}$.

   (a) Prove that CP^n is a compact Hausdorff space.

   (b) Prove that CP^1 is homeomorphic to S^2.

   (c) Regard S^{2n+1} as the set of vectors $z = (z_0, \ldots, z_n) \in \mathbb{R}^{n+1}$ such that $|z_0|^2 + \cdots + |z_n|^2 = 1$. Show that the natural projection $\pi : S^{2n+1} \to CP^n$ is onto and that $\pi^{-1}(x)$ is homeomorphic to S^1 for all $x \in CP^n$. Show that each $x \in CP^n$ has an open neighborhood U such that $\pi^{-1}(U)$ is homeomorphic to the product space $U \times S^1$.

Remark: Intuitively speaking, S^{2n+1} is a "twisted product" of CP^n and S^1. We say that S^{2n+1} is a *fiber space* over CP^n, with fiber S^1.

Homotopy Theory

THREE

Many questions that arise naturally in topology are difficult to handle if only the basic definitions and their immediate consequences are used. It is certainly natural to inquire whether two given topological spaces are homeomorphic, and yet this question is hard to answer even for familiar spaces. Intuition suggests that the punctured plane $\mathbb{R}^2\backslash\{(0,0)\}$ is not homeomorphic to the plane $\mathbb{R}^2$, but a precise proof of this fact using only the foundational material of Chapter II is difficult. A successful approach to such problems involves a principle that at first description may seem unduly abstract. It is the principle of associating with topological objects certain algebraic objects, thus converting topological problems into algebraic problems, which, one hopes, are more tractable. This idea has developed into a whole branch of mathematics, *algebraic topology*. The purpose of this chapter is to develop an important concrete instance of the general process.

We begin in Section 1 by introducing the appropriate algebraic concept, that of a *group*. Sections 2 through 6 form a unit in which the fundamental group of a topological space is introduced and the fundamental group of the circle is computed from a geometric point of view using covering spaces. Several elementary applications are given, such as the Brouwer Fixed Point Theorem in dimension two. In Sections 7 and 8 we discuss homotopic maps and give an alternative computation of the fundamental group of the circle, this time from an analytic point of view. Sections 9 and 10 include deeper applications of the ideas introduced in this chapter. In Section 9 we prove that every vector field on the 2-sphere has a zero, and in Section 10 we prove the deceptively difficult Jordan Curve Theorem, which states that every simple closed curve in the plane has an "inside" and an "outside."

1. GROUPS

A *group* is a set G, and a mapping $(a,b) \to a * b$ of $G \times G$ into G, that satisfy the following axioms:

(1.1) *Associativity of Multiplication:* $a * (b * c) = (a * b) * c$ for all a, b, $c \in G$.

(1.2) *Existence of an Identity:* There exists an element $e \in G$ such that $e * a = a * e = a$ for all $a \in G$.

(1.3) *Existence of Inverse:* For each element $a \in G$, there exists an element $b \in G$ such that $a * b = b * a = e$.

Note that the identity of a group is unique. Indeed, if e and e' both satisfy (1.2), then $e = e * e' = e'$.

If both b and b' are elements of G that satisfy (1.3), then

$$b = b * e = b * (a * b') = (b * a) * b' = e * b' = b'.$$

Consequently the element b satisfying (1.3) is unique. It is called the *inverse of a* and denoted by a^{-1}, so that

$$a * a^{-1} = a^{-1} * a = e, \qquad a \in G.$$

By (1.2), $e^{-1} = e$.

The group G is *commutative*, or *abelian*, if $a * b = b * a$ for all $a, b \in G$. Examples of groups abound. Here is a list of a few common groups:

(i) The trivial group consisting of only one element, the identity, with the only possible operation, is obviously a commutative group.

(ii) The set $\mathbb{Z}$ of integers, with the operation of addition, forms an abelian group. The integer 0 is the identity of $\mathbb{Z}$, and the inverse of m is $-m$.

(iii) The set $\mathbb{R}$ of real numbers and the set $\mathbb{C}$ of complex numbers, with the operation of addition, are commutative groups.

(iv) Any real or complex vector space, with the operation of addition, is a commutative group. In particular, $\mathbb{R}^n$ and $\mathbb{C}^n$ are commutative groups.

(v) The set of nonzero real numbers, with the operation of multiplication, is a commutative group. The identity is 1, and the inverse of s is $1/s$.

(vi) The unit circle S^1 in the complex plane, with the operation of multiplication, is a commutative group. The identity is 1, and the inverse of $a \in S^1$ is the complex conjugate $\bar{a}$ of a.

(vii) The set $GL(n,\mathbb{R})$ of invertible $n \times n$ matrices with real entries, with matrix multiplication as the operation, forms a group. If we allow complex entries, we obtain a group $GL(n,\mathbb{C})$. These groups are called *general linear groups*. For $n \geq 2$, they are not commutative.

(viii) The set S_n of permutations of n objects, with composition as the operation, forms a group called the *symmetric group*. The group S_n has $n!$ elements. If $n \geq 3$, S_n is not commutative.

(ix) The set $\mathbb{Z}_n$ of congruence classes of integers modulo n, with the operation of addition, forms a commutative group of n elements.

A subset H of a group G is a *subgroup* of G if H forms a group when endowed with the multiplication inherited from G. In order that a subset H of G be a subgroup of G, it is necessary and sufficient that H be "closed" under multiplication and taking inverses, that is, that

(1.4) $\qquad a * b \in H \qquad$ whenever $a \in H$ and $b \in H$, and

(1.5) $\qquad a^{-1} \in H \qquad$ whenever $a \in H$.

For example, $\mathbb{Z}$ is a subgroup of $\mathbb{R}$ and $\mathbb{R}$ is a subgroup of $\mathbb{C}$ (operation of addition). Subspaces of a vector space are in particular subgroups of the vector space. However, a subgroup of a vector space need not be a subspace, as indicated by the example $\mathbb{Z} \subset \mathbb{R}$.

Now let G and H be two groups. A function f from G to H is a *homomorphism* if

$$f(a * b) = f(a) * f(b), \qquad \text{all } a, b \in G. \tag{1.6}$$

Here the product $a * b$ is taken in G and the product $f(a) * f(b)$ is taken in H. For instance, a map h from $\mathbb{R}$ (operation of addition) to the nonzero real numbers (operation of multiplication) is a homomorphism if and only if

$$h(s + t) = h(s)h(t), \qquad s,t \in \mathbb{R}.$$

An example of such a map is given by

$$h(t) = e^t, \qquad t \in \mathbb{R}.$$

Some other examples of homomorphisms are as follows:

(i) The map $f : \mathbb{Z} \to \mathbb{Z}$ defined by

$$f(m) = 5m, \qquad m \in \mathbb{Z}.$$

(ii) The map $f : \mathbb{R} \to S^1$ defined by

$$f(s) = e^{is}, \qquad s \in \mathbb{R}.$$

(iii) The map $f : \mathbb{Z} \to \mathbb{Z}_n$ defined so that $f(m)$ is the congruence class of m (modulo n).

(iv) The map $f : \mathbb{R} \to GL(2,\mathbb{R})$ defined by

$$f(t) = \begin{pmatrix} 1 & t \\ 0 & 1 \end{pmatrix}, \qquad t \in \mathbb{R}.$$

(v) If H is a subgroup of G, then the inclusion map $H \hookrightarrow G$ is a homomorphism.

Let G and H be groups and let $f : G \to H$ be a homomorphism. If e is the identity of G and e' is the identity of H, then $f(e) = e'$. This follows from the chain of identities

$$e' = f(e) * f(e)^{-1} = f(e * e) * f(e)^{-1} = f(e) * f(e) * f(e)^{-1} = f(e).$$

(Here we ignore the order in which the operations are performed since they are associative.) From the identity $e' = f(e) = f(aa^{-1}) = f(a)f(a^{-1})$, we conclude that

$$f(a^{-1}) = f(a)^{-1}, \qquad a \in G.$$

If K is another group and both $f : G \to H$ and $g : H \to K$ are homomorphisms, then the composition $g \circ f : G \to K$ is also a homomorphism. Indeed,

$$\begin{aligned}(g \circ f)(a * b) &= g(f(a * b)) = g(f(a) * f(b)) = g(f(a)) * g(f(b)) \\ &= [(g \circ f)(a)] * [(g \circ f)(b)]\end{aligned}$$

whenever $a,b \in G$.

A homomorphism $f : G \to H$ is an *isomorphism* if f is one-to-one and onto. If there is an isomorphism of G and H, then G and H are said to be *isomorphic*, written $G \cong H$. The composition of two isomorphisms is an isomorphism. Furthermore, the inverse of an isomorphism is an isomorphism. Indeed, if $x,y \in H$, say $x = f(a)$ and $y = f(b)$, then $f(a * b) = x * y$, so that

$$f^{-1}(x * y) = a * b = f^{-1}(x) * f^{-1}(y).$$

One familiar isomorphism is the map $f : \mathbb{R}^2 \to \mathbb{C}$ given by

$$f((x,y)) = x + iy, \qquad (x,y) \in \mathbb{R}^2.$$

Another example of an isomorphism is the function f from $\mathbb{R}$ to the group $\mathbb{R}_+$ of positive real numbers, with the operation of multiplication, given by

$$f(t) = e^t, \qquad t \in \mathbb{R}.$$

EXERCISES

1. Prove that any two groups with one element are isomorphic. Prove that any two groups with two elements are isomorphic. Prove that any two groups with three elements are isomorphic.
2. Show that for any fixed positive integer n, there are at most a finite number of nonisomorphic groups of order n. *Hint:* If $\alpha_1, \ldots, \alpha_n$ are the elements of the group, then the group is determined by knowing all the products $\alpha_i * \alpha_j$, $1 \leq i \leq j \leq n$.
3. Find the number of nonisomorphic groups of order 4.
4. Prove that if f is a homomorphism from a group G to a group H, then $f(G)$ is a subgroup of H. Prove that if e is the identity of H, then $f^{-1}(e)$ is a subgroup of G.
5. Let G and H be groups. Define an operation in the Cartesian product $G \times H$ by

$$(a_1,b_1) * (a_2,b_2) = (a_1 * a_2, b_1 * b_2), \qquad a_1,a_2 \in G;\ b_1,b_2 \in H.$$

Show that endowed with this operation, $G \times H$ becomes a group. Show that $G \times H$ is commutative if and only if G and H are commutative.

Remark. $G \times H$ is the *direct product* of G and H. If G and H are abelian and if the operations in G and H are denoted by "+", then it is customary to refer to the direct product of G and H as the *direct sum* of G and H and denote it by $G \oplus H$. For example, the group $\mathbb{R} \oplus \mathbb{R}$ is isomorphic to $\mathbb{R}^2$ and $\mathbb{Z} \oplus \mathbb{Z}$ is isomorphic to the subgroup of integral lattice points in $\mathbb{R}^2$.

2. HOMOTOPIC PATHS

Let X be a topological space and let $a,b \in X$. Recall that a path in X from a to b is a map $\gamma : [0,1] \to X$ such that $\gamma(0) = a$ and $\gamma(1) = b$. Two paths γ_0 and γ_1 from a to b are *homotopic with endpoints fixed,* written $\gamma_0 \simeq \gamma_1$ rel$\{0,1\}$, if there is a map

$$F : [0,1] \times [0,1] \to X$$

such that

$$F(s,0) = \gamma_0(s), \qquad 0 \le s \le 1,$$

$$F(s,1) = \gamma_1(s), \qquad 0 \le s \le 1,$$

$$F(0,t) = a, \qquad 0 \le t \le 1,$$

$$F(1,t) = b, \qquad 0 \le t \le 1.$$

The map F is referred to as a *homotopy* of γ_0 and γ_1. For each $t \in [0,1]$, the map $\gamma_t : [0,1] \to X$ defined by

$$\gamma_t(s) = F(s,t), \qquad 0 \le s \le 1, \tag{2.1}$$

is then a path from a to b. The variable t is to be regarded as a parameter, and the paths γ_t move "continuously" with t. As the parameter t moves from 0 to 1, the path γ_0 is continuously deformed to the path γ_1 through the γ_t's. Often we shall refer to $\{\gamma_t\}_{0 \le t \le 1}$ as a homotopy of γ_0 and γ_1, understanding that the homotopy F can be recovered from the γ_t's by (2.1). The situation may be represented by a square, representing the parameter space, with left edge labeled "a" to indicate that the restriction of the homotopy F to the left edge is the constant path at a, with bottom edge labeled "γ_0" to indicate that the restriction of F to the bottom edge coincides with γ_0, etc. (see the diagram).

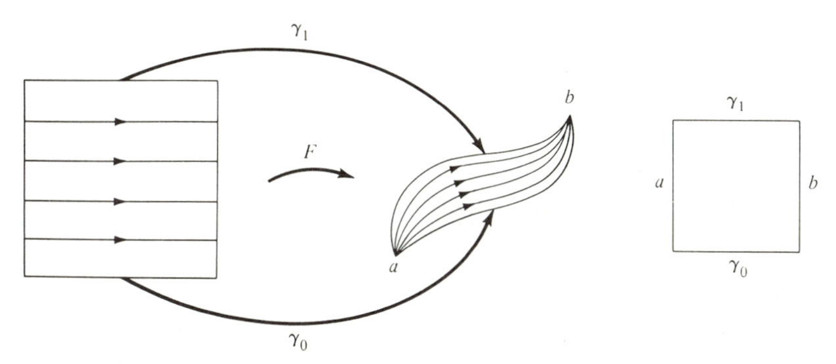

A subset X of $\mathbb{R}^n$ is *convex* if, whenever x and y belong to X, then the straight-line interval joining x to y lies in X. In other words, X is convex if, whenever $x,y \in X$, then $tx + (1 - t)y \in X$ for $0 \le t \le 1$. The following theorem provides the fundamental example of homotopic paths.

2.1 Theorem: Let X be a convex subset of $\mathbb{R}^n$ and let $a,b \in X$. Then any two paths in X from a to b are homotopic in X with endpoints fixed.

Proof: Let γ_0 and γ_1 be paths in X from a to b. The homotopy $\{\gamma_t\}$ is defined so that for each fixed s, $\gamma_t(s)$ moves from $\gamma_0(s)$ to $\gamma_1(s)$ along the straight-line interval joining the points. An explicit homotopy is given by defining

$$F(s,t) = \gamma_t(s) = (1 - t)\gamma_0(s) + t\gamma_1(s), \qquad 0 \le s,t \le 1.$$

One checks that each γ_t is a path in X from a to b and that the γ_t move continuously with t, that is, the homotopy F is continuous. □

The remainder of this section is devoted to a series of elementary lemmas, which will form the foundation for the discussion of the fundamental group in Section 3. We fix a topological space X.

2.2 Lemma: The relation $\gamma \simeq \alpha \operatorname{rel}\{0,1\}$ is an equivalence relation on the set of paths in X from a to b.

Proof: A homotopy of γ to γ is given by $\gamma_t = \gamma$, $0 \le t \le 1$, so that the relation is reflexive.

Suppose that $\gamma_0 \simeq \gamma_1 \operatorname{rel}\{0,1\}$ and let $\{\gamma_t\}$ be a homotopy of γ_0 and γ_1. Then by changing the direction of the parameter, we obtain a homotopy $t \to \gamma_{1-t}$, $0 \le t \le 1$, of γ_1 to γ_0. In other words,

$$F(s,t) = \gamma_{1-t}(s), \qquad 0 \le s,t \le 1,$$

is a homotopy of γ_1 and γ_0. Hence the relation is symmetric.

Suppose that $\gamma_0 \simeq \gamma_1 \operatorname{rel}\{0,1\}$ and $\gamma_1 \simeq \gamma_2 \operatorname{rel}\{0,1\}$. Let $\{\gamma_t\}_{0 \le t \le 1}$ be the homotopy of γ_0 and γ_1 and let $\{\alpha_t\}_{0 \le t \le 1}$ be the homotopy of γ_1 to γ_2, so that $\alpha_0 = \gamma_1$ and $\alpha_1 = \gamma_2$. Set

$$\beta_t = \begin{cases} \gamma_{2t}, & 0 \le t \le 1/2, \\ \alpha_{2t-1}, & 1/2 \le t \le 1. \end{cases}$$

Since $\gamma_1 = \alpha_0$, $\beta_{1/2}$ is well defined. The homotopy β_t corresponds to deforming γ_0 to γ_1 at double speed and then deforming γ_1 to γ_2 at double speed. The map $(s,t) \to \beta_t(s)$ is continuous, so that $\{\beta_t\}$ is indeed a homotopy of γ_0 and γ_2, and the relation is transitive. □

The equivalence classes of paths in X from a to b modulo this equivalence relation are called the *homotopy classes* of paths from a to b. The homotopy class of a path γ is denoted by $[\gamma]$. Then $[\gamma_0] = [\gamma_1]$ means that $\gamma_0 \simeq \gamma_1 \operatorname{rel}\{0,1\}$, that is, that γ_0 and γ_1 are homotopic with endpoints fixed.

The next lemma shows that any reparametrization of a path lies in the same homotopy class.

2.3 Lemma: Let γ be a path in X from a to b. Let ρ be any map from $[0,1]$ to $[0,1]$ such that $\rho(0) = 0$ and $\rho(1) = 1$. Then $[\gamma \circ \rho] = [\gamma]$.

Proof: Note that $\gamma \circ \rho$ is a path from a to b. It is obtained by running along γ at a varying speed, with perhaps some backsliding. A homotopy of γ to $\gamma \circ \rho$ is given explicitly by

$$\gamma_t(s) = \gamma((1 - t)s + t\rho(s)), \qquad 0 \le s,t \le 1.$$

This homotopy is precisely the composition of γ and the homotopy $\alpha_t(s) = (1 - t)s + t\rho(s)$ of ρ, regarded as a path in $[0,1]$ from 0 to 1, and the identity path $\alpha_0(s) = s$ from 0 to 1. In particular, γ_t is well defined and moves continuously with t. □

Now let $a,b,c \in X$, let α be a path from a to b, and let β be a path from b to c. As in Section II.9, we define a path $\alpha\beta$ from a to c by

$$(\alpha\beta)(s) = \begin{cases} \alpha(2s), & 0 \leq s \leq 1/2, \\ \beta(2s - 1), & 1/2 \leq s \leq 1. \end{cases}$$

In other words, $\alpha\beta$ is obtained by running first along α at double speed and then along β at double speed.

2.4 Lemma: Let $a,b,c \in X$, let α_0 and α_1 be paths from a to b, and let β_0 and β_1 be paths from b to c. If $[\alpha_0] = [\alpha_1]$ and $[\beta_0] = [\beta_1]$, then $[\alpha_0\beta_0] = [\alpha_1\beta_1]$.

Proof: If $\{\alpha_t\}_{0\leq t\leq 1}$ is a homotopy of α_0 and α_1 and $\{\beta_t\}_{0\leq t\leq 1}$ is a homotopy of β_0 and β_1, then $\{\alpha_t\beta_t\}_{0\leq t\leq 1}$ is a homotopy of $\alpha_0\beta_0$ and $\alpha_1\beta_1$. □

Lemma 2.4 allows us to define under certain circumstances the products of homotopy classes. If α is a path from a to b and β is a path from b to c, we define

$$[\alpha][\beta] = [\alpha\beta]$$

By Lemma 2.4, the class $[\alpha\beta]$ does not depend on the representatives of $[\alpha]$ and $[\beta]$ used to define it.

Consider three paths α, β, and γ such that the terminal point of α coincides with the initial point of β and the terminal point of β coincides with the initial point of γ. Then $(\alpha\beta)\gamma$ and $\alpha(\beta\gamma)$ are both defined, and they are given by

$$((\alpha\beta)\gamma)(s) = \begin{cases} \alpha(4s), & 0 \leq s \leq 1/4, \\ \beta(4s - 1), & 1/4 \leq s \leq 1/2, \\ \gamma(2s - 1), & 1/2 \leq s \leq 1, \end{cases}$$

and

$$(\alpha(\beta\gamma))(s) = \begin{cases} \alpha(2s), & 0 \leq s \leq 1/2, \\ \beta(4s - 2), & 1/2 \leq s \leq 3/4, \\ \gamma(4s - 3), & 3/4 \leq s \leq 1. \end{cases}$$

It is easy to give examples of α, β, and γ such that $(\alpha\beta)\gamma \neq \alpha(\beta\gamma)$. In other words, multiplication of paths (when defined) is not an associative operation. However, when we pass to homotopy classes, we do obtain an associative operation.

2.5 Lemma: Let α, β, and γ be paths in X as above, so that $(\alpha\beta)\gamma$ and $\alpha(\beta\gamma)$ are defined. Then

$$(\alpha\beta)\gamma \simeq \alpha(\beta\gamma) \text{ rel}\{0,1\}.$$

In other words,

$$([\alpha][\beta])[\gamma] = [\alpha]([\beta][\gamma]).$$

Proof: Comparing the formulas above for the respective products, we find that

$$(\alpha(\beta\gamma))(s) = ((\alpha\beta)\gamma)(\rho(s)), \qquad 0 \le s \le 1,$$

where

$$\rho(s) = \begin{cases} s/2, & 0 \le s \le 1/2, \\ s - 1/4, & 1/2 \le s \le 3/4, \\ 2s - 1, & 3/4 \le s \le 1. \end{cases}$$

In other words, $\alpha(\beta\gamma)$ is a reparametrization of $(\alpha\beta)\gamma$, so that, by Lemma 2.3, the paths are homotopic with endpoints fixed. □

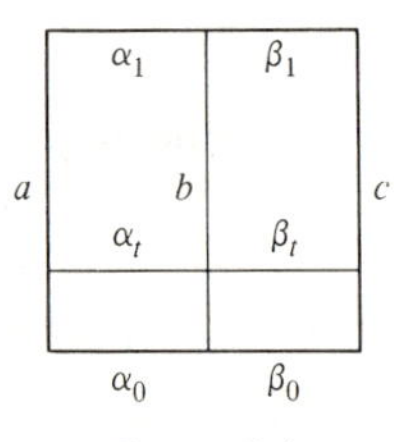

Lemma 2.4

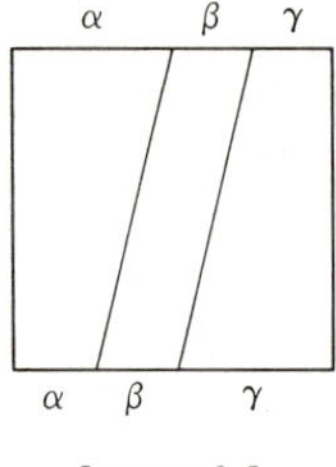

Lemma 2.5

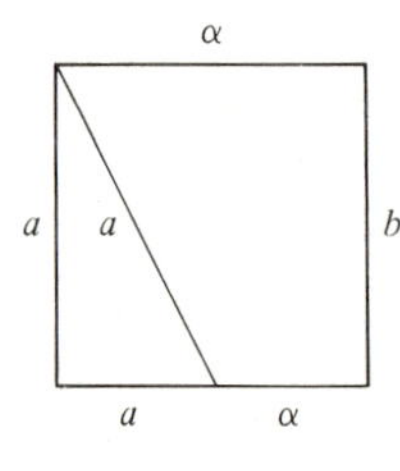

Lemma 2.6

The constant path at the point b will also be denoted by b. It is defined by $b(s) = b$, $0 \le s \le 1$.

2.6 Lemma: If α is a path from a to b, then

$$a\alpha \simeq \alpha \simeq \alpha b \text{ rel}\{0,1\}.$$

In other words,

$$[a][\alpha] = [\alpha] = [\alpha][b].$$

Proof: The path $a\alpha$ is given by

$$(a\alpha)(s) = \begin{cases} a, & 0 \le s \le 1/2, \\ \alpha(2s - 1), & 1/2 \le s \le 1. \end{cases}$$

If

$$\rho(s) = \begin{cases} 0, & 0 \le s \le 1/2, \\ 2s - 1, & 1/2 \le s \le 1, \end{cases}$$

then $a\alpha = \alpha \circ \rho$. By Lemma 2.3, $a\alpha \simeq \alpha$ rel$\{0,1\}$. Similarly, $\alpha b \simeq \alpha$ rel$\{0,1\}$. □

Now let α be a path in X from a to b. As in Section II.9, we define a path α^{-1} from b to a by

$$(\alpha^{-1})(s) = \alpha(1 - s), \qquad 0 \leq s \leq 1,$$

so that α^{-1} corresponds to running backwards along α. In particular, for the constant path b, we have $b^{-1} = b$.

2.7 Lemma: Let α_0 and α_1 be paths in X from a to b. If $[\alpha_0] = [\alpha_1]$, then $[\alpha_0{}^{-1}] = [\alpha_1{}^{-1}]$.

Proof: If $\{\alpha_t\}_{0 \leq t \leq 1}$ is a homotopy of α_0 and α_1, then $\{\alpha_t{}^{-1}\}_{0 \leq t \leq 1}$ is a homotopy of α_1^{-1} and α_0^{-1}. □

Lemma 2.7 allows us to define the inverse of the homotopy class of a path α by

$$[\alpha]^{-1} = [\alpha^{-1}].$$

By the lemma, the homotopy class defining $[\alpha]^{-1}$ does not depend on the choice of the representative of the homotopy class of α.

The final preparatory lemma asserts that the path obtained by running along α at double speed and then returning backwards along α at double speed is homotopic to a constant path. Note that in a homotopy of a product path, the intermediate stopping points are not required to be held fixed.

2.8 Lemma: Let α be a path in X from a to b. Then $\alpha\alpha^{-1}$ is homotopic to the constant path at a with endpoints fixed. In other words,

$$[\alpha][\alpha]^{-1} = [a].$$

Proof: For $0 \leq t \leq 1$, let γ_t be the path described by running along α at double speed until time $t/2$, then resting until time $1 - t/2$, then returning along α at double speed. The explicit formula for γ_t is

$$\gamma_t(s) = \begin{cases} \alpha(2s), & 0 \leq s \leq t/2, \\ \alpha(t), & t/2 \leq s \leq 1 - t/2, \\ \alpha(2 - 2s), & 1 - t/2 \leq s \leq 1. \end{cases}$$

Then γ_0 is the constant path at a and γ_1 and $\alpha\alpha^{-1}$. It is easy to check that the γ_t move continuously with t, so that they form a homotopy. □

EXERCISES

1. Suppose that $(\alpha\beta)\gamma = \alpha(\beta\gamma)$ for any three paths α, β, and γ in X for which the product is defined. Show that each path component of X consists of a single point.
2. Let X be path-connected and let $b \in X$. Show that every path in X is homotopic with endpoints fixed to a path passing through b.

3. Let D be an open subset in $\mathbb{R}^n$, let α be a path in D from x to y, and set

$$d = \inf\{|\alpha(s) - w| : w \in \partial D, 0 \leq s \leq 1\}.$$

Show that if β is any path in D from x to y such that $|\beta(s) - \alpha(s)| < d$, $0 \leq s \leq 1$, then β is homotopic to α with endpoints fixed.

4. A path α in $\mathbb{R}^n$ is *polygonal* if there is a subdivision $0 = s_0 < s_1 < \cdots < s_m = 1$ of the unit interval such that α maps each interval $[s_{j-1}, s_j]$ onto the straight-line segment from $\alpha(s_{j-1})$ to $\alpha(s_j)$. Show that every path in an open subset D of $\mathbb{R}^n$ is homotopic in D with endpoints fixed to a polygonal path.

5. Show that any path in S^n is homotopic with endpoints fixed to a polygonal path on S^n, where "polygonal" is now interpreted to mean that the path is formed from arcs lying on great circles of S^n.

6. Let (X,d) be a compact metric space and let $a,b \in X$. Let $\mathcal{P}$ be the set of all paths in X from a to b with the metric

$$\rho(\alpha,\beta) = \sup\{d(\alpha(s),\beta(s)) : 0 \leq s \leq 1\}.$$

Show that two paths $\alpha,\beta \in \mathcal{P}$ are homotopic with endpoints fixed if and only if α and β lie in the same path component of $\mathcal{P}$.

3. THE FUNDAMENTAL GROUP

A *pointed space* is a pair (X,b) consisting of a topological space X and a point $b \in X$. The point b is referred to as the *base point* of X. A *loop* in X based at b is a path in X that begins and ends at b. The product of any two loops based at b is well defined, so that the product of the homotopy classes of any two such loops is well defined. Let $\pi_1(X,b)$ denote the set of homotopy classes of loops based at b, together with the multiplication of homotopy classes defined in Section 2.

3.1 Theorem: The set $\pi_1(X,b)$ of homotopy classes of loops based at b, with the operation of multiplication of homotopy classes, is a group.

Proof: Lemma 2.5 shows that the multiplication in $\pi_1(X,b)$ is associative. Lemma 2.6 shows that the homotopy class $[b]$ of the constant map b is an identity for $\pi_1(X,b)$. Lemma 2.8 shows that the homotopy class $[\alpha]^{-1}$ of α^{-1} is an inverse for the homotopy class of α. □

The group $\pi_1(X,b)$ is called the *fundamental group of X based at b*. If $\pi_1(X,b)$ consists of only the identity, we say that $\pi_1(X,b)$ is *trivial* and write $\pi_1(X,b) \cong 0$. Thus $\pi_1(X,b)$ is trivial if and only if every loop based at b is homotopic with endpoints fixed to the constant loop at b. Theorem 2.1 then yields the following result.

3.2 Theorem: If X is a convex subset of $\mathbb{R}^n$ and $b \in X$, then $\pi_1(X,b) \cong 0$.

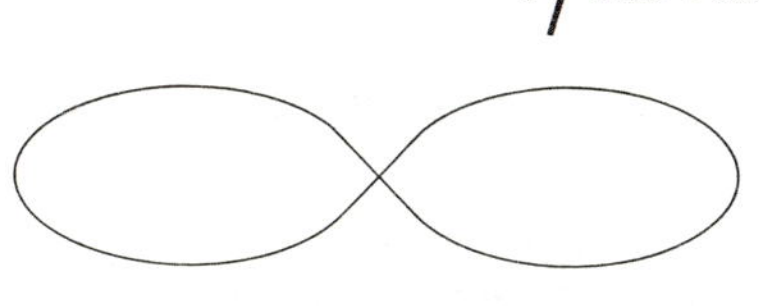

Reclining figure eight

It turns out that the fundamental group of a space need not be commutative. For instance, if X is a "figure eight," then $\pi_1(X,b)$ is isomorphic to a certain non-commutative group, namely, the free group on two generators. This theorem is due to Van Kampen, and a proof will be indicated in the exercises in Section 5. Meanwhile, it will take us some effort to prove that there is a space X such that $\pi_1(X,b)$ is not trivial. The simplest space with a nontrivial fundamental group is the circle S^1, and it will be shown in Section 5 that

$$\pi_1(S^1,1) \cong \mathbb{Z}.$$

Now the question arises of how $\pi_1(X,b)$ depends on the base point b. Any loop based at b lies within the path component of b in X. Therefore if $c \in X$ lies in a path component different from b, then $\pi_1(X,c)$ is in no way related to $\pi_1(X,b)$. However, if c lies in the same path component as b, then $\pi_1(X,c)$ is isomorphic to $\pi_1(X,b)$. In fact, the following is true.

3.3 Theorem: Let $b,c \in X$ and suppose α to be a path in X from b to c. Then for each loop γ based at c, the homotopy class

$$\alpha_*([\gamma]) = [\alpha][\gamma][\alpha]^{-1}$$

is defined and α_* is an isomorphism of $\pi_1(X,c)$ and $\pi_1(X,b)$.

Proof: Since the initial and terminal points of the paths match up appropriately, α_* is well defined. Since the multiplication of homotopy classes is associative, the parentheses indicating the order of multiplication are omitted.

Suppose that β and γ are both loops at c. From Lemmas 2.6 and 2.8 we obtain

$$[\beta][\gamma] = [\beta][c][\gamma] = [\beta][\alpha]^{-1}[\alpha][\gamma].$$

Consequently

$$\begin{aligned}\alpha_*([\beta][\gamma]) &= [\alpha][\beta][\gamma][\alpha]^{-1} \\ &= [\alpha][\beta][\alpha]^{-1}[\alpha][\gamma][\alpha]^{-1} \\ &= \alpha_*([\beta])\alpha_*([\gamma]).\end{aligned}$$

Hence α_* is a homomorphism.

Suppose next that the loops β and γ at c satisfy $\alpha_*([\beta]) = \alpha_*([\gamma])$. Then

$$[\beta] = [\alpha]^{-1}[\alpha][\beta][\alpha]^{-1}[\alpha] = [\alpha]^{-1}[\alpha][\gamma][\alpha]^{-1}[\alpha] = [\gamma].$$

Hence α_* is one-to-one.

Finally, let λ be a loop based at b. Then $\beta = (\alpha^{-1}\lambda)\alpha$ is a loop based at c that satisfies

$$\alpha_*([\beta]) = [\alpha][\alpha]^{-1}[\lambda][\alpha][\alpha]^{-1} = [\lambda].$$

Consequently α_* is onto and is an isomorphism. □

If X is path-connected, then the various groups $\pi_1(X,b)$, $b \in X$, are all isomorphic. Thus the isomorphism type of $\pi_1(X,b)$ is well defined; it is called the *fundamental group* of X, denoted by $\pi_1(X)$. The situation here is slightly peculiar linguistically in that the fundamental group $\pi_1(X)$ is not equal to any of the $\pi_1(X,b)$ but is rather the (common) isomorphism class of all the distinct but isomorphic groups $\pi_1(X,b)$, $b \in X$. In the following development, the assertions to be made about $\pi_1(X)$ will be such assertions as "$\pi_1(X)$ is isomorphic to G," where G is some concretely described group. This assertion has the precise meaning that, for all $b \in X$, $\pi_1(X,b)$ is isomorphic to G; the isomorphism holds for all b in X if it holds for any one b in X since all $\pi_1(X,b)$ are isomorphic.

A space X is *simply connected* if X is path-connected and $\pi_1(X)$ is trivial. Since any convex subset of $\mathbb{R}^n$ is path-connected, Theorem 3.2 shows that a convex subset of $\mathbb{R}^n$ is simply connected.

There is another way of viewing loops in X based at b, which is convenient for some purposes. Let S^1 denote the unit circle in $\mathbb{R}^2 \cong \mathbb{C}$. Then each map

$$f : S^1 \to X$$

satisfying $f(1) = b$ determines a loop α based at b via the formula

$$\alpha(s) = f(e^{2\pi i s}), \qquad 0 \le s \le 1.$$

Conversely, any loop α based at b arises from a map $f : S^1 \to X$ satisfying $f(1) = b$. The point is that the exponential map $s \to e^{2\pi i s}$ is a homeomorphism from the interval $[0,1]$, with the endpoints identified, to the circle S^1.

If the loops α_0 and α_1 based at b are homotopic with endpoints fixed, then the corresponding maps $f_0, f_1 : S^1 \to X$ are homotopic relative to the base points $1 \in S^1$ and $b \in X$ in the sense that there are maps $\{f_t\}_{0 \le t \le 1}$ from S^1 to X such that $f_t(1) = b$, $0 \le t \le 1$, and the f_t move continuously with t. In other words, there is a map

$$F : S^1 \times [0,1] \to X$$

such that

$$F(z,0) = f_0(z), \qquad z \in S^1,$$

$$F(z,1) = f_1(z), \qquad z \in S^1,$$

$$F(1,t) = b, \qquad 0 \le t \le 1.$$

Conversely, if f_0 and f_1 are homotopic as above, then $(s,t) \to F(e^{2\pi i s},t)$ is a homotopy of α_0 and α_1 with endpoints fixed.

With this identification in mind, we shall think of maps $f : (S^1,1) \to (X,b)$ also as loops in X based at $f(1)$.

EXERCISES

1. Prove that if $n \geq 2$, then S^n is simply connected. *Hint:* Use Exercise 2.5 to show that every loop in S^n is homotopic to a loop that does not cover all of S^n.
2. Prove that if there are simply connected open subsets U and V of X such that $U \cup V = X$ and $U \cap V$ is nonempty and path-connected, then X is simply connected.
3. A space X is *contractible to a point* $x_0 \in X$ *with* x_0 *held fixed* if there is a map $F : X \times [0,1] \to X$ such that

$$F(x,0) = x_0, \qquad x \in X,$$

$$F(x,1) = x, \qquad x \in X,$$

$$F(x_0,t) = x_0, \qquad 0 \leq t \leq 1.$$

Show that such a space is simply connected.

4. Let X be the *comb space*, that is, the compact subset of $\mathbb{R}^2$ consisting of the horizontal interval $\{(x,0) : 0 \leq x \leq 1\}$ and the closed vertical intervals of unit length with lower endpoints at $(0,0)$ and at $(0,1/n)$, $1 \leq n < \infty$.

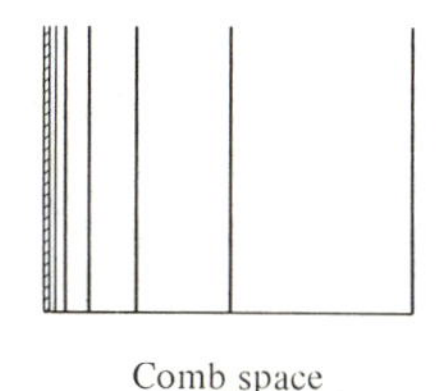

Comb space

Show that X is contractible to $(0,0)$ with $(0,0)$ held fixed. Show that X is not contractible to $(0,1)$ with $(0,1)$ held fixed.

5. A subset W of $\mathbb{R}^n$ is *star-shaped* with respect to a point $w \in W$ if, whenever $y \in W$, then the straight-line segment from w to y is contained in W.
   (a) Show that a subset W of $\mathbb{R}^n$ is convex if and only if it is star-shaped with respect to each of its points.
   (b) Give an example of a star-shaped set that is not convex.
   (c) Show that a star-shaped set in $\mathbb{R}^n$ is contractible to a point.
   (d) Show that a star-shaped set in $\mathbb{R}^n$ is simply connected.
6. Let (X,x_0) and (Y,y_0) be pointed spaces. Show that $\pi_1(X \times Y,(x_0y_0))$ is isomorphic to the direct product $\pi_1(X,x_0) \times \pi_1(Y,y_0)$.
7. Prove that the product of simply connected spaces is simply connected.
8. Prove that if $n \geq 3$, then $\mathbb{R}^n\backslash\{0\}$ is simply connected.
9. Let X be a path-connected topological space and let $b,c \in X$. Let B^2 be the closed unit ball in $\mathbb{R}^2$, with boundary circle S^1. Show that the following are equivalent.
   (a) X is simply connected.
   (b) Any two paths from b to c are homotopic with endpoints fixed.
   (c) Every map $f : S^1 \to X$ extends to a map $F : B^2 \to X$.

4. INDUCED HOMOMORPHISMS

Let (X,b) and (Y,c) be pointed topological spaces. A *map* $f : (Y,c) \to (X,b)$ is a continuous function f from Y to X that satisfies $f(c) = b$. We aim to show that any such map induces a homomorphism f_* from $\pi_1(Y,c)$ to $\pi_1(X,b)$. Despite the notational similarity, f_* is unrelated to the induced map α_* of the preceding section. The map f_* will be defined in the obvious way, by making the path γ in Y correspond to the path $f\circ\gamma$ in X. That this correspondence respects homotopy classes is the content of the following elementary but useful lemma.

4.1 Lemma: Let X and Y be topological spaces and let $f : Y \to X$ be a map. Let α_0 and α_1 be paths in Y that are homotopic with endpoints fixed. Then $f\circ\alpha_0$ and $f\circ\alpha_1$ are paths in X that are homotopic with endpoints fixed.

Proof: If $\{\alpha_t\}_{0\leq t\leq 1}$ is the homotopy of α_0 to α_1, then the paths $\{f\circ\alpha_t\}_{0\leq t\leq 1}$ form a homotopy of $f\circ\alpha_0$ and $f\circ\alpha_1$. □

The following corollary to Lemma 4.1 and Theorem 2.1 is sufficiently useful to merit a separate statement.

4.2 Corollary: Let Y be a convex subset of $\mathbb{R}^n$, let $y_0,y_1 \in Y$, and let f be a map from Y to X. If α and β are paths in Y from y_0 to y_1, then $f\circ\alpha$ is homotopic to $f\circ\beta$ with endpoints fixed.

Suppose now that $c_0,c_1 \in Y$ and that f is a map from Y to X. If α is a path in Y from c_0 to c_1, then $f_*([\alpha])$ is defined to be the homotopy class of paths in X from $f(c_0)$ to $f(c_1)$ that includes $f\circ\alpha$:

$$f_*([\alpha]) = [f\circ\alpha].$$

By Lemma 4.1, this definition does not depend on the choice of the representative of the homotopy class $[\alpha]$. If the path product $\alpha\beta$ is defined, then $f\circ(\alpha\beta) = (f\circ\alpha)(f\circ\beta)$, so that

$$f_*([\alpha][\beta]) = f_*([\alpha])f_*([\beta]). \tag{4.1}$$

Since the inverse of the path $f\circ\alpha$ is $f\circ(\alpha^{-1})$,

$$f_*([\alpha]^{-1}) = f_*([\alpha])^{-1}.$$

Finally,

$$f_*([c]) = [f(c)]$$

whenever c is a constant path in Y.

4.3 Theorem: Let (X,b) and (Y,c) be pointed topological spaces and let $f : (Y,c) \to (X,b)$ be a map. Then f_* is a homomorphism of $\pi_1(Y,c)$ and $\pi_1(X,b)$. If, furthermore, (W,d) is a pointed topological space and $g : (W,d) \to (Y,c)$ is a map, then

$$(f \circ g)_* = f_* \circ g_*.$$

Finally, if $X = Y$, $b = c$, and f is the identity map, then f_* is the identity isomorphism.

Proof: That f_* is a homomorphism follows from (4.1). The other statements follow directly from the definitions. □

4.4 Corollary: If $f : Y \to X$ is a homeomorphism and if $c \in Y$ and $b = f(c)$, then f_* is an isomorphism of $\pi_1(Y,c)$ and $\pi_1(X,b)$.

Proof: Since $f \circ f^{-1}$ and $f^{-1} \circ f$ are the identity maps of X and Y, respectively, $f_* \circ (f^{-1})_*$ and $(f^{-1})_* \circ f_*$ are the identity isomorphisms of $\pi_1(X,b)$ and $\pi_1(Y,c)$, respectively. Since $(f^{-1})_* \circ f_*$ is one-to-one, so is f_*. Since $f_* \circ (f^{-1})_*$ is onto, so is f_*. Hence f_* is an isomorphism. □

With Corollary 4.4, we have attained the goal indicated at the beginning of the chapter, namely, we have associated to each (pointed) topological space an algebraic object, its fundamental group, in such a way that if two spaces are homeomorphic (via a base-point-preserving homeomorphism), then the algebraic objects are isomorphic. Thus, for instance, two spaces could be shown to be nonhomeomorphic by showing that their fundamental groups were not isomorphic. To give this idea any real significance, it is necessary to be able to compute the fundamental groups of whatever topological spaces one wishes to study. In the following section, we shall show how to determine the fundamental group of the circle S^1—it turns out to be (isomorphic to) the group $\mathbb{Z}$ of integers with addition as the group operation. We focus attention first on the circle S^1 because it is obviously the simplest space containing a closed curve without self-intersections. The method used to compute $\pi_1(S^1)$ will serve to compute the fundamental group of many other topological spaces. Some applications of the fundamental group to fixed-point theorems and other related results will be presented in Section 6 and later sections.

EXERCISES

1. Show that simple connectivity is a topological property.
2. A subspace A of a topological space X is a *retract* of X if there is a map $f : X \to A$ such that $f(y) = y$ for all $y \in A$. The map f is called a *retraction* of X onto A. Show that the unit sphere S^n in $\mathbb{R}^{n+1}$ is a retract of $\mathbb{R}^{n+1} \setminus \{0\}$.
3. Let f be a retraction of X onto A and let $x_0 \in A$. Let $j : A \hookrightarrow X$ be the inclusion map. Prove the following:
   (a) $j_* : \pi_1(A,x_0) \to \pi_1(X,x_0)$ is one-to-one.
   (b) $f_* : \pi_1(X,x_0) \to \pi_1(A,x_0)$ is onto.
   (c) If X is simply connected, then A is simply connected.

5. COVERING SPACES

We wish now to compute the fundamental group of the circle S^1. This will be accomplished with the aid of the exponential map $p : \mathbb{R} \to S^1$, defined by

$$(5.1) \qquad p(t) = e^{2\pi it}, \qquad t \in \mathbb{R}.$$

Since the line of proof will be quite general, we axiomatize those properties of the spaces $\mathbb{R}$ and S^1 and the exponential map p that will be needed.

Let E and X be topological spaces and let $p : E \to X$ be a map. An open subset U of X is *evenly covered* by p if the inverse image $p^{-1}(U)$ is a union of disjoint open subsets of E, each of which is mapped homeomorphically by p onto U. The map p is a *covering map* if p maps E onto X and if each $x \in X$ has an open neighborhood that is evenly covered by p. In this case, E is a *covering space* over X.

If $x \in X$, the set $p^{-1}(x)$ is called the *fiber* over x. It is evidently a discrete subspace of E. According to the definition, each $x \in X$ has an open neighborhood U such that $p^{-1}(U)$ is homeomorphic to $p^{-1}(x) \times U$. The subsets of $p^{-1}(U)$ that are mapped homeomorphically onto U are called the *sheets* of $p^{-1}(U)$. If U is connected, then the sheets of $p^{-1}(U)$ coincide with the connected components of $p^{-1}(U)$.

The exponential map $p : \mathbb{R} \to S^1$ is a covering map. Indeed, let $e^{2\pi it_0} \in S^1$, fix $0 < \varepsilon < 1/2$, and let $U = \{e^{2\pi it} : |t - t_0| < \varepsilon\}$. Then $p^{-1}(U)$ is the disjoint union of the intervals $(m + t_0 - \varepsilon, m + t_0 + \varepsilon)$, m an integer, and each of these intervals is mapped homeomorphically by p onto U. For the purposes of visualizing the covering map, it is convenient to think of $\mathbb{R}$ as the helix $\{(\cos(2\pi t), \sin(2\pi t), t) : -\infty < t < \infty\}$ in $\mathbb{R}^3$ by identifying $t \in \mathbb{R}$ with the corresponding point on the helix. The covering map p then projects the helix onto the circle S^1 in the x,y-plane.

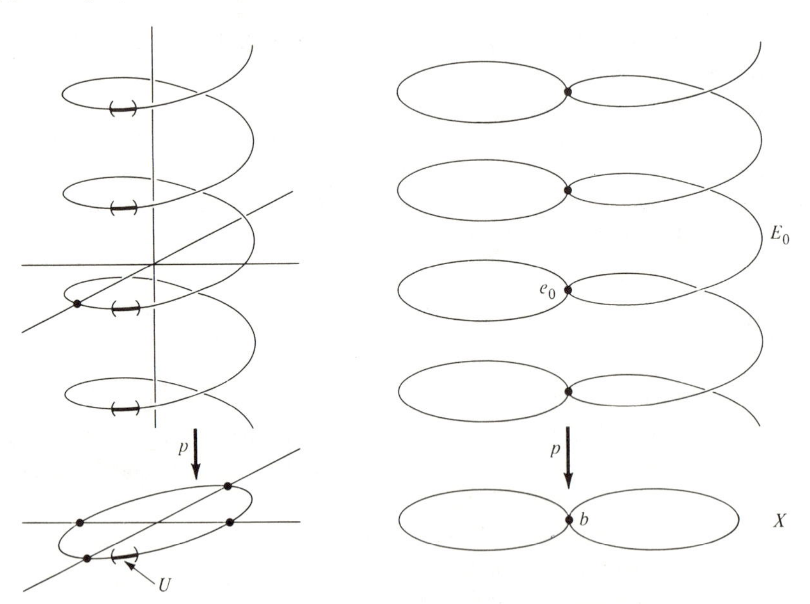

As a second example, let X be a figure eight, consisting of two loops touching at one point b. Let E_0 be the space obtained by unwinding one of the loops, so that

E_0 is a helix with loops attached as indicated in the figure above. If we unwind the loop of E_0 touching the base point e_0 of E_0, we obtain another covering space E over X, as suggested by the figure appearing towards the end of this section. The covering map is the composition of two covering maps, one from E to E_0, the other from E_0 to X.

As another example, consider the n-dimensional projective space P^n, defined in Exercise II.13.8. Recall that P^n is the quotient space obtained from the unit sphere S^n in $\mathbb{R}^{n+1}$ by identifying antipodal points. Let $p : S^n \to P^n$ be the quotient map. If $x_0 \in S^n$, $\varepsilon > 0$ is small, and $V = \{x \in S^n : |x - x_0| < \varepsilon\}$, then $U = p(V)$ is an open subset of P^n and $p^{-1}(U)$ is the disjoint union of V and $-V$. Since both V and $-V$ are mapped homeomorphically by p onto U, p is a covering map. In this case, every fiber consists of two points.

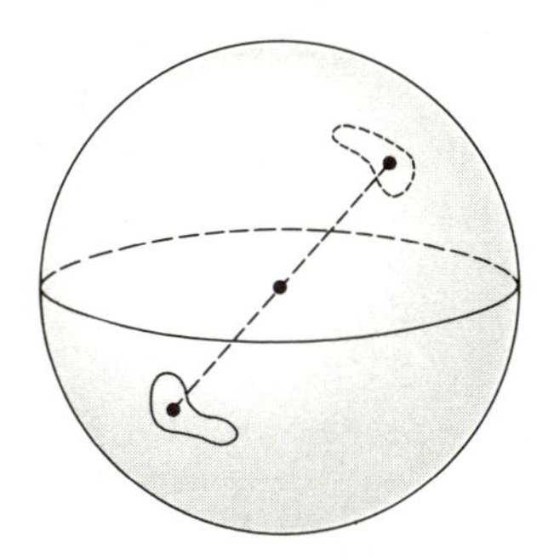

Products of covering spaces are covering spaces. For instance, if $T^n = S^1 \times \cdots \times S^1$ is the n-dimensional torus (product of n circles), then the exponential map $p : \mathbb{R}^n \to T^n$, defined by

$$p(x_1, \ldots, x_n) = (e^{2\pi i x_1}, \ldots, e^{2\pi i x_n}),$$

is a covering map.

Now let $p : E \to X$ be a covering map, let Y be a topological space, and let $f : Y \to X$ be a map. For later developments it will be important to determine whether there is a map $g : Y \to E$ such that $p \circ g = f$. Such a map g is called a *lift* of f. The situation may be represented schematically by the diagram

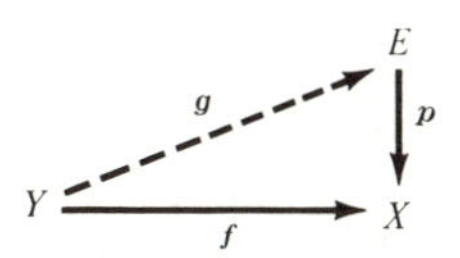

The relation $p \circ g = f$ means that the image of a point $y \in Y$ is the same whether f is applied directly (the "low road") or g is applied followed by the projection p (the "high road"). If $p \circ g = f$, the diagram is said to *commute*.

The uniqueness of lifts will be handled by the following lemma.

5.1 Lemma: Let $p : E \to X$ be a covering map and let Y be a connected topological space. Let $f : Y \to X$ be a map and let $g,h : Y \to E$ be two lifts of f. If $g(y) = h(y)$ for some point $y \in Y$, then $g = h$.

Proof: Let

$$S = \{y \in Y : g(y) = h(y)\},$$

$$T = \{y \in Y : g(y) \neq h(y)\}.$$

Then $S \cup T = Y$ and $S \cap T = \varnothing$. We must show that either $S = Y$ or $T = Y$. Since Y is connected, it suffices to show that S and T are each open.

Let $y \in Y$ and let U be an open neighborhood of $f(y)$ that is evenly covered. Let V and W be sheets of $p^{-1}(U)$ provided by the definition of covering map, such that $g(y) \in V$ and $h(y) \in W$. If $g(y) = h(y)$, then $V = W$, whereas if $g(y) \neq h(y)$, then V and W are disjoint.

Since g and h are continuous at y, there is an open neighborhood N of y such that $g(N) \subset V$ and $h(N) \subset W$. If $y \in T$, then $V \cap W = \varnothing$, so that $g(z) \neq h(z)$ for all $z \in N$, and $N \subset T$. It follows that T is open. On the other hand, if $y \in S$, then $V = W$. Furthermore, for each $z \in N$, $g(z)$ must be that unique point $v \in V$ such that $p(v) = f(z)$ and $h(z)$ must be that unique point $w \in W$ such that $p(w) = f(z)$. Since $V = W$, we conclude that $v = w$, so that $g = h$ on N, $N \subset S$, and S is also open. □

5.2 Theorem *(Path Lifting Theorem):* Let $p : E \to X$ be a covering map, let $\gamma : [0,1] \to X$ be a path, and let $e_0 \in E$ satisfy $p(e_0) = \gamma(0)$. Then there exists a unique path $\alpha : [0,1] \to E$ such that $\alpha(0) = e_0$ and $p \circ \alpha = \gamma$.

Proof: For each $x \in X$, choose an open neighborhood $U(x)$ of x that is evenly covered by p. The open sets $\gamma^{-1}(U(x))$, $x \in X$, form an open cover of $[0,1]$. Since $[0,1]$ is compact, we can find $0 = s_0 < s_1 < \cdots < s_m = 1$ and evenly covered open sets $U_1, \ldots, U_m$ such that $\gamma^{-1}(U_j)$ includes $[s_{j-1}, s_j]$, $1 \leq j \leq m$. In other words,

$$\gamma([s_{j-1}, s_j]) \subset U_j, \qquad 1 \leq j \leq m.$$

The lift is performed now in m steps, as follows.

Since $p(e_0) = \gamma(0) \in U_1$, there is an open neighborhood V_1 of e_0 that is mapped homeomorphically by p onto U_1. Define α on the interval $[0, s_1]$ such that $\alpha(s)$ is the unique point of V_1 covering $\gamma(s)$. In other words, set

$$\alpha = (p|_{V_1})^{-1} \circ \gamma \qquad \text{on } [0, s_1].$$

Then $\alpha(0) = e_0$ and $p \circ \alpha = \gamma$ on $[0, s_1]$.

Now perform the same procedure, with $e_0 = \alpha(0)$ replaced by $\alpha(s_1)$ and U_1 replaced by U_2, to extend α to the interval $[s_1, s_2]$. After m steps, we shall have lifted the entire path α.

The uniqueness of the lifted path α follows from Lemma 5.1. □

Actually, one can lift a family of paths depending continuously on a parameter, so that the lifted paths also depend continuously on the parameter. A version of this principle, which will suffice for our purposes, is as follows.

5.3 Theorem: Let $p : E \to X$ be a covering map, let $F : [0,1] \times [0,1] \to X$ be a map, and let $e_0 \in E$ satisfy $p(e_0) = F(0,0)$. Then there exists a unique lift $G : [0,1] \times [0,1] \to E$ of F such that $G(0,0) = e_0$.

Proof: The uniqueness again follows from Lemma 5.1.

According to the Path Lifting Theorem, there is a unique path $t \to e_t$, $0 \le t \le 1$, in E such that $p(e_t) = F(0,t)$, $0 \le t \le 1$. By the same theorem, there exists for each t a unique path $s \to G(s,t)$, $0 \le s \le 1$, such that $G(0,t) = e_t$ and $p(G(s,t)) = F(s,t)$, $0 \le s \le 1$. This defines the lift G of F. We must show that G is continuous on $[0,1] \times [0,1]$.

Let $\gamma : [0,1] \to X$ be the path defined by

$$\gamma(s) = F(s,0), \qquad 0 \le s \le 1.$$

Consider the construction of the lift $\alpha(s) = G(s,0)$ given in the proof of Theorem 5.2 and retain the notation of that proof. Since $\gamma^{-1}(U_j)$ is an open neighborhood of $[s_{j-1},s_j]$, $F^{-1}(U_j)$ includes $[s_{j-1},s_j] \times [0,\varepsilon]$ for some small $\varepsilon > 0$. Consequently there exists $\varepsilon > 0$ such that

$$F([s_{j-1},s_j] \times [0,\varepsilon]) \subset U_j, \qquad 1 \le j \le m.$$

Since $e_0 \in V_1$, we can assume also that the initial points e_t belong to V_1 for $0 \le t \le \varepsilon$ since they depend continuously on t. Now, the procedure for constructing the lift shows that

$$G = (p|_{V_1})^{-1} \circ F \qquad \text{on } [0,s_1] \times [0,\varepsilon].$$

In particular, G is continuous on $[0,s_1] \times [0,\varepsilon]$. Proceeding in this fashion, we find that G is continuous on each rectangle $[s_{j-1},s_j] \times [0,\varepsilon]$, $1 \le j \le m$, so that G is continuous on $[0,1] \times [0,\varepsilon]$.

The same proof shows that for each $t_0 \in (0,1]$, there exists $\varepsilon > 0$ such that G is continuous on the rectangle $[0,1] \times [t_0 - \varepsilon, t_0 + \varepsilon]$. (Replace $t_0 + \varepsilon$ by t_0 if $t_0 = 1$.) Consequently G is continuous on $[0,1] \times [0,1]$. □

Now let (E,e) and (X,b) be pointed spaces and let $p : (E,e) \to (X,b)$ be a covering map, that is, $p : E \to X$ is a covering map satisfying $p(e) = b$. Let $\gamma : [0,1] \to X$ be a loop based at b. By the Path Lifting Theorem, there is a unique lift $\alpha : [0,1] \to E$ of γ that satisfies $\alpha(0) = e$. The lift α need not be a loop since it need not terminate at e. However, the terminal point $\alpha(1)$ of α satisfies $p(\alpha(1)) = \gamma(1) = b$, so that $\alpha(1)$ lies in the fiber $p^{-1}(b)$ over b.

Suppose now that γ_1 is another loop in X based at b such that γ_1 is homotopic to γ with endpoints fixed. Let $\{\gamma_t\}_{0 \le t \le 1}$ be the homotopy, so that $\gamma_0 = \gamma$. Applying Theorem 5.3 to $F(s,t) = \gamma_t(s)$, we obtain a map $G : [0,1] \times [0,1] \to E$ such that $G(0,0) = e$ and $p(G(s,t)) = \gamma_t(s)$, $0 \le s,t \le 1$. Then the path α_t in E defined by $\alpha_t(s) = G(s,t)$, $0 \le t \le 1$, is a lift of γ_t and $\alpha_0 = \alpha$. We claim that the α_t's all start at e. To see this, observe that the map $t \to G(0,t)$ is the unique lift to E, starting at e, of the constant path at b. Hence the lift coincides with the constant path at e and $e = G(0,t) = \alpha_t(0)$, $0 \le t \le 1$. Similarly, the map $t \to G(1,t)$ is the unique lift to

E, starting at $G(1,0) = \alpha(1)$, of the constant path at b. Hence $\alpha(1) = G(1,t) = \alpha_t(1)$ for $0 \leq t \leq 1$ and all of the paths α_t terminate at $\alpha(1)$. In particular, the lift α_1 of γ_1 to E starts at e and terminates at $\alpha(1)$.

We conclude that the terminal point $\alpha(1)$ is the same for all loops in the same homotopy class of γ. This allows us to define a function

$$\Phi : \pi_1(X,b) \to p^{-1}(b),$$

so that $\Phi([\gamma])$ is the terminal point of the lift of γ to E that starts at e.

As an application, consider the figure eight X and its covering space E given in the illustration. The loop $\alpha\beta$ in X lifts to a path in E that begins at e and terminates

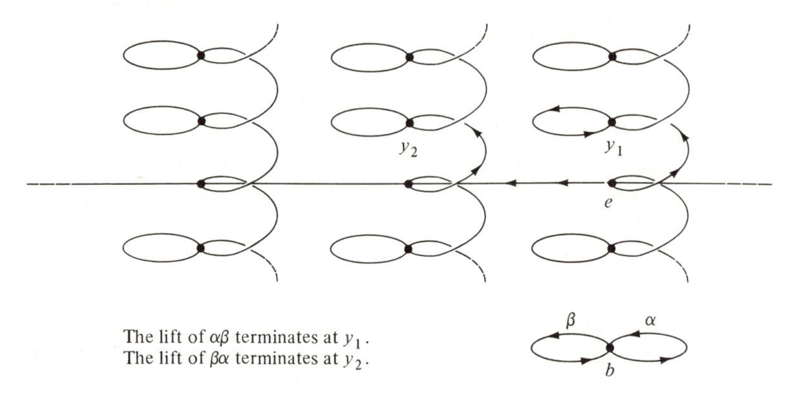

The lift of $\alpha\beta$ terminates at y_1.
The lift of $\beta\alpha$ terminates at y_2.

at y_1. On the other hand, the loop $\beta\alpha$ in X lifts to a path in E that begins at e and terminates at y_2. Since $y_1 \neq y_2$, $\alpha\beta$ is not homotopic to $\beta\alpha$ and

$$[\alpha][\beta] \neq [\beta][\alpha].$$

Thus we arrive at the striking discovery that the fundamental group of the figure eight is not commutative. By applying the idea used above to more elaborate covering spaces, it is possible to prove Van Kampen's Theorem, that the fundamental group of the figure eight is the group called the *free group on two generators* (Exercise 6).

5.4 Theorem: Let $p : (E,e) \to (X,b)$ be a covering map and suppose that E is simply connected. Then Φ is a one-to-one correspondence of $\pi_1(X,b)$ and the fiber $p^{-1}(b)$.

Proof: Suppose that $y \in p^{-1}(b)$. Let α be a path in E from e to y and set $\gamma = p \circ \alpha$. Then α is the lift of γ to E that starts at e, so that $\Phi([\gamma]) = \alpha(1) = y$. Hence the function Φ is onto.

Suppose that γ_0 and γ_1 are loops in X based at b such that $\Phi([\gamma_0]) = \Phi([\gamma_1])$. Let α_0 and α_1 be lifts of γ_0 and γ_1, respectively, that start at e. Then α_0 and α_1 have the same terminal point, so that the path $\alpha_0\alpha_1^{-1}$ in E is a loop based at e. Since E is simply connected, there is a homotopy $F : [0,1] \times [0,1] \to E$ of $\alpha_0\alpha_1^{-1}$ to the point e. Then $p \circ F : [0,1] \times [0,1] \to X$ is a homotopy of the loop $\gamma_0\gamma_1^{-1}$ to the point b. Hence $[\gamma_0][\gamma_1]^{-1} = [b]$ and $[\gamma_0] = [\gamma_1]$. It follows that Φ is one-to-one. □

As a corollary of the proof, we obtain the following.

5.5 Corollary: Let $p : (E,e) \to (X,b)$ be a covering map and suppose that E is simply connected. For each point $y \in p^{-1}(b)$, let α_y be a path in E from e to y and let $\gamma_y = p \circ \alpha_y$ be a loop in X based at b. If γ is any loop in X based at b, then there is a unique loop γ_y such that γ is homotopic to γ_y with endpoints fixed.

As an example, we apply Theorem 5.4 to the covering map

$$p : S^n \to P^n$$

discussed earlier in this section. Let e be the north pole of S^n and let $b = p(e)$. Then $p^{-1}(b)$ consists of two points, the north and south poles of S^n. If $n \geq 2$, then S^n is simply connected (Exercise 3.1). Consequently Theorem 5.4 applies and $\pi_1(P^n,b)$ has exactly two elements. One element is the identity, and the other element is the homotopy class of $p \circ \alpha$, where α is any path on S^n from the north pole to the south pole. Since any group with two elements is isomorphic to $\mathbb{Z}_2$, we obtain

$$\pi_1(P^n) \cong \mathbb{Z}_2, \qquad n \geq 2.$$

Now we return to the prototypical covering map $p : (\mathbb{R},0) \to (S^1,1)$ given by (5.1). Since $p^{-1}(1)$ coincides with the subset $\mathbb{Z}$ of $\mathbb{R}$ consisting of the integers, the elements of the fundamental group $\pi_1(S^1,1)$ are in one-to-one correspondence with the integers. We wish to show that this correspondence is a group isomorphism. For this, we consider in detail the procedure by which a loop in S^1 determines an integer.

Let γ be a loop in S^1 based at 1. Then the lift of γ is a map

$$h : [0,1] \to \mathbb{R}$$

that satisfies

$$e^{2\pi i h(s)} = \gamma(s), \qquad 0 \leq s \leq 1,$$
$$h(0) = 0.$$

The terminal point $h(1)$ of h is then an integer, which is called the *index of* γ and denoted by $\operatorname{ind}(\gamma)$. Thus $\operatorname{ind}(\gamma)$ is the element of $p^{-1}(0)$ associated with $[\gamma]$ by Theorem 5.4.

5.6 Theorem: Two loops α and β in S^1 based at 1 are in the same homotopy class if and only if they have the same index. The correspondence

$$[\alpha] \to \operatorname{ind}(\alpha)$$

is an isomorphism of $\pi_1(S^1,1)$ and the integers $\mathbb{Z}$.

Proof: The only item that remains to be proved is that the correspondence is a homomorphism. To verify this, it suffices to show that

$$\operatorname{ind}(\alpha_1\alpha_2) = \operatorname{ind}(\alpha_1) + \operatorname{ind}(\alpha_2)$$

whenever α_1 and α_2 are both loops in S^1 based at 1.

Choose maps $h_1, h_2 : [0,1] \to \mathbb{R}$ so that $h_1(0) = h_2(0) = 0$ and

$$\alpha_j(s) = e^{2\pi i h_j(s)}, \qquad 0 \leq s \leq 1; j = 1,2.$$

Define

$$h(s) = \begin{cases} h_1(2s), & 0 \le s \le 1/2, \\ h_1(1) + h_2(2s - 1), & 1/2 \le s \le 1. \end{cases}$$

Then $h : [0,1] \to \mathbb{R}$ is continuous, $h(0) = 0$, and

$$(\alpha_1\alpha_2)(s) = e^{2\pi i h(s)}, \qquad 0 \le s \le 1.$$

Consequently

$$\text{ind}(\alpha_1\alpha_2) = h(1) = h_1(1) + h_2(1) = \text{ind}(\alpha_1) + \text{ind}(\alpha_2). \ \square$$

In Chapter IV we shall generalize the notion of index to mappings of the n-sphere. In that context, the integer associated with a map will be called the *degree of the map*. Thus $\text{ind}(\gamma)$ coincides with the degree $\deg(\gamma)$ of γ, to be defined in Chapter IV. It is a historical accident that the terminology used for curves on the plane differs from that used in higher-dimensional topology.

EXERCISES

1. For m an integer, let α_m be the loop in S^1 defined by

$$\alpha_m(s) = e^{2\pi i m s}, \qquad 0 \le s \le 1.$$

Show that every loop in S^1 based at 1 is homotopic with endpoints fixed to precisely one of the loops α_m.

2. Suppose that $p : (E,e) \to (X,b)$ is a covering map and that E is simply connected. Suppose furthermore that E and X are groups with identities e and b, respectively, and that p is a homomorphism. Suppose finally that for each fixed $y \in E$, the group multiplication $z \to y * z$ is a continuous function on E. Prove that the fiber $p^{-1}(b)$ is a subgroup of E, and that

$$\pi_1(X,b) \cong p^{-1}(b).$$

3. Show that the exponential map $p : \mathbb{R}^n \to T^n$, defined earlier in this section, is a covering map of $\mathbb{R}^n$ onto the n-torus T^n. Show that

$$\pi_1(T^n) \cong \mathbb{Z} \oplus \cdots \oplus \mathbb{Z} \qquad (n \text{ summands}).$$

For each n-tuple $(m_1,\ldots,m_n) \in \mathbb{Z} \oplus \cdots \oplus \mathbb{Z}$, define explicitly a loop in T^n based at $(1,\ldots,1)$ in the corresponding homotopy class.

4. Show that the map $p : \mathbb{C} \to \mathbb{C}\backslash\{0\}$, defined by

$$p(z) = e^z, \qquad z \in \mathbb{C},$$

is a covering map. What is $\pi_1(\mathbb{C}\backslash\{0\})$?

5. Show that the restriction of the map p of Exercise 4 to the horizontal strip $E = \{x + iy : c < y < d\}$ is a covering map of E over the open annulus $\{w : e^c < |w| < e^d\}$. What is the fundamental group of the open annulus? What is the fundamental group of a closed annulus?

6. Let X be the figure-eight space, and let α and β be the loops in X indicated in the figure and discussion preceding Theorem 5.4. Show that every element of $\pi_1(X)$ can be expressed uniquely as a finite product
$$[\alpha]^{m_1}\,[\beta]^{m_2}\,[\alpha]^{m_3}\cdots,$$
where $m_1, m_2, \ldots$ are integers and $m_j \neq 0$ for $j \geq 2$ ($m_1 = 0$ is allowed). *Note:* This proves Van Kampen's Theorem, that $\pi_1(X)$ is the free group with generators $[\alpha]$ and $[\beta]$. For the uniqueness assertion, construct an appropriate covering space of X.
7. A topological space Y is *locally path-connected* if for each $y \in Y$ and neighborhood U of y, there exists a neighborhood V of y such that every point of V can be joined to y by a path in U. Let (Y,c) be a pointed topological space such that Y is locally path-connected and simply connected, let $p : (E,e) \to (X,b)$ be a covering map.
   (a) Show that every map $f : (Y,c) \to (X,b)$ can be uniquely lifted to a map $g : (Y,c) \to (E,b)$.
   (b) Suppose in addition that X is locally path-connected and E is simply connected. Show that if f is a covering map, then g is a homeomorphism of Y and E. (In other words, a simply connected covering space of a locally path-connected space is unique.)
8. Let $p : E \to X$ be a covering map. A map $f : E \to E$ is a *covering transformation* if $p \circ f = f$.
   (a) Show that with the operation of composition, the covering transformations form a group.
   (b) Show that if X is locally path-connected and E is simply connected, then the group of covering transformations is isomorphic to the fundamental group of X.
9. What are the covering transformations of S^2 over P^2? of $\mathbb{R}$ over S^1?
10. Let X be a path-connected topological space such that every $x \in X$ has an open neighborhood that is simply connected. Fix $b \in X$ and let E be the set of all pairs $(x,[\gamma])$, where $x \in X$ and γ is a path in X from b to x. For each simply connected open subset U of X and each path γ in X starting at b and terminating at some point $\gamma(1) \in U$, define $W(U,\gamma)$ to be the set of all pairs $(x,[\gamma][\alpha])$ in E such that $x \in U$ and α is a path in U from $\gamma(1)$ to x. Prove the following.
    (a) The sets $W(U,\gamma)$ form a base for a topology for E.
    (b) The natural projection $(x,[\gamma]) \to x$ of E onto X is a covering map.
    (c) E is simply connected.

Remark: By Exercise 7, the space E is essentially unique. It is called the *universal covering space* of X.

11. Let X be the quotient space obtained from the union of circles $\{x^2 + y^2 = 1,\ z = 1/n\}$ in $\mathbb{R}^3$, for $1 \leq n \leq \infty$, by identifying the set $\{(1,0,1/n) : 1 \leq n \leq \infty\}$ to a point b. Show that if $p : (E,e) \to (X,b)$ is a covering map, then E is not simply connected. (Thus X has no universal covering space. Why does Exercise 10 not apply?)

6. SOME APPLICATIONS OF THE INDEX

Recall that B^n is the closed unit ball in $\mathbb{R}^n$. It will be convenient to identify $\mathbb{R}^2$ and $\mathbb{C}$ so that B^2 becomes the closed unit disc in the complex plane, with boundary circle S^1. The applications to be presented in this section are based on the following theorem.

6.1 Theorem: Let f be a map from B^2 to S^1 that satisfies $f(1) = 1$. Then the loop α, defined by

$$\alpha(s) = f(e^{2\pi is}), \qquad 0 \leq s \leq 1,$$

has index zero:

$$\text{ind}(\alpha) = 0.$$

Proof: Define a loop β in B^2 by

$$\beta(s) = e^{2\pi is}, \qquad 0 \leq s \leq 1.$$

Then $\alpha = f \circ \beta$. Since B^2 is convex, Corollary 4.2 shows that α is homotopic with endpoints fixed to the constant loop in S^1 at 1. By Theorem 5.6, $\text{ind}(\alpha) = 0$. □

The first application asserts that S^1 is not a retract of B^2.

6.2 Theorem: There is no map f of B^2 onto S^1 such that $f(z) = z$ for all $z \in S^1$.

Proof: If there were such a map f, then the loop $s \rightarrow e^{2\pi is}$, $0 \leq s \leq 1$, in S^1 would have index zero, by Theorem 6.1. However, the index of the loop is 1. □

There is another proof of Theorem 6.2, one which illustrates more clearly the method of algebraic topology (sometimes referred to as *algebraic arrowology*), that is, the converting of topological problems to problems in algebra.

Alternate Proof: Let $j : S^1 \hookrightarrow B^2$ be the inclusion map. By Theorem 4.3, the mappings in the following commutative diagram

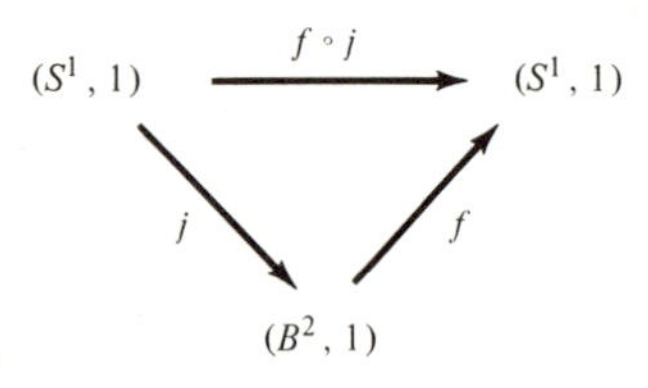

generate the following commutative diagram of group homomorphisms:

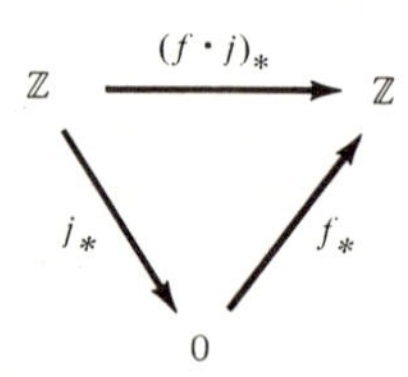

Since $f \circ j$ is the identity map of S^1, $(f \circ j)_* = f_* \circ j_*$ is the identity isomorphism of $\mathbb{Z} \cong \pi_1(S^1,1)$. However, since $\pi_1(B^2,1) = 0$, both f_* and j_* are the zero homomorphisms. Again we have reached a contradiction. □

A point $x \in X$ is a *fixed point* of a map $f : X \to X$ if $f(x) = x$. In general, one does not expect a map to have a fixed point. For instance, the antipodal map $z \to -z$ of S^n onto S^n has no fixed points. In contrast, the celebrated Brouwer Fixed Point Theorem asserts that every map $f : B^n \to B^n$ has a fixed point. We prove this theorem in the special case where $n = 2$. The general case is treated in Chapter IV.

6.3 Theorem: Any map $f : B^2 \to B^2$ has a fixed point.

Proof: Suppose that f has no fixed point. For each $z \in B^2$, let $g(z)$ be the point of

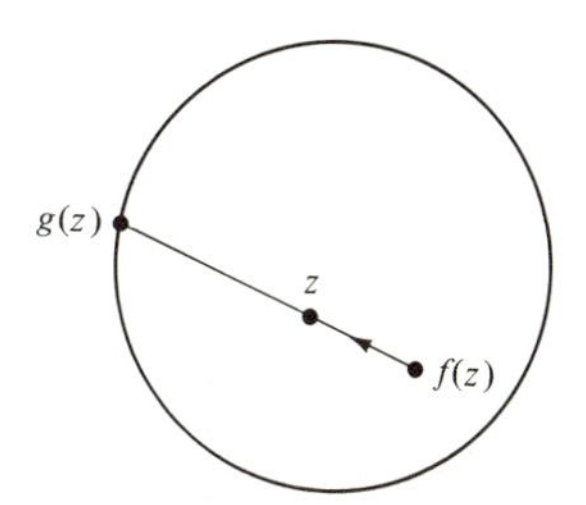

S^1 at which the ray issuing from $f(z)$ and passing through z leaves B^2. Then g is a continuous map from B^2 to S^1 (the proof is left as an exercise). Since $g(z) = z$ if $z \in S^1$, we obtain a contradiction to Theorem 6.2. □

The next application involves a preliminary reduction and then a more detailed analysis of the behavior of the index. Lurking in the background is projective space.

6.4 Theorem *(Borsuk-Ulam Theorem):* Let f be a map from S^2 to $\mathbb{R}^2$. Then there exist antipodal points w and $-w$ in S^2 such that $f(w) = f(-w)$.

Proof: Define a map $g : S^2 \to \mathbb{R}^2$ by

$$g(w) = f(w) - f(-w), \qquad w \in S^2.$$

We must show that g vanishes at some point of S^2. The proof will depend only on the following property of g:

$$(6.1) \qquad g(-w) = -g(w), \qquad w \in S^2.$$

Consider the map $h : B^2 \to \mathbb{R}^2$ defined by

$$h(x,y) = g(x,y,\sqrt{1 - x^2 - y^2}\,), \qquad (x,y) \in B^2.$$

Here the nonnegative value of the square root is taken, so that h is obtained from g by flattening the top half of S^2 onto the closed disc B^2. From (6.1) we have

$$(6.2) \qquad h(-z) = -h(z), \qquad z \in S^1.$$

It suffices to show that any map h from B^2 to $\mathbb{R}^2$ satisfying (6.2) vanishes at some point of B^2.

Suppose that such an h does not vanish on B^2. Then

$$\varphi(z) = \frac{h(z)}{|h(z)|}\frac{|h(1)|}{h(1)}, \qquad z \in B^2,$$

defines a map $\varphi : B^2 \to S^1$ which satisfies

$$\varphi(-z) = -\varphi(z), \qquad z \in S^1, \tag{6.3}$$

$$\varphi(1) = 1. \tag{6.4}$$

By Theorem 6.1, the path

$$\alpha(s) = \varphi(e^{2\pi is}), \qquad 0 \leq s \leq 1,$$

has index zero. We aim to obtain a contradiction by showing that the index of α is odd.

Choose $k : [0,1] \to \mathbb{R}$ such that

$$\varphi(e^{2\pi is}) = e^{2\pi ik(s)}, \qquad 0 \leq s \leq 1,$$

$$k(0) = 0.$$

Then $\text{ind}(\alpha) = k(1)$. The condition (6.3) shows that

$$\begin{aligned}\exp[2\pi ik(s + 1/2)] &= -\exp[2\pi ik(s)] \\ &= \exp[2\pi i(k(s) + 1/2)], \qquad 0 \leq s \leq 1/2.\end{aligned}$$

For each fixed $s \in [0,1/2]$, the number

$$k(s + 1/2) - k(s) - 1/2 \tag{6.5}$$

is then an integer. Since the function defined by (6.5) depends continuously on s and has discrete range, it is constant—equal to the integer m, say—so that

$$k(s + 1/2) - k(s) = m + 1/2, \qquad 0 \leq s \leq 1/2.$$

Then

$$\begin{aligned}\text{ind}(\alpha) &= k(1) = k(1) - k(1/2) + k(1/2) - k(0) \\ &= m + 1/2 + m + 1/2 = 2m + 1,\end{aligned}$$

an odd integer. □

6.5 "Corollary": At any given instant of time, there are two antipodal points on the surface of the earth at which the temperature and the wind speed are the same.

The next result concerns the division of volumes by planes. It derives its picturesque name from its interpretation as the assertion that it is possible, with a single knife stroke, to divide two pieces of bread and a piece of ham each into equal halves, no matter how irregular the three pieces or how askew their relative locations.

6.6 Theorem *(Ham Sandwich Theorem):* Let U, V, and W be three bounded connected open subsets of $\mathbb{R}^3$. Then there is a plane in $\mathbb{R}^3$ that divides each of the sets into two pieces of equal volume.

Proof: We can assume that each of the sets U, V, and W is nonempty. Otherwise the proof is easier.

Let $w \in S^2$ and let L be the straight line passing through 0 and w. Then there is a unique point lying on L such that the plane through the point and perpendicular to L divides U in half by volume. The existence of the point requires elementary facts about volumes of sets in $\mathbb{R}^3$, together with the finiteness of the volume of U. The uniqueness of the point requires in addition that U be nonempty and connected. Intuitively one moves a point p forward along L and observes that the volume $h(p)$ of the part of U behind the plane through p perpendicular to L is a continuous function of p that increases from 0 to the volume of U. The point required is the point p at which $h(p)$ is exactly half the volume of U. We denote this point by $g_0(w)w$, so that g_0 becomes a function from S^2 to $\mathbb{R}$. It is easy to check that $g_0(w)$ depends continuously on w. Furthermore, from the construction it is clear that

$$g_0(-w) = -g_0(w), \qquad w \in S^2. \tag{6.6}$$

Maps g_1 and g_2 from S^2 to $\mathbb{R}$ are defined similarly, replacing U by V and W, respectively. We must show that there exists $w \in S^2$ such that $g_0(w) = g_1(w) = g_2(w)$.

Define a map $f : S^2 \to \mathbb{R}^2$ by

$$f(w) = (g_0(w) - g_1(w), g_0(w) - g_2(w)), \qquad w \in S^2.$$

It suffices to show that f vanishes at some point of S^2.

By the Borsuk-Ulam Theorem, there exists $w_0 \in S^2$ such that

$$f(-w_0) = f(w_0). \tag{6.7}$$

From (6.6) and the corresponding identities for g_1 and g_2, we obtain

$$f(-w) = -f(w), \qquad w \in S^2. \tag{6.8}$$

From (6.7) and (6.8) it follows that $f(w_0) = 0$. □

EXERCISES

1. Prove in detail that the function g appearing in the proof of the Brouwer Fixed Point Theorem is continuous.
2. Prove in detail that the functions g_0, g_1, and g_2 appearing in the proof of the Ham Sandwich Theorem are continuous.
3. Let U, V, and W be open balls in $\mathbb{R}^3$. Describe the plane in $\mathbb{R}^3$ which divides each of U, V, and W in half by volume. When is the plane unique?
4. If U and V are bounded connected open subsets of $\mathbb{R}^3$ and $w \in \mathbb{R}^3$, then prove that there exists a plane in $\mathbb{R}^3$ passing through w and dividing U and V in half by volume.
5. A topological space X has the *fixed-point property* if every map from X to X has a fixed point. Show that the fixed-point property is a topological property.
6. Recall (from Exercise 4.2) that a subspace Y of X is a *retract* of X if there is a map $f : X \to Y$ such that f is the identity on Y. Show that the unit ball

B^n in $\mathbb{R}^n$ has the fixed-point property if and only if its boundary S^{n-1} is not a retract of B^n. (It turns out that B^n has the fixed-point property. See Section IV.2.)

7. Prove that if f is a map from S^1 to $\mathbb{R}$, then there exist antipodal points of S^1 at which f has equal values.
8. Prove that if U and V are bounded connected open subsets of $\mathbb{R}^2$, then there exists a straight line that divides each of U and V in half by area. (This is the *Pancake Theorem*.)

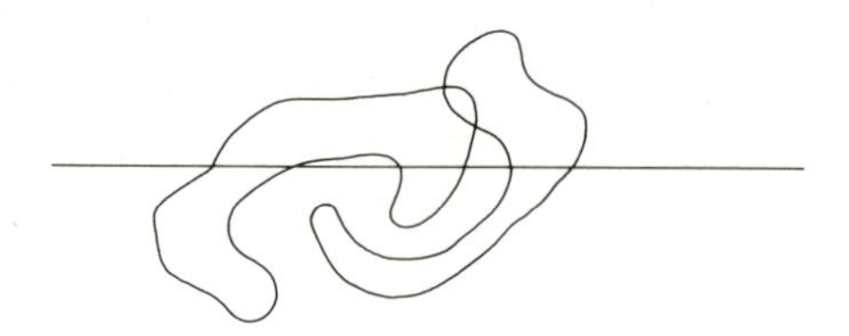

9. Let S be the circle in $\mathbb{R}^3$ centered at (1,0,0), with radius 1, and contained in the plane $\{y = 0\}$. Let T be the circle in $\mathbb{R}^3$ centered at (0,0,0), with radius 1, and contained in the plane $\{z = 0\}$. Show that S and T are *linked*, that is, there does not exist a map $f : B^2 \to \mathbb{R}^3 \backslash T$ such that the restriction of f to S^1 maps S^1 homeomorphically onto S. (Intuitively, every membrane with boundary S must be punctured somewhere by T.) *Hint:* Show first that S is a retract of $\mathbb{R}^3 \backslash T$. A retraction is given by first projecting $\mathbb{R}^3 \backslash T$ circularly onto the punctured half-space $\{(x,0,z) : x \geq 0\} \backslash (1,0,0)$ and then projecting radially onto the circle S.

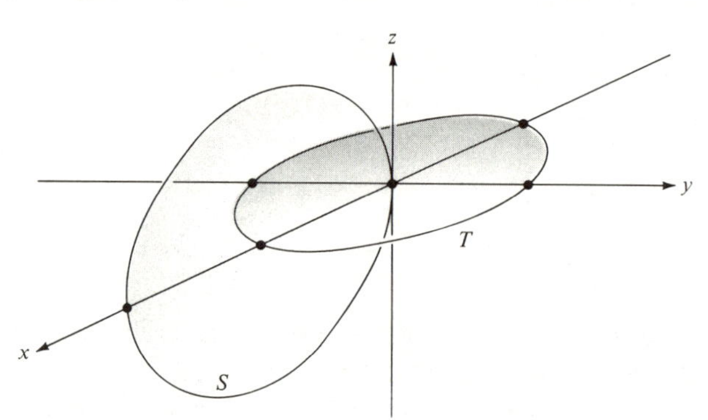

7. HOMOTOPIC MAPS

Let X and Y be topological spaces and let A be a subset of Y. Let f and g be two maps from Y to X such that $f = g$ on A. Then *f is homotopic to g relative to A*, written

$$f \simeq g \text{ rel} A,$$

if there is a map

$$F : Y \times [0,1] \to X$$

such that

$$F(y,0) = f(y), \qquad y \in Y,$$

$$F(y,1) = g(y), \qquad y \in Y,$$

$$F(a,t) = f(a) = g(a), \qquad a \in A,\ 0 \leq t \leq 1.$$

The map F is called a *homotopy* of f and g. The situation is depicted graphically in the figure for the case where $X = \mathbb{R}$. The various contour curves indicated in the figure are the graphs of the slice functions f_t defined by $f_t(y) = F(y,t)$, which are deformed continuously from $f_0 = f$ to $f_1 = g$.

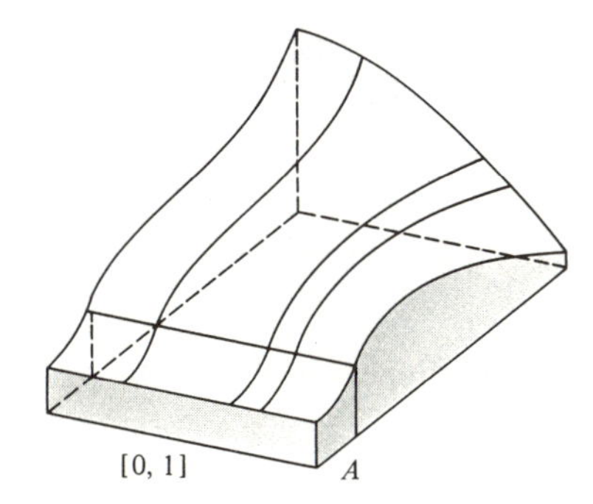

The case $Y = [0,1]$ and $A = \{0,1\}$ specializes to the definition of homotopic paths with endpoints fixed, given in Section 2. The proof given in Section 2 can be extended in the obvious way to show that being homotopic rel$A$ is an equivalence relation among maps from Y to X which coincide with a given map on A.

In this section, we shall be primarily interested in the case in which A is empty. In this case, we simply say that f is *homotopic* to g and write $f \simeq g$.

7.1 Theorem: Let F be a homotopy of two maps f and g from Y to X. Let $y_0 \in Y$, $x_0 = f(y_0)$, and $x_1 = g(y_0)$. Define a path α from x_0 to x_1 by

$$\alpha(t) = F(y_0,t), \qquad 0 \leq t \leq 1.$$

Then

$$g_* = \alpha_* \circ f_*,$$

that is, the following diagram commutes:

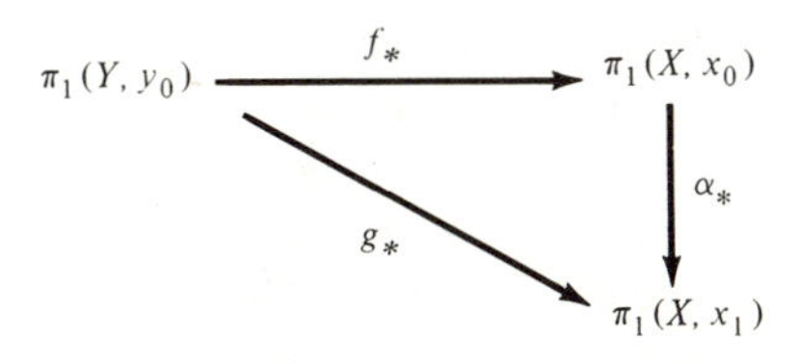

In particular, f_* is an isomorphism if and only if g_* is an isomorphism.

Proof: Recall that α_* was defined in Section 3 and g_* and f_* were defined in Section 4. Since α_* is an isomorphism, the final statement of the theorem follows from the commutativity of the diagram.

Let $\gamma : [0,1] \to Y$ be a loop based at y_0. We must show that $\alpha_*(f_*([\gamma])) = g_*([\gamma])$. For this, it suffices to show that $[\alpha^{-1}][f\circ\gamma][\alpha] = [g\circ\gamma]$.

Consider the map

$$G : [0,1] \times [0,1] \to X$$

defined by

$$G(s,t) = F(\gamma(s),t), \qquad 0 \le s,t \le 1.$$

Let β_1, β_2, β_3, and β_4 be the four paths in the unit square indicated by the following figure:

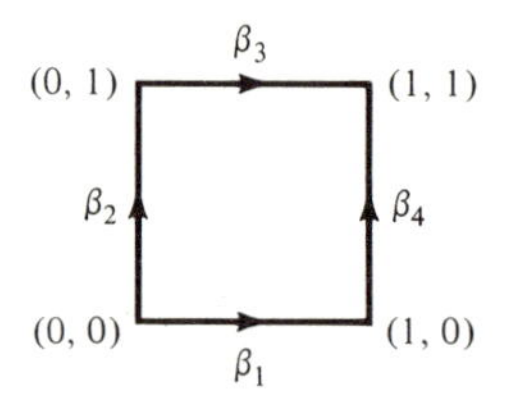

Assume the paths are parametrized linearly, so that, for instance, $\beta_1(s) = (s,0)$, $0 \le s \le 1$. Then

$$G\circ\beta_1 = f\circ\gamma,$$

$$G\circ\beta_2 = \alpha,$$

$$G\circ\beta_3 = g\circ\gamma,$$

$$G\circ\beta_4 = \alpha.$$

Since the square is convex, the paths β_3 and $(\beta_2^{-1}\beta_1)\beta_4$ from $(0,1)$ to $(1,1)$ are homotopic with endpoints fixed. By Lemma 4.1 or Corollary 4.2,

$$G\circ\beta_3 \simeq G\circ((\beta_2^{-1}\beta_1)\beta_4) \text{ rel}\{0,1\}.$$

Consequently

$$[g\circ\gamma] = [(\alpha^{-1}(f\circ\gamma))\alpha] = [\alpha^{-1}][f\circ\gamma][\alpha],$$

as required. □

A map $f : Y \to X$ is a *homotopy equivalence* of Y and X if there exists a map $g : X \to Y$ such that $f\circ g$ is homotopic to the identity map of X and $g\circ f$ is homotopic to the identity map of Y. The map g is a *homotopy inverse* of f. Evidently g is a homotopy equivalence of X and Y, with homotopy inverse f.

Two spaces X and Y are *homotopically equivalent* if there is a homotopy equivalence of Y and X. It is easy to check that the relation of being homotopically equivalent is an equivalence relation. The remarks above show that the relation is symmetric. Transitivity follows from the fact that the composition of two homotopy equivalences is a homotopy equivalence (when defined). The details of this verification are left as an exercise. Reflexivity follows from the fact that the identity map of a topological space is a homotopy equivalence. Indeed, if f is any homeomorphism, then f is a homotopy equivalence with homotopy inverse f^{-1}.

As an example, we note that the inclusion map

$$f : S^1 \hookrightarrow \mathbb{C}\backslash\{0\}$$

is a homotopy equivalence of the circle S^1 and the punctured plane, and a homotopy inverse for f is given by the map $g : \mathbb{C}\backslash\{0\} \to S^1$ defined by

$$g(z) = z/|z|, \qquad z \in \mathbb{C}\backslash\{0\}.$$

The composition $f \circ g$ is the map $z \to z/|z|$ of $\mathbb{C}\backslash\{0\}$ into itself. A homotopy of $f \circ g$ and the identity map of $\mathbb{C}\backslash\{0\}$ is given by

$$F(z,t) = z/|z|^t, \qquad z \in \mathbb{C}\backslash\{0\},\ 0 \leq t \leq 1.$$

This homotopy pulls the various points of the punctured plane to the unit circle along

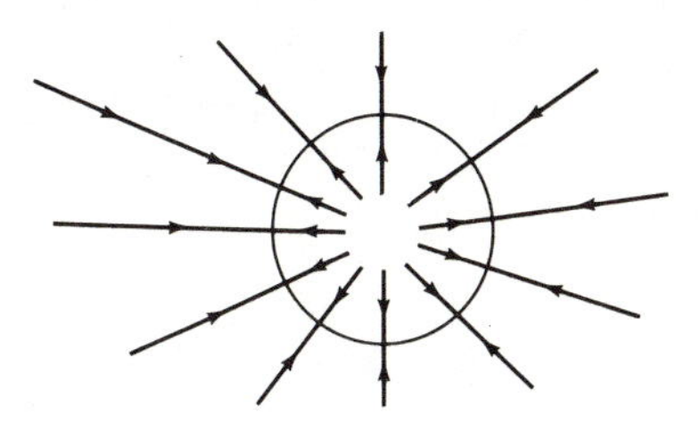

rays, as indicated in the figure. The map $g \circ f$ is already the identity of S^1. Hence f is indeed a homotopy equivalence of the circle and the punctured plane, as asserted.

7.2 Theorem: Let $f : Y \to X$ be a homotopy equivalence, let $y_0 \in Y$, and let $x_0 = f(y_0)$. Then f_* is an isomorphism of $\pi_1(Y,y_0)$ and $\pi_1(X,x_0)$.

Proof: Let g be a homotopy inverse for f and let $y_1 = g(x_0)$. Now the identity map of Y induces the identity isomorphism of $\pi_1(Y,y_0)$. Applying Theorem 7.1 to $g \circ f$ and the identity map of Y, we find that $(f \circ g)_* = f_* \circ g_*$ is an isomorphism. It follows that f_* is one-to-one and that g_* is onto. Since the roles of f and g are interchangeable, f_* is also onto. Consequently f_* is an isomorphism. □

One consequence of Theorem 7.2 is that the homotopy equivalence $S^1 \hookrightarrow \mathbb{C}\backslash\{0\}$ induces an isomorphism,

$$\pi_1(\mathbb{C}\backslash\{0\}) \cong \mathbb{Z}.$$

The isomorphism can be made explicit by tracing through the steps in the proof of Theorem 7.2 and its predecessors. The integer corresponding to a loop γ in $\mathbb{C}\backslash\{0\}$ based at 1 is the index of the loop $\alpha = \gamma/|\gamma|$ in S^1.

A topological space X is *contractible* if X is homotopically equivalent to a space consisting of a single point. This occurs if and only if there is a point $x_0 \in X$ and a map $F : X \times [0,1] \rightarrow X$ such that

$$F(x,0) = x_0, \qquad x \in X,$$

$$F(x,1) = x, \qquad x \in X.$$

Note that is it not required that x_0 be held fixed by the homotopy F.

Any convex subset X of $\mathbb{R}^n$ is contractible. Indeed, fix $x_0 \in X$ and define $F : X \times [0,1] \rightarrow X$ by

$$F(x,t) = tx + (1 - t)x_0, \qquad x \in X, 0 \leq t \leq 1.$$

Then F is a homotopy of the identity and the constant map at x_0.

7.3 Theorem: A contractible space is simply connected.

Proof: Suppose that X is contractible and F is a homotopy of the identity map and the constant map at $x_0 \in X$. For each fixed $x \in X$, the map $t \rightarrow F(x,t)$ is a path in X joining x_0 and x. Consequently X is path-connected. By Theorem 7.2, $\pi_1(X,x_0)$ is isomorphic to the fundamental group of the one-point space $\{x_0\}$, which is trivial. □

EXERCISES

1. Prove in detail that the relation of being homotopically equivalent is transitive.
2. Show that if Y is contractible, then $X \times Y$ is homotopically equivalent to X.
3. Prove that any annulus in $\mathbb{R}^2$ is homotopically equivalent to S^1.
4. Prove that the punctured open disc $\{z : 0 < |z| < 1\}$ in $\mathbb{C}$ is homotopically equivalent to S^1.
5. Prove that $\mathbb{R}^{n+1}\backslash\{0\}$ is homotopically equivalent to S^n.
6. Prove that the doubly punctured plane is homotopically equivalent to the figure eight.
7. Prove that if Y is contractible, then any two maps from X to Y are homotopic.
8. Prove that if X is contractible and Y is path-connected, then any two maps from X to Y are homotopic.
9. Let (X,b) be a pointed topological space, let α_0 and α_1 be loops in X based at b, and let $f_j(e^{2\pi is}) = \alpha_j(s)$, $0 \leq s \leq 1$, be the corresponding maps of S^1 into X. Show that f_0 is homotopic to f_1 if and only if there exists a loop β in X based at b such that
$$[\alpha_1] = [\beta]^{-1}[\alpha_0][\beta].$$
10. If X is path-connected and $\pi_1(X)$ is commutative, show that there is a one-to-one correspondence between elements of $\pi_1(X)$ and homotopy classes of maps of S^1 into X.

8. MAPS INTO THE PUNCTURED PLANE

In this section, we aim to classify the homotopy classes of maps from a circle to the punctured plane. Such a classification can be obtained by using the description of the fundamental group $\pi_1(S^1)$ given in Section 5. We shall develop the results independently of Section 5, though, and as a dividend we shall obtain an alternative proof that $\pi_1(S^1)$ is isomorphic to the integers. Whereas the material involving covering spaces has a topological flavor, our approach now is more analytic in nature. A familiarity with complex functions and in particular with the logarithm function will be assumed.

Let X be a topological space. We wish to consider maps from X to the punctured complex plane $\mathbb{C}\backslash\{0\}$. In other words, the objects under consideration are continuous complex-valued functions on X which do not vanish at any point of X. Such functions form a group, with pointwise multiplication as the operation:

$$(fg)(x) = f(x)g(x), \qquad x \in X.$$

The constant function 1 serves as the identity of the group, and the inverse of f is the function $1/f$.

A map $f : X \to \mathbb{C}\backslash\{0\}$ is an *exponential* if $f = e^g$ for some map $g : X \to \mathbb{C}$. Since $1/e^g = e^{-g}$ and $e^g e^h = e^{g+h}$, the exponentials form a subgroup of the maps from X to $\mathbb{C}\backslash\{0\}$.

Every continuous function f from X to the positive real numbers is an exponential. In this case, f can be expressed as e^g, where $g = \log f$ is the usual natural logarithm function composed with f. More generally, the following is true.

8.1 Lemma: Suppose $f : X \to \mathbb{C}\backslash\{0\}$ is a map that omits the negative real axis (that is, $f(X) \cap (-\infty,0) = \varnothing$). Then f is an exponential.

Proof: Recall that for $z = re^{i\theta} \in \mathbb{C}\backslash(-\infty,0]$, with $-\pi < \theta < \pi$, the *principal value* of $\log z$ is defined to be the complex number

$$\operatorname{Log} z = \log r + i\theta.$$

In other words,

$$\operatorname{Log} z = \log|z| + i \arg z,$$

where the value of the argument $\arg z$ is chosen to satisfy

$$-\pi < \arg z < \pi.$$

Then $\operatorname{Log} z$ depends continuously on z and $e^{\operatorname{Log} z} = z$. If $g = \operatorname{Log} f$, then g is well-defined and continuous and $e^g = f$. □

The following result is related to Rouché's Theorem from complex function theory. Note the symmetry between f and g; the estimate (8.1) remains the same if f and g are interchanged.

8.2 Theorem: Let f and g be maps from X to $\mathbb{C}$ which satisfy

$$|f(x) - g(x)| < |f(x)| + |g(x)|, \qquad x \in X. \tag{8.1}$$

Then g/f and f/g are exponentials. In particular, f is an exponential if and only if g is.

Proof: First observe that the strict inequality in (8.1) implies that neither f nor g can vanish at any point of X. Dividing (8.1) by $f(x)$, we obtain

$$|1 - g(x)/f(x)| < 1 + |g(x)/f(x)|, \qquad x \in X.$$

It follows that g/f cannot assume negative real values. By Lemma 8.1, g/f is an exponential, and then so is f/g. The final statement of the theorem is a consequence of the fact that the product of exponentials is an exponential. □

8.3 Theorem: Let X be a compact metric space and let f and g be maps from X to $\mathbb{C}\backslash\{0\}$. Then f and g are homotopic if and only if f/g is an exponential.

Proof: Suppose first that f/g is an exponential, say $f/g = e^h$. Then

$$F(x,t) = g(x)e^{th(x)}, \qquad x \in X, 0 \leq t \leq 1,$$

defines a homotopy F of f and g.

Conversely, suppose that f and g are homotopic. Let F be a homotopy of f and g, so that

$$F(x,0) = f(x), \qquad x \in X,$$

$$F(x,1) = g(x), \qquad x \in X.$$

Since $X \times [0,1]$ is compact, the continuous positive function $|F|$ on $X \times [0,1]$ attains its minimum value, and that minimum is positive:

$$(8.2) \qquad 0 < b = \inf\{|F(x,t)| : x \in X, 0 \leq t \leq 1\}.$$

By Theorem I.6.3, F is uniformly continuous on $X \times [0,1]$. Consequently there exists $\delta > 0$ such that whenever $|s - t| < \delta$, then

$$(8.3) \qquad |F(x,s) - F(x,t)| < b, \qquad \text{all } x \in X.$$

Now choose an integer $n > 1/\delta$ and consider the maps $f_j : X \to \mathbb{C}\backslash\{0\}$ defined by

$$f_j(x) = F(x,j/n), \qquad x \in X, 0 \leq j \leq n.$$

Then $f_0 = f$ and $f_n = g$. Furthermore, (8.2) and (8.3) show that

$$|f_j(x) - f_{j-1}(x)| < b \leq |f_j(x)|, \qquad x \in X, 1 \leq j \leq n.$$

By Theorem 8.2, each f_{j-1}/f_j is an exponential. Consequently

$$f/g = (f_0/f_1)(f_1/f_2) \cdots (f_{n-1}/f_n)$$

is an exponential. □

Since any constant map is an exponential, we obtain the following corollaries.

8.4 Corollary: Let X be a compact metric space and let f be a map from X to $\mathbb{C}\backslash\{0\}$. Then f is an exponential if and only if f is homotopic to a constant map.

8.5 Corollary: Let X be a compact contractible space. Then every map f from X to $\mathbb{C}\backslash\{0\}$ is an exponential.

Proof: Let $F : X \times [0,1] \to X$ be the homotopy of the identity map of X and a constant map x_0. Then $f \circ F$ is a homotopy of f to the constant map $f(x_0)$. By Corollary 8.4, f is an exponential. □

Now we restrict our attention to maps of S^1 into the punctured plane. We wish to assign to any such map an index that corresponds to the number of times the function wraps around the origin. The definition is made as follows.

Let $f : S^1 \to \mathbb{C}\backslash\{0\}$ be a map. Consider the map $\theta \to f(e^{i\theta})$ of $[0,2\pi]$ into $\mathbb{C}\backslash\{0\}$. Since the interval $[0,2\pi]$ is contractible, Theorem 8.5 shows that the map is an exponential, that is, that there is a map

$$g : [0,2\pi] \to \mathbb{C}\backslash\{0\}$$

such that

$$f(e^{i\theta}) = e^{g(\theta)}, \qquad 0 \le \theta \le 2\pi. \tag{8.4}$$

Suppose $g_1 : [0,2\pi] \to \mathbb{C}$ is another map that satisfies (8.4). Then $e^{g(\theta)-g_1(\theta)} = 1$, $0 \le \theta \le 2\pi$, so that the map $g - g_1$ assumes only values which are integral multiples of $2\pi i$. Since the range of $g - g_1$ is discrete, $g - g_1$ is constant. The number $g(2\pi) - g(0)$ is then independent of the choice of g satisfying (8.4). Consequently the number

$$\operatorname{ind}(f) = [g(2\pi) - g(0)]/2\pi i \tag{8.5}$$

is well defined, where g is any map satisfying (8.4). This number $\operatorname{ind}(f)$ is called the *index* of the map $f : S^1 \to \mathbb{C}\backslash\{0\}$. Since $e^{g(2\pi)} = f(e^{2\pi i}) = f(1) = e^{g(0)}$, $g(2\pi) - g(0)$ is an integral multiple of $2\pi i$ and $\operatorname{ind}(f)$ is an integer.

As an example, fix an integer m and consider the function $f(z) = z^m$, regarded as a map from S^1 to $\mathbb{C}\backslash\{0\}$. The formula $f(e^{i\theta}) = e^{im\theta}$ shows that we may write f in the form (8.4), for $g(\theta) = im\theta$. The index of f given by (8.5) is then simply m:

$$\operatorname{ind}(z^m) = m, \qquad m \in \mathbb{Z}.$$

8.6 Theorem: The index function, defined on maps from S^1 to $\mathbb{C}\backslash\{0\}$, has the following properties:

(i) $\operatorname{ind}(f_1 f_2) = \operatorname{ind}(f_1) + \operatorname{ind}(f_2)$.

(ii) $\operatorname{ind}(f) = 0$ if and only if f is an exponential.

(iii) $\operatorname{ind}(f) = \operatorname{ind}(f/|f|)$.

(iv) If $f : S^1 \to S^1$ is a map satisfying $f(1) = 1$, then $\text{ind}(f)$ coincides with the index of the loop α defined by $\alpha(s) = f(e^{2\pi is})$, $0 \le s \le 1$.

Proof: Suppose $f_1(e^{i\theta}) = e^{g_1(\theta)}$ and $f_2(e^{i\theta}) = e^{g_2(\theta)}$, $0 \le \theta \le 2\pi$. Then

$$(f_1f_2)(e^{i\theta}) = e^{g_1(\theta)+g_2(\theta)},$$

so that

$$\begin{aligned}\text{ind}(f_1f_2) &= [g_1(2\pi) + g_2(2\pi) - (g_1(0) + g_2(0))]/2\pi i \\ &= \text{ind}(f_1) + \text{ind}(f_2).\end{aligned}$$

That proves (i).

Suppose that f is an exponential, say

$$f(e^{i\theta}) = e^{h(e^{i\theta})}, \ 0 \le \theta \le 2\pi.$$

If $g(\theta) = h(e^{i\theta})$, then g satisfies (8.4). Since $g(2\pi) = g(0)$, (8.5) shows that $\text{ind}(f) = 0$. Conversely, suppose that $\text{ind}(f) = 0$. Write $f(e^{i\theta}) = e^{ig(\theta)}$, $0 \le \theta \le 2\pi$. Then $g(0) = g(2\pi)$, so that the function $h : S^1 \to \mathbb{C}$ obtained by setting $h(e^{i\theta}) = g(\theta)$, $0 \le \theta \le 2\pi$, is well defined and continuous. Since $f = e^h$, f is an exponential. That proves (ii).

Since $|f|$ is a positive function, $|f|$ is an exponential. By (i), $\text{ind}(f) = \text{ind}(f/|f|) + \text{ind}(|f|)$, and by (ii), $\text{ind}(|f|) = 0$. That yields (iii).

Now let f and α be as in (iv). Choose $h : [0,1] \to \mathbb{R}$ such that $h(0) = 0$ and $\alpha(s) = e^{2\pi ih(s)}$, $0 \le s \le 1$. Then $\text{ind}(\alpha) = h(1)$. Define $g : [0,2\pi] \to \mathbb{C}$ by

$$g(\theta) = 2\pi ih(\theta/2\pi), \qquad 0 \le \theta \le 2\pi.$$

Then g satisfies (8.4), so that

$$\text{ind}(f) = [g(2\pi) - g(0)]/2\pi i = h(1),$$

and this coincides with $\text{ind}(\alpha)$. □

8.7 Theorem: The following are equivalent for maps f and g from S^1 to $\mathbb{C}\backslash\{0\}$:

(i) f is homotopic to g.

(ii) $\text{ind}(f) = \text{ind}(g)$.

(iii) f/g is an exponential.

Proof: The equivalence of (i) and (iii) is the special case of Theorem 8.3 in which $X = S^1$.

If f/g is an exponential, then by (ii) of Theorem 8.6, $\text{ind}(f/g) = 0$. Writing $f = g \cdot (f/g)$ and using the additivity of the index, we obtain

$$\text{ind}(f) = \text{ind}(g) + \text{ind}(f/g) = \text{ind}(g).$$

Conversely, if $\text{ind}(f) = \text{ind}(g)$, then $\text{ind}(f/g) = 0$, so that by (ii) of Theorem 8.6, f/g is an exponential. □

Since we have computed the index of z^m to be m, we obtain immediately the following corollary.

8.8 Corollary: Each map f from S^1 to $\mathbb{C}\backslash\{0\}$ is homotopic to precisely one of the maps z^m, $m \in \mathbb{Z}$, namely, the map corresponding to $m = \operatorname{ind}(f)$.

Now the isomorphism $\pi_1(S^1,1) \cong \mathbb{Z}$ can be obtained as follows. Since the maps z^m are not homotopic, the loops $\alpha_m : [0,1] \to S^1$ based at 1, defined by

$$\alpha_m(s) = e^{2\pi ims}, \qquad 0 \le s \le 1,$$

cannot be homotopic with endpoints fixed (see the remarks at the end of Section 3). On the other hand, let $\alpha : [0,1] \to S^1$ be an arbitrary loop based at 1. Define $f: S^1 \to S^1$ by $f(e^{2\pi is}) = \alpha(s)$, $0 \le s \le 1$, and let $m = \operatorname{ind}(f)$. By Theorem 8.7, f/z^m is an exponential, say

$$f(e^{2\pi is})/e^{2\pi ims} = e^{h(e^{2\pi is})}, \qquad 0 \le s \le 1.$$

Then

$$F(s,t) = e^{th(e^{2\pi is})}, \qquad 0 \le s,t \le 1,$$

is a homotopy of the loop α and the loop α_m with endpoints fixed. Consequently the correspondence $[\alpha_m] \to m$ is a one-to-one correspondence of $\pi_1(S^1,1)$ and $\mathbb{Z}$. One checks that the path product $\alpha_m\alpha_n$ corresponds to a map from S^1 to S^1 of index $m + n$, so that $\alpha_m\alpha_n$ is homotopic to α_{m+n}. Consequently the correspondence is a group isomorphism.

As an application of these results, we give a topological proof of the Fundamental Theorem of Algebra. One version of that theorem is as follows.

8.9 Theorem: A polynomial

$$p(z) = z^n + a_{n-1}z^{n-1} + \cdots + a_1z + a_0$$

of degree $n \ge 1$ and with complex coefficients $a_0, \ldots, a_{n-1}$ has a root in the complex plane.

Proof: Choose R so large that

$$\left| \frac{a_{n-1}}{R} w^{n-1} + \cdots + \frac{a_1}{R^{n-1}} + \frac{a_0}{R^n} \right| < 1, \qquad |w| \le 1.$$

This inequality is valid, for instance, whenever $R > |a_{n-1}| + \cdots + |a_0| + 1$. Define a map g from the closed unit disc B^2 to $\mathbb{C}$ by

$$g(w) = \frac{p(Rw)}{R^n} = w^n + \frac{a_{n-1}}{R} w^{n-1} + \cdots + \frac{a_0}{R^n}, \qquad |w| \le 1.$$

The estimate above shows that

$$|g(w) - w^n| < 1 = |w^n|, \qquad |w| = 1.$$

By Theorem 8.2, the restriction of w^n/g to the unit circle is an exponential. Now w^n is not an exponential since its index is $n \geq 1$. Hence $g = w^n \cdot (g/w^n)$ is not an exponential. Corollary 8.5 shows that g has a zero on B^2, and hence p has a complex root. □

EXERCISES

1. Let X be compact and let $f : X \to \mathbb{C}\backslash\{0\}$ be an exponential. Show that there exists $\varepsilon > 0$ such that every map $g : X \to \mathbb{C}\backslash\{0\}$ which satisfies $|f(x) - g(x)| < \varepsilon$, $x \in X$, is an exponential.
2. Let X be a locally compact Hausdorff space. Show that two maps f and g from X to $\mathbb{C}\backslash\{0\}$ are homotopic if and only if f/g is an exponential.
   Hint: Consider first the case in which X is compact.
3. Let X be a locally compact Hausdorff space that is contractible. Show that any map from X to $\mathbb{C}\backslash\{0\}$ is an exponential.
4. Let f be a map from S^1 to $\mathbb{C}\backslash\{0\}$. Show that there exists $\varepsilon > 0$ such that any map $g : S^1 \to \mathbb{C}\backslash\{0\}$ satisfying $|g(z) - f(z)| < \varepsilon$, $z \in S^1$, has the same index as f.
5. Suppose that f and g are maps from S^1 to S^1 such that f and g do not assume antipodal values at any point of S^1. Show the $\text{ind}(f) = \text{ind}(g)$.
6. If $n \geq 2$, show that every map $f : S^n \to \mathbb{C}\backslash\{0\}$ is an exponential.
7. If $n \geq 2$, show that every map $f : P^n \to \mathbb{C}\backslash\{0\}$ is an exponential. (Note, however, that P^n is not simply connected: $\pi_1(P^n) \cong \mathbb{Z}_2$ for $n \geq 2$. Can you offer an explanation for this state of affairs?)
8. Classify the homotopy classes of maps from the figure eight to $\mathbb{C}\backslash\{0\}$.

9. VECTOR FIELDS

The main idea of this section, the idea of a vector field, has many important applications not only in various parts of mathematics but also in physics. We shall introduce vector fields and prove that every field on the unit sphere S^2 in $\mathbb{R}^3$ has a zero. One colorful way of expressing this theorem is that it is impossible to comb a hairy billiard ball.

A *vector field* F on a subset S of $\mathbb{R}^n$ is simply a continuous function F from S to $\mathbb{R}^n$. Vector fields were encountered in Section I.8 in connection with the Cauchy-Picard Existence Theorem (I.8.4). Vector fields on $\mathbb{R}^2$ are often represented graphically by sketching the vector $F(x)$, with its tail at x, for typical points x. Some examples of vector fields on $\mathbb{R}^2$ are as follows:

(i) If S is any subset of $\mathbb{R}^n$ and if $q_0 \in \mathbb{R}^n$ is fixed, then function $F(p) = q_0$, $p \in S$, is a vector field on S. Such vector fields are called *constant vector fields*.

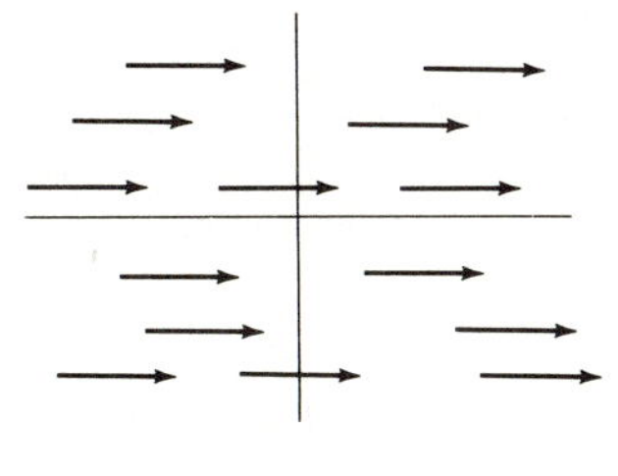

$F(x, y) = (1, 0)$

(ii) If U is any open subset of $\mathbb{R}^n$ and if f is a continuous real-valued function on U, all of whose first-order partial derivatives exist and are continuous, then the gradient ∇f of f, defined by

$$(\nabla f)(p) = \left(\frac{\partial f}{\partial x_1}(p), \ldots, \frac{\partial f}{\partial x_n}(p)\right), \qquad p \in U,$$

is a vector field on U.

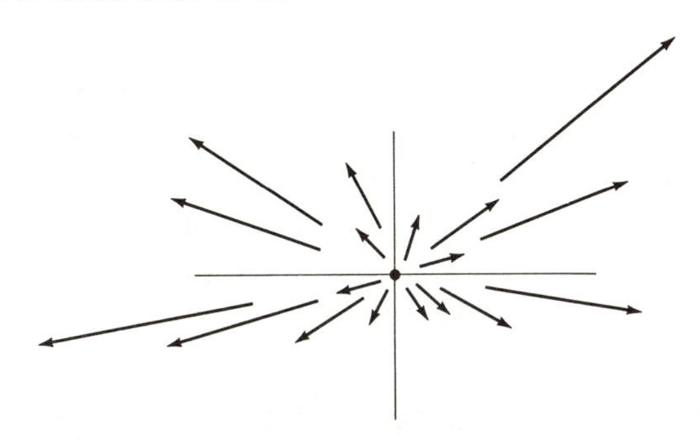

$F(x, y) = \nabla(x^2 + y^2) = (2x, 2y)$

(iii) The function $(x,y) \to (-y,x)$ is a vector field on $\mathbb{R}^2$ which represents the velocity at each point of $\mathbb{R}^2$ when $\mathbb{R}^2$ is rotated in the counterclockwise direction at a uniform rate.

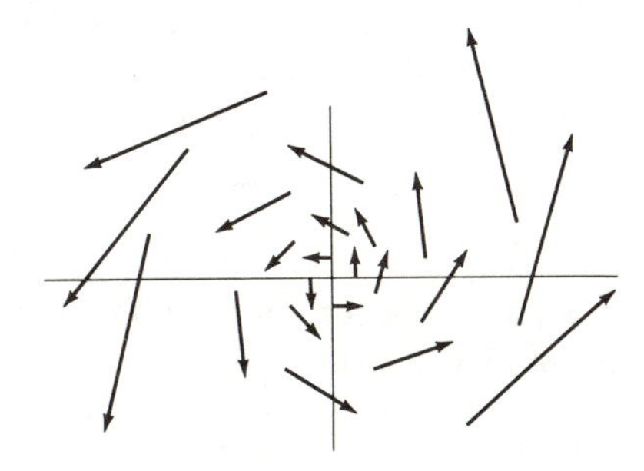

$F(x, y) = (-y, x)$

(iv) If U is any open subset of $\mathbb{R}^3$ in which a fluid is circulating in a reasonably regular fashion, then the function assigning to each $p \in U$ the velocity vector of the fluid particles at that point is a vector field on U.

As these illustrations show, one thinks of a vector field as a continuous assignment of a vector to each point. Here one makes use of the idea, familiar in $\mathbb{R}^2$ and $\mathbb{R}^3$, that the points of $\mathbb{R}^n$ may also be thought of as vectors. Of course, this viewpoint is just a change of intuitive concept. Whether one considers the n-tuples which make up $\mathbb{R}^n$ as points or vectors is, aside from the intuitive significance, only a matter of terminology.

A point $p \in S$ is a *zero* of a vector field F on S if $F(p) = 0$. The vector fields depicted in examples (ii) and (iii) above each have only one zero, the origin. While the vector field F itself behaves quite well at a zero, the direction of F (the function $F/|F|$) can behave quite wildly near a zero of F; for this reason, the zeros of F are referred to in the older literature as "singularities" of F.

It is important to observe that there is a notion of index available for vector fields in $\mathbb{R}^2$. Indeed, if F is any nonzero vector field on a subset S of $\mathbb{R}^2$ and if $e^{i\theta} \to \gamma(e^{i\theta})$, $0 \leq \theta \leq 2\pi$, is a loop in S, then the *index of F around* γ is defined to be the index of $F \circ \gamma$, regarded as a map from the circle to the nonzero complex numbers, as defined in the preceding section. The index of F around γ counts the number of times that $F(p)$ winds around the origin as p moves around γ and is thus referred to also as the *winding number* of F around γ.

Our work in the preceding section leads to a simple criterion for a vector field on $\mathbb{R}^2$ to have a zero.

9.1 Lemma: Let F be a vector field on the closed disc $\{x^2 + y^2 \leq R^2\}$ that does not vanish in the boundary circle $\{x^2 + y^2 = R^2\}$. Suppose that the index of F around the loop $e^{i\theta} \to Re^{i\theta}$ is not zero. Then F has a zero at some interior point of the disc.

Proof: Suppose F has no zeros on the closed disc. We must show that the map $F_R : e^{i\theta} \to F(Re^{i\theta})$, from the circle to the punctured plane, has index zero. For this, observe that the maps F_r defined by

$$F_r(e^{i\theta}) = F(re^{i\theta}), \qquad 0 \leq \theta \leq 2\pi,\ 0 \leq r \leq R,$$

provide a homotopy of F_R and the constant map F_0. Since the index of a constant map is zero, we conclude from Theorem 8.7 that the index of F_R is zero. □

Recall that the unit sphere in $\mathbb{R}^3$ is defined by

$$S^2 = \{(x,y,z) : x^2 + y^2 + z^2 = 1\}.$$

A vector field F on S^2 is *tangent to* S^2 if for each $p \in S^2$, $F(p)$ is orthogonal to p. If F_1, F_2, and F_3 are the coordinate functions of F, so that $F = (F_1, F_2, F_3)$, then the condition for F to be tangent to S^2 is that

$$(9.1) \qquad xF_1(x,y,z) + yF_2(x,y,z) + zF_3(x,y,z) = 0, \qquad (x,y,z) \in S^2.$$

For example, the only constant vector field tangent to S^2 is the vector field 0 (Exercise 2). The vector field

$$F(x,y,z) = (-y,x,0), \qquad (x,y,z) \in S^2,$$

is tangent to S^2. It points from west to east everywhere except at the north and south poles $(0,0,\pm 1)$, which are zeros of the vector field. As another example, consider the vector field

$$G(x,y,z) = (xz,yz,z^2 - 1), \qquad (x,y,z) \in S^2.$$

One checks that condition (9.1) is satisfied, so that G is tangent to S^2. The vector field G points from north to south except at the north and south poles, which are zeros of G.

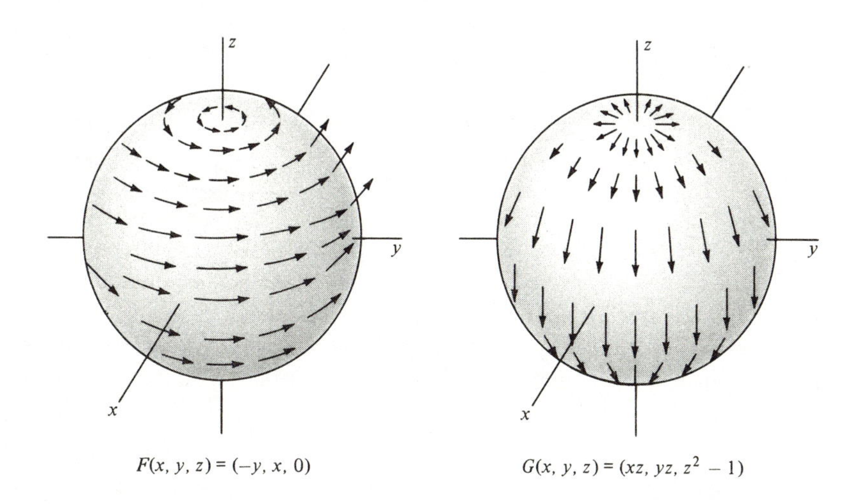

$F(x, y, z) = (-y, x, 0)$ $\qquad$ $G(x, y, z) = (xz, yz, z^2 - 1)$

The velocity vector field of a fluid flowing on the surface of a sphere is tangent to the sphere. For instance, regarding S^2 as the conventional spherical realization of the earth's surface, we obtain a vector field tangent to S^2 by assigning to any point the vector expressing the magnitude and direction of the wind at that point, at some fixed time.

The result we are aiming to prove in this section is the following.

9.2 Theorem: Every tangent vector field to S^2 has a zero.

In other words, there is always a point on the surface of the earth at which the wind is not blowing. Before undertaking the formal proof of Theorem 9.2, we outline the idea behind it.

Suppose the vector field F does not have a zero at the north pole, labeled N in the next figure, and consider a small circle Γ that is centered at the north pole and inside of which the vector field is approximately constant. If one were an observer standing at the north pole and regarding the surface of the earth nearby as flat, one would compute the index of F around Γ to be zero. However, if one were an observer standing at the south pole and studying the vector field on a map obtained by projecting the sphere stereographically from the north pole into the plane, one would compute that the winding number of the vector field around the circle corresponding to Γ is $+2$. Hence F vanishes somewhere on the map, by Lemma 9.1.

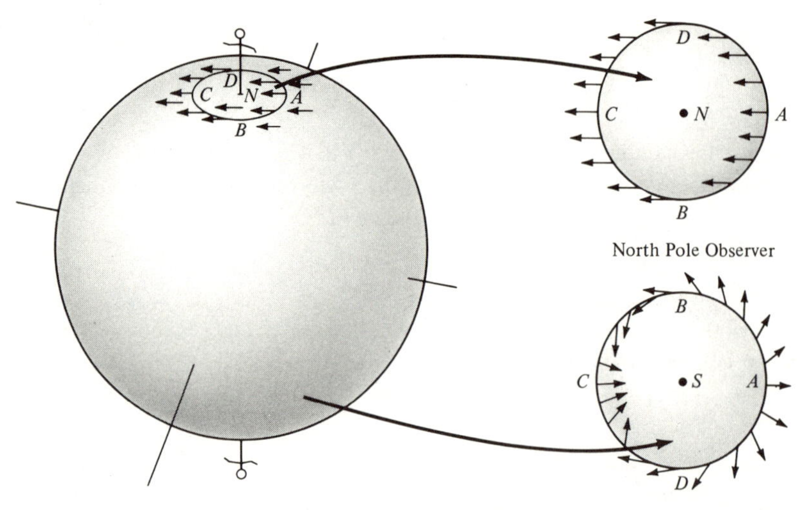

To convert this idea into a detailed proof, the crucial step is to associate with a vector field on $S^2\backslash\{N\}$ another on $\mathbb{R}^2$, corresponding to the vector field on the south pole observer's map. This requires some explicit computations with the stereographic projection.

Proof of Theorem 9.2: Let $N = (0,0,1)$ be the north pole of the sphere. Define the stereographic projection

$$\sigma : S^2\backslash\{N\} \to \mathbb{R}^2$$

by

$$\sigma(x,y,z) = (x/(1 - z), y/(1 - z)).$$

Evidently σ is continuous, and a direct computation shows that σ has a continuous inverse, given by

$$\sigma^{-1}(u,v) = \left(\frac{2u}{u^2 + v^2 + 1}, \frac{2v}{u^2 + v^2 + 1}, \frac{u^2 + v^2 - 1}{u^2 + v^2 + 1}\right), \qquad (u,v) \in \mathbb{R}^2.$$

In particular, σ maps $S^2\backslash\{N\}$ homeomorphically onto $\mathbb{R}^2$.

Suppose that $F = (F_1, F_2, F_3)$ is a tangent vector field on S^2. To F we associate a vector field G on $\mathbb{R}^2$ by

$$G(u,v) = ((1 - z)F_1 + xF_3,(1 - z)F_2 + yF_3) \tag{9.2}$$

where

$$(u,v) = \sigma(x,y,z). \tag{9.3}$$

Clearly G is continuous.

Suppose $G(u,v) = 0$ for some $(u,v) \in \mathbb{R}^2$. Then with (x,y,z) as in (9.3), we obtain

$$(9.4) \qquad \begin{aligned} (1 - z)F_1 + xF_3 &= 0, \\ (1 - z)F_2 + yF_3 &= 0. \end{aligned}$$

Multiplying the first equation by x, the second by y, adding, and utilizing the relations

$$xF_1 + yF_2 = -zF_3,$$

$$x^2 + y^2 = 1 - z^2,$$

we obtain $(1 - z)(-zF_3) + (1 - z^2)F_3 = 0$, which simplifies to $(1 - z)F_3 = 0$. Since $z \neq 1$, we have $F_3 = 0$, and (9.4) then yields $F_1 = F_2 = 0$, so that $F(x,y,z) = 0$. Conversely, if $F(x,y,z) = 0$, then the definition shows that $G(u,v) = 0$. Thus p is a zero of F if and only if $\sigma(p)$ is a zero of G.

Consider the particular case of the tangent vector field $\tilde{F}$ on S^2 defined by

$$\tilde{F}(x,y,z) = (z,0,-x).$$

The corresponding vector field $\tilde{G}$ on $\mathbb{R}^2$ is given by

$$\tilde{G}(u,v) = (z - z^2 - x^2, \; - xy).$$

Substituting for x, y, and z and expressing u and v in polar coordinates, we obtain

$$\tilde{G}(r\cos\theta, r\sin\theta) = \frac{-2r^2}{(r^2 + 1)^2}\left(\frac{1}{r^2} + \cos(2\theta), \sin(2\theta)\right).$$

If $r > 1$, then $\tilde{G}$ does not vanish and we can consider the index of $\tilde{G}$ around the circle $e^{i\theta} \to re^{i\theta}$. This is the same as the index of the map $e^{i\theta} \to \left(\frac{1}{r^2} + \cos(2\theta), \sin(2\theta)\right)$. By Theorem 8.2, this index coincides with the index of the map $e^{i\theta} \to (\cos(2\theta), \sin(2\theta))$, which is $+2$.

Now suppose that F is a vector field on S^2, such that

$$(9.5) \qquad F(0,0,1) = (1,0,0) = \tilde{F}(0,0,1).$$

Let G be the corresponding vector field on $\mathbb{R}^2$. Then $F \cdot \tilde{F} = F_1\tilde{F}_1 + F_2\tilde{F}_2 + F_3\tilde{F}_3$ has value 1 at (0,0,1). By continuity, $F \cdot \tilde{F} > 0$ in a neighborhood of U of (0,0,1). It follows that the convex combinations $tF + (1 - t)\tilde{F}$ have no zeros in U for $0 \leq t \leq 1$; otherwise we obtain a contradiction by taking the scalar product with $\tilde{F}$. Since $tG + (1 - t)\tilde{G}$ is the vector field on $\mathbb{R}^2$ corresponding to $tF + (1 - t)\tilde{F}$, we see that $tG + (1 - t)\tilde{G}$ has no zeros on $\sigma(U)$. Choose R so large that $(u,v) \in U$ for $r^2 = u^2 + v^2 \geq R$. The family $tG + (1 - t)\tilde{G}$, $0 \leq t \leq 1$, gives a homotopy of the maps $e^{i\theta} \to G(R\cos\theta, R\sin\theta)$ and $e^{i\theta} \to \tilde{G}(R\cos\theta, R\sin\theta)$ in the punctured plane $\mathbb{R}^2\backslash\{0\}$. We have observed that the index of the latter map is $+2$; consequently the index of the former map is also $+2$. By Lemma 9.1, G has a zero inside the disc $\{u^2 + v^2 \leq R^2\}$. Hence F has a zero.

Now let F be an arbitrary vector field on S^2. We claim that F has a zero. If $F(0,0,1) = 0$, we are done. Otherwise let T be a rotation of $\mathbb{R}^3$ such that $T(F(0,0,1)) = (\alpha,0,0)$, $\alpha \in \mathbb{R}$, $\alpha \neq 0$, and $T(0,0,1) = (0,0,1)$. According to the result just obtained, the vector field $(1/\alpha)T \circ F \circ T^{-1}$ has a zero. It follows that F has a zero. □

EXERCISES

1. Sketch a vector field on S^2 that has precisely one zero.
2. Show that the only constant vector field tangent to S^2 is identically zero.
3. Let F be a continuous function from S^2 to $\mathbb{R}^3$. Show that there is a point $p \in S^2$ at which $F(p)$ is normal to S^2, that is, $F(p)$ is a multiple of p.
4. Let φ be a continuously differentiable map from $\mathbb{R}^n$ to $\mathbb{R}^m$. Then φ induces a transformation φ^* from vector fields on subsets of $\mathbb{R}^n$ to vector fields on $\mathbb{R}^m$ as follows. For a vector field $F = (F_1,\ldots,F_n)$ on $\mathbb{R}^n$, define $\varphi^*F = ((\varphi^*F)_1,\ldots,(\varphi^*F)_m)$ on $\mathbb{R}^m$ by the formula

$$(\varphi^*F)_j(\varphi(p)) = \sum_{k=1}^{n} \frac{\partial \varphi_j}{\partial x_k}(p)F_k(p), \qquad 1 \leq j \leq m.$$

Show that if $\sigma : S^2\backslash N \to \mathbb{R}^2$ is the stereographic projection, if $F = (F_1,F_2,F_3)$ is a vector field on S^2, and if G is the vector field on $\mathbb{R}^2$ defined by (9.2), then

$$(\sigma^*F)(u,v) = (1 - z)^{-2}G(u,v),$$

where $(u,v) = \sigma(x,y,z)$. In particular, σ^*F and G point in the same direction.

5. A *closed curve* (= closed path = loop) in the complex plane $\mathbb{C}$ is a continuous mapping γ of S^1 into $\mathbb{C}$. If $p \notin \gamma(S^1)$, then the *winding number* of the closed curve γ around p is the index of the map $e^{i\theta} \to \gamma(e^{i\theta}) - p$, $0 \leq \theta \leq 2\pi$. Show that the winding numbers of γ around p are identical for all p in a fixed connected component of $\mathbb{C}\backslash\gamma(S^1)$. Show that the winding number of γ around p is zero for p in the unbounded connected component of $\mathbb{C}\backslash\gamma(S^1)$.
6. Suppose that $F : U \to \mathbb{R}^2$ is a continuous vector field on an open subset U of $\mathbb{R}^2$. Let p be an isolated zero of F, i.e., suppose that there is a disc $\overline{B(p;\varepsilon)} = \{q : |q - p| \leq \varepsilon\}$ contained in U such that $F(q) \neq 0$ for $q \in \overline{B(p;\varepsilon)}$, $q \neq p$, while $F(p) = 0$.
   (a) Show that the index of F around γ is the same for any loop γ in $B(p;\varepsilon)\backslash\{p\}$ with winding number 1 around p. This common value is called the *index of F at p*.
   (b) Express the index of F around γ in terms of the index of F at p, when γ is a loop in $B(p;\varepsilon)\backslash\{p\}$ with winding number m around p. Prove your answer.
   (c) What are the indices at the origin of the vector fields depicted in the figures discussed at the beginning of this section?
   (d) If F has index m at an isolated zero p, what is the index of $-F$ at p?
7. Suppose that $F : \overline{B(0;1)} \to \mathbb{R}^2$ is a continuous vector field on the closed unit disc $\overline{B(0;1)} = \{(x,y) : x^2 + y^2 \leq 1\}$ in $\mathbb{R}^2$ with only a finite number of zeros, all of which lie in the open disc $B(0;1)$. Show that the index of F around the positively traversed unit circle is the sum of the indices of F at its zeros. *Hint:* Suppose first that F has only two zeros, say at $(-\frac{1}{2},0)$

and $(\frac{1}{2},0)$. Show that the index of F around the unit circle is the sum of the indices of F around appropriate semicircular loops separating the zeros.

8. Let $F : S^2 \to \mathbb{R}^2$ be a vector field on the sphere with only a finite number of zeros, such that $F \neq 0$ near the north pole N. Let $\sigma : S^2\backslash\{N\} \to \mathbb{R}^2$ be the stereographic projection. Show that the index of $F\circ\sigma^{-1}$ around a large circular loop $e^{i\theta} \to (R \cos \theta, R \sin \theta)$, $0 \leq \theta \leq 2\pi$, is $+2$. (Thus with reasonable definitions, the sum of the indices of a vector field on S^2, with only isolated zeros, is $+2$. Check this for the vector fields depicted in the figures in the discussion after Lemma 9.1.)

10. THE JORDAN CURVE THEOREM

The purpose of this section is to prove that if Γ is a subset of $\mathbb{R}^2$ that is homeomorphic to the unit circle S^1, then $\mathbb{R}^2\backslash\Gamma$ has two connected components. This result is certainly intuitively plausible. In fact, one assumes it in everyday life almost without thought. The children who form a ring with hands clasped never seriously doubt that a child inside the ring will have to cross the ring to get outside even if the ring is not precisely circular. The very use of the words "inside" and "outside" in this setting already implies an idea of disconnection. Even in mathematics, the idea was for a long time taken for granted. In the geometry of the Greeks, it is always assumed without proof that a simple closed (polygonal) curve divides the plane into two parts and that a curve from a point of one part to a point of the other must intersect the simple closed curve. Despite its intuitive appeal, this theorem is not easy to prove.

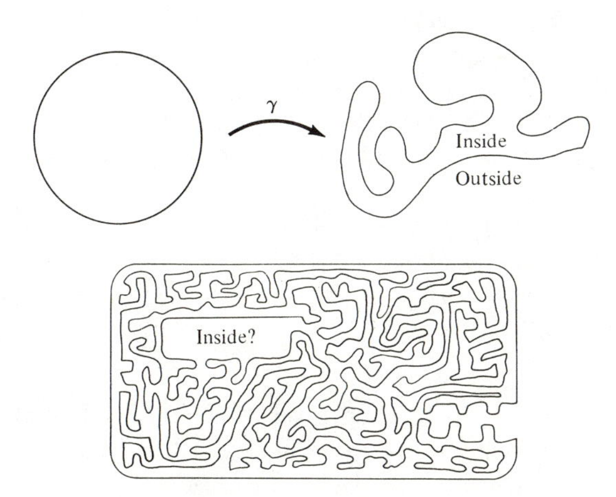

A *simple closed curve* (or *Jordan curve*) in a topological space X is a one-to-one continuous mapping γ of S^1 into X. If X is a Hausdorff space, then the mapping is necessarily a homeomorphism onto its image,

$$\Gamma = \gamma(S^1).$$

The image Γ is often referred to also as a *simple closed curve*.

The theorem we aim to prove in this section is stated formally as follows.

10.1 Theorem *(Jordan Curve Theorem):* If γ is a simple closed curve in $\mathbb{R}^2$ with image Γ, then $\mathbb{R}^2\backslash\Gamma$ has precisely two connected components.

The exterior of any disc including Γ is included in a single connected component of $\mathbb{R}^2\backslash\Gamma$, called the *outside* of Γ. The remaining components of $\mathbb{R}^2\backslash\Gamma$, if any, are bounded. Our task is to prove that there is precisely one bounded component of $\mathbb{R}^2\backslash\Gamma$, the *inside* of Γ.

There are various reformulations of the Jordan Curve Theorem. Two of them are as follows.

10.2 Theorem: If γ is a simple closed curve in S^2 with image Γ, then $S^2\backslash\Gamma$ consists of precisely two connected components.

10.3 Theorem: Let $h : \mathbb{R} \to \mathbb{R}^2$ be a one-to-one continuous map such that $|h(t)| \to \infty$ as $|t| \to \infty$. Then $\mathbb{R}^2\backslash h(\mathbb{R})$ consists of precisely two connected components.

Observe that Theorem 10.3 implies Theorem 10.2. To see this, we regard the sphere S^2 as the one-point compactification $\mathbb{R}^2 \cup \{\infty\}$ of the plane $\mathbb{R}^2$ and we regard the circle S^1 as the one-point compactification $\mathbb{R}^1 \cup \{\infty'\}$ of the real line $\mathbb{R}$. Suppose that γ is a simple closed curve in S^2, with image Γ. By rotating the sphere, we may arrange that $\gamma(\infty') = \infty$. Then the restriction h of γ to $\mathbb{R}$ satisfies the hypothesis of Theorem 10.3. Since $\mathbb{R}^2\backslash h(\mathbb{R}) = S^2\backslash\Gamma$, Theorem 10.3 shows that $S^2\backslash\Gamma$ consists of precisely two connected components. Thus Theorem 10.2 holds.

Note that Theorem 10.2 implies the Jordan Curve Theorem. Indeed, a simple closed curve in $\mathbb{R}^2$ can be regarded as a simple closed curve in $S^2 = \mathbb{R}^2 \cup \{\infty\}$. If $S^2\backslash\Gamma$ consists of two connected components U and V, and say $\infty \in U$, then $\mathbb{R}^2\backslash\Gamma$ has precisely two connected components, namely, $U\backslash\{\infty\}$ and V.

Thus to establish the Jordan Curve Theorem, it suffices to prove Theorem 10.3. Our efforts through the remainder of this section will be devoted to the proof of Theorem 10.3.

Note that the theorem becomes less "geometrically obvious" as the complexity of the simple closed curve increases as the figure above indicates.

That there are at least two connected components of $\mathbb{R}^2\backslash h(\mathbb{R})$ will depend upon the following lemma (cf. Exercise 3.2).

10.4 Lemma: Let U and V be open subsets of a path-connected space X that cover X. If U and V are simply connected and $U \cap V$ is path-connected, then X is simply connected.

Proof: Let $p \in U \cap V$ and let $\gamma : [0,1] \to X$ be a loop in X based at p. Then $\gamma^{-1}(X\backslash U)$ is a compact subset of the open subset $\gamma^{-1}(V)$ of $[0,1]$. A moment's reflection shows there are a finite number of disjoint closed subintervals $I_j = [s_j,t_j]$, $1 \leq j \leq N$, of $[0,1]$ such that the I_j's cover $\gamma^{-1}(X\backslash U)$, and $\gamma(I_j) \subset V$ for $1 \leq j \leq N$. In particular, $\gamma(s_j)$ and $\gamma(t_j)$ belong to $U \cap V$ for $1 \leq j \leq N$. By hypothesis, there is a path $\alpha_j : I_j \to U \cap V$ from $\gamma(s_j)$ to $\gamma(t_j)$. Let α be the path in U defined so that $\alpha(s)$

$= \alpha_j(s)$ for $s \in I_j$, $1 \leq j \leq N$, and $\alpha(s) = \gamma(s)$ for $s \notin \cup I_j$. Since V is simply connected, the path α_j is homotopic to the restriction path of γ to I_j. Combining these homotopies in the obvious way, we see that α is homotopic to γ. Now the path α lies in U and U is simply connected, so that α is homotopic to a point. It follows that γ is homotopic to a point and X is simply connected. □

The proof that $\mathbb{R}^2 \backslash h(\mathbb{R})$ has at most two components depends on a more careful analysis of a situation more general than that of Lemma 10.4.

Suppose that X is a path-connected space that can be represented as the union of two (connected) simply connected open subsets U and V. What can be said about $\pi_1(X)$? It turns out that $\pi_1(X)$ is the free group with number of generators one less than the number of components of $U \cap V$. For our purposes, we need only the following piece of information.

10.5 Lemma: Let X be a connected, locally path-connected space and let U and V be simply connected open subsets of X which cover X. If $U \cap V$ has three or more distinct path components, then $\pi_1(X)$ is not abelian.

Proof: The proof depends on the construction of an appropriate covering space unwinding two independent loops in X. While it appears complicated, the reader will upon reflection recognize that the construction corresponds completely to the mattress-spring figure preceding Theorem 5.4.

Let W_1 and W_2 be distinct path components of $U \cap V$ and let $W_0 = (U \cap V) \backslash (W_1 \cup W_2)$. Then W_0, W_1, and W_2 are disjoint open subsets of X whose union is $U \cap V$. The hypothesis guarantees that the W_j's are nonempty.

Let U_{mn} and V_{mn} be disjoint copies of U and V, respectively, $m,n \in Z$. An element of U_{mn} will be denoted by a quadruple (x,U,m,n). Let Y denote the union of the U_{mn}'s and the V_{mn}'s. Define an equivalence relation "$\sim$" on Y so that

$$(x,U,j,k) \sim (y,V,m,n)$$

if and only if $x = y \in U \cap V$ and one of the following conditions holds:

(i) $x = y \in W_0$, $\quad j = m,\ k = n$,

(ii) $x = y \in W_1$, $\quad j = m + 1,\ k = n$,

(iii) $x = y \in W_2$, $\quad j = m \neq 0,\ k = n$,

(iv) $x = y \in W_2$, $\quad j = m = 0,\ k = n + 1$.

Thus conditions (i) through (iv) give a prescription for pasting the copies of U to the copies of V. Let E denote the quotient space of Y obtained from these equivalence relations, endowed with the quotient topology. The natural projections of the U_{mn}'s and V_{mn}'s onto the first coordinates determine a natural projection π of E onto X. The projection π is continuous, and it is easily checked to be a covering mapping.

Fix $p \in U$. For $j = 1,2$ let α_j be a loop in X that starts at p, proceeds through U to W_0, continues in V to W_j, and returns to p in U. One checks that the lift of α_1 to E starting at $(p,U,0,0)$ travels through U_{00}, thence to V_{00}, then back through U_{10}, to terminate at $(p,U,1,0)$. Similarly, the lift of α_2 to E starting at $(p,U,1,0)$ terminates at $(p,U,1,0)$, so that the lift of $\alpha_1\alpha_2$ to E starting at $(p,U,0,0)$ terminates at $(p,U,1,0)$.

On the other hand, the lift of $\alpha_2\alpha_1$ to E starting at $(p,U,0,0)$ terminates at $(p,U,1,1)$. Since the lifts of $\alpha_1\alpha_2$ and $\alpha_2\alpha_1$ terminate at different points of E, $\alpha_1\alpha_2$ is not homotopic to $\alpha_2\alpha_1$ (cf. the remarks preceding Theorem 5.4). Hence $\pi_1(X)$ is not abelian. □

10.6 Lemma: Let T be a proper closed subset of $\mathbb{R}^2$ and let Q be the open subset of $\mathbb{R}^3$ defined by

$$Q = \mathbb{R}^3 \backslash \{(x,y,0) : (x,y) \in T\}.$$

If $\mathbb{R}^2\backslash T$ is connected, then Q is simply connected. If $\mathbb{R}^2\backslash T$ has at least three connected components, then $\pi_1(Q)$ is not abelian.

Proof: Since $\mathbb{R}^2\backslash T$ is not empty, Q is connected. Define subsets U and V of Q by

$$U = \{(x,y,z) : z > 0\} \cup \{(x,y,z) : z > -1,\ (x,y) \notin T\},$$

$$V = \{(x,y,z) : z < 0\} \cup \{(x,y,z) : z < 1,\ (x,y) \notin T\}.$$

Then U and V are connected open subsets of Q that cover Q.

We claim that U and V are simply connected. It is enough to see that U is simply connected since U is homeomorphic to V under the reflection $(x,y,z) \to (x,y,-z)$. Let $\gamma : [0,1] \to U$ be a loop with $\gamma(0) = \gamma(1) = (0,0,1)$. Define $H : [0,1] \times [0,1] \to U$ by $H(t,s) = \gamma(t) + (0,0,s - s \cdot z(t))$, where $z(t)$ is defined to be the z-coordinate of $\gamma(t)$ if this coordinate is negative and to be 0 if the z-coordinate is positive. The plus sign means vector addition in $\mathbb{R}^3$. It is easy to check that the image of H is a subset of U. Moreover, H is continuous, $H(t,0) = \gamma(t)$, and $H(t,1) \subset \{(x,y,z) : z \geq 1\}$. Since $\{(x,y,z) : z \geq 1\}$ is a convex subset of U containing $(0,0,1)$, the loop $t \to H(t,1)$ is homotopic in U to the constant loop at $(0,0,1)$. Therefore γ is homotopic in U to the constant loop at $(0,0,1)$. Hence U is simply connected, and so is V.

Next observe that

$$U \cap V = \{(x,y,z) : (x,y) \in T,\ -1 < z < 1\}$$

is homeomorphic to $(\mathbb{R}^2\backslash T) \times (-1,1)$. If $\mathbb{R}^2\backslash T$ is connected, then $U \cap V$ is connected, and Lemma 10.4 shows that Q is simply connected. On the other hand, if $\mathbb{R}^2\backslash T$ has at least three components, then so does $U \cap V$, and Lemma 10.5 shows that $\pi_1(Q)$ is not abelian. □

10.7 Lemma: Let $h : \mathbb{R} \to \mathbb{R}^2$ be a one-to-one continuous map such that $|h(t)| \to \infty$ as $|t| \to \infty$. Let $\iota : \mathbb{R}^2 \to \mathbb{R}^3$ be the cannonical embedding, defined by

$$\iota(x,y) = (x,y,0), \qquad (x,y) \in \mathbb{R}^2.$$

Then there is a homeomorphism F of $\mathbb{R}^3$ such that

$$(F \circ \iota \circ h)(t) = (0,0,t), \qquad t \in \mathbb{R}.$$

Proof: Define $g : h(\mathbb{R}) \to \mathbb{R}$ by $g(h(t)) = t$, $t \in \mathbb{R}$. One checks that g is continuous on $h(\mathbb{R})$. By the Tietze Extension Theorem (II.5.4), there is a continuous function

$G : \mathbb{R}^2 \to \mathbb{R}$ such that $G(p) = g(p)$ for all $p \in h(\mathbb{R})$. Let h_1 and h_2 be the coordinate functions of h, so that $h(t) = (h_1(t), h_2(t))$, $t \in \mathbb{R}$. Then G satisfies

$$(10.1) \qquad G(h_1(t), h_2(t)) = t, \qquad t \in \mathbb{R}.$$

Define $F_1 : \mathbb{R}^3 \to \mathbb{R}^3$ by

$$F_1(x,y,z) = (x,y,G(x,y) + z).$$

Since F_1 has continuous inverse $(x,y,z) \to (x,y,z - G(x,y))$, F_1 is a homeomorphism. Define $F_2 : \mathbb{R}^3 \to \mathbb{R}^3$ by

$$F_2(x,y,z) = (x - h_1(z), y - h_2(z), z).$$

Since F_2 has continuous inverse $(x,y,z) \to (x + h_1(z), y + h_2(z), z)$, F_2 is also a homeomorphism.

Set $F = F_2 \circ F_1$, a homeomorphism of $\mathbb{R}^3$. Then using (10.1) we obtain

$$\begin{aligned}(F \circ \iota \circ h)(t) &= F_2(F_1(h_1(t), h_2(t),0)) \\ &= F_2(h_1(t), h_2(t), t) \\ &= t. \ \square\end{aligned}$$

Proof of Theorem 10.3: First note that $h(\mathbb{R}) \neq \mathbb{R}^2$. Indeed, $\mathbb{R}$ has the property that the removal of any of its points leaves a disconnected space, whereas $\mathbb{R}^2$ does not have this property. Since this property is preserved under homeomorphisms, $\mathbb{R}$ and $\mathbb{R}^2$ are not homeomorphic. The condition on h ensures that h is a homeomorphism of $\mathbb{R}$ onto its range. Hence the range of h cannot coincide with $\mathbb{R}^2$.

Consider the set Q of Lemma 10.6 for $T = h(\mathbb{R})$. The homeomorphism F of Lemma 10.7 maps Q homeomorphically onto the set

$$\mathbb{R}^3 \backslash \{(0,0,t) : -\infty < t < \infty\} \cong (\mathbb{R}^2 \backslash \{0\}) \times \mathbb{R}.$$

It is easy to check that the fundamental group of this space is $\mathbb{Z}$. Indeed, $\mathbb{R}$ is contractible, so that $(\mathbb{R}^2 \backslash \{0\}) \times \mathbb{R}$ is homotopically equivalent to the punctured plane (Exercise 7.2) which has fundamental group $\mathbb{Z}$. It follows that

$$\pi_1(Q) \cong \mathbb{Z}.$$

Since $\pi_1(Q)$ is nonzero and abelian, Lemma 10.6 shows that $\mathbb{R}^2 \backslash h(\mathbb{R})$ consists of precisely two connected components. $\square$

Thus Theorem 10.3 is proved and, as remarked earlier, this also establishes the Jordan Curve Theorem.

The Jordan Curve Theorem was first stated formally by C. Jordan in 1893, in his *Cours d'Analyse*. The proof he gave was incomplete, and many false proofs of the theorem have been subsequently proposed. The first correct proof of the theorem was given by O. Veblen in 1905, and the proof we have given was discovered by P. Doyle in 1968. An alternate proof is laid out in Exercises 5 through 10.

We mention in passing the Schoenflies Theorem, which contains more information than the Jordan Curve Theorem, though it does not extend to higher dimensions. The Schoenflies Theorem asserts that if $\gamma : S^1 \to \mathbb{R}^2$ is a simple closed curve, then there is a homeomorphism H: $\mathbb{R}^2 \to \mathbb{R}^2$ such that $H \circ \gamma$ is the identity mapping of S^1. The homeomorphism H then maps Γ onto the unit circle, so that topologically a simple closed curve in $\mathbb{R}^2$ is indistinguishable from the unit circle. In particular, each connected component of $\mathbb{R}^2 \backslash \Gamma$ has boundary Γ (Exercise 4).

EXERCISES

1. Let X be a connected, locally path-connected space and let U and V be simply connected open subsets of X which cover X. Show that if $U \cap V$ consists of two connected components with disjoint closures, then $\pi_1(X) \cong \mathbb{Z}$.
2. Let $h : [0,\infty) \to \mathbb{R}^2$ be a one-to-one continuous map such that $|h(t)| \to \infty$ as $t \to \infty$. Show that $\mathbb{R}^2 \backslash h([0,\infty))$ is connected. *Hint:* Follow the proof of Theorem 10.3 and use Exercise 1.
3. A *simple arc* in a space X is the homeomorphic image in X of the unit interval $[0,1]$. Show that the complement of a simple arc in $\mathbb{R}^2$ or in S^2 is connected.
4. If γ is a simple closed curve in $\mathbb{R}^2$ with image Γ, then show that each component of $\mathbb{R}^2 \backslash \Gamma$ has boundary Γ. *Hint:* Use Exercise 3 or Exercise 6.
5. Let E be a compact subset of the complex plane $\mathbb{C}$ and let $p \in \mathbb{C} \backslash E$. Prove that p lies in the unbounded connected component of $\mathbb{C} \backslash E$ if and only if $z - p$ is an exponential on E.
6. Using Exercise 5, show that if E is a compact contractible subset of $\mathbb{C}$, then $\mathbb{C} \backslash E$ is connected. (This generalizes Exercise 3.)
7. Let E be a compact subset of $\mathbb{C}$ and suppose $p_1, \ldots, p_m$ lie in different bounded components of $\mathbb{C} \backslash E$. Let $q_1, \ldots, q_m$ be integers. Prove that the function

$$f = (z - p_1)^{q_1} \cdots (z - p_m)^{q_m}$$

is an exponential on E if and only if $q_1 = \cdots = q_m = 0$. *Hint:* Let U be the connected component of $\mathbb{C} \backslash E$ containing p_1. Write $f = e^h$ on E, extend h continuously to $\mathbb{C}$, and consider the function g defined to be f on U and e^h on $\mathbb{C} \backslash U$.

8. Using Exercise 7, show that the complement of a simple closed curve in $\mathbb{C}$ has at most two connected components.
9. Using Exercises 4, 6, and 8, show that the Jordan Curve Theorem is valid, providing that Γ contains a straight-line segment. *Hint:* By considering the index of $\gamma - p$, show that $z - p$ cannot be an exponential for p lying on both sides of the straight-line segment in Γ.
10. Prove the Jordan Curve Theorem by introducing a straight line segment in $\mathbb{C} \backslash \Gamma$ between two appropriate points of Γ and applying Exercise 9 to the resulting simple closed curves.

11. A straight line L in $\mathbb{R}^2$ is *topologically transversal* to a simple closed curve Γ in $\mathbb{R}^2$ if for every $p \in \Gamma \cap L$, there is an $\varepsilon > 0$ and a homeomorphism of $B(p;\varepsilon)$ onto $B((0,0);1)$ mapping $B(p;\varepsilon) \cap \Gamma$ onto the vertical interval $\{(0,t) : -1 < t < 1\}$ and mapping $B(p;\varepsilon) \cap L$ onto the horizontal interval $\{(s,0) : -1 < s < 1\}$. Informally, Γ can be straightened out near p without bending L.

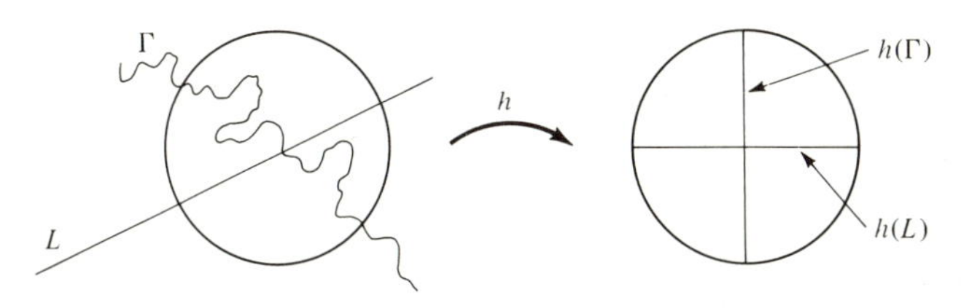

(a) Prove that if L is topologically transversal to Γ, then the number of points in $\Gamma \cap L$ is finite and even. *Hint:* Use Exercise 4.

(b) Prove that if $q \in L\backslash\Gamma$, then q is in the bounded component of $\mathbb{R}^2\backslash\Gamma$ if and only if the number of points in $L \cap \Gamma$ on each side of q is odd.

(c) Formulate and prove a similar result for a closed half-line.

Higher Dimensional Homotopy

FOUR

Some of the consequences of the fact that $\pi_1(S^1)$ is not the trivial group—for example, the Brouwer Fixed Point Theorem (Theorem III.6.3)— could be easily generalized to higher dimensions if one knew that the identity map of the n-dimensional sphere to itself was not homotopic to a constant mapping. The purpose of this chapter is to pursue in some detail this idea of generalizing to higher dimensions some of the results of Chapter III on the homotopy of paths and on the fundamental group.

In the first section, the appropriate n-dimensional generalization of the fundamental group is defined. This generalization associates to each topological space X a sequence of groups $\pi_1(X), \pi_2(X), \ldots$. The group $\pi_1(X)$ is just the fundamental group as previously defined. The group $\pi_n(X)$, the *nth homotopy group* of X, measures the existence of homotopically distinct continuous maps of the n-sphere into X, in much the same sense that $\pi_1(X)$ measures the existence of homotopically distinct closed curves.

Section 2 is devoted to a proof of the fact that the n-sphere S^n is not contractible, that is, that the identity map of the n-sphere is not homotopic to the constant map. There is no known proof of this fact that is both reasonably simple and free of appeal to concepts not strictly belonging to topology. The proof given in Section 2 buys simplicity by using calculus methods, specifically Stokes' Theorem. The use of such methods plays an important role in much of modern topology. The reader is by no means detouring far from topics of topological interest in studying this proof, even though the initial impression might be that it is a *deus ex machina*.

In the third section, we introduce some topics that are closely related to the development of homology theory, an important part of modern topology. We introduce a class of geometric objects called *affine simplexes* and a procedure for subdividing them called *barycentric subdivision*. Using this subdivision procedure, we prove that continuous functions can be approximated by functions of a particularly well-behaved kind called *piecewise linear functions.* These results are applied in Section 4 to obtain another proof of the noncontractibility of S^n.

In Section 5, the important idea of the degree of a map from S^n to S^n is described and developed in outline. This idea is a generalization of the concept of index of a map from S^1 to S^1 introduced in Chapter III; the expression "index" is used in the special case where $n = 1$ because of historical precedent in the subject of complex analysis. Applications of the idea are discussed, and some indications of further general directions in the development of topology are given.

1. HIGHER HOMOTOPY GROUPS

Let X be a topological space and fix a base point $b \in X$. We have defined $\pi_1(X,b)$ to be the set of homotopy classes of mappings α from the unit interval $I = [0,1]$ to X that satisfy $\alpha(0) = \alpha(1) = b$. This definition generalizes to higher dimensions in a natural manner.

The *n-cube*, consisting of all n-tuples $(s_1,\ldots,s_n) \in \mathbb{R}^n$ such that $0 \leq s_j \leq 1$, $1 \leq j \leq n$, will be denoted by I^n. Its boundary ∂I^n consists of all such n-tuples for which at least one of the components s_j is 0 or 1.

An *n-cube at b in X* is a continuous function $\alpha : I^n \to X$ such that $\alpha(s) = b$ for all $s \in \partial I^n$. Two n-cubes α and β at b in X are *homotopic* if there is a continuous family α_t, $0 \leq t \leq 1$, of n-cubes at b in X such that $\alpha_0 = \alpha$ and $\alpha_1 = \beta$. By "continuous family," we mean simply that the map $(s,t) \to \alpha_t(s)$ of $I^n \times I$ to X is continuous. This notion of "homotopic" is the same as the notion of "homotopic relative to ∂I^n" introduced in Section III.7, and it is denoted by

$$\alpha \simeq \beta \ \mathrm{rel}\,\partial I^n.$$

As noted in Section III.7, the relation is an equivalence relation. The set of all n-cubes at b in X homotopic to a given α is called the *homotopy class* of α, denoted by $[\alpha]$. The collection of such homotopy classes is denoted by $\pi_n(X,b)$.

We wish to endow $\pi_n(X,b)$ with a group structure. For this purpose, we define the product of two n-cubes α and β at b in X by

$$(\alpha\beta)(s_1,\ldots,s_n) = \begin{cases} \alpha(2s_1,s_2,\ldots,s_n), & 0 \leq s_1 \leq \frac{1}{2}, \\ \beta(2s_1 - 1,s_2,\ldots,s_n), & \frac{1}{2} \leq s_1 \leq 1. \end{cases}$$

Since both α and β map ∂I^n to b, this composite map patches up continuously at the interface of the defining regions. Moreover, $\alpha\beta$ maps ∂I^n to b, so that $\alpha\beta$ is also an n-cube at b in X.

In the case where $n = 2$, one can conveniently represent the product $\alpha\beta$ by a diagram:

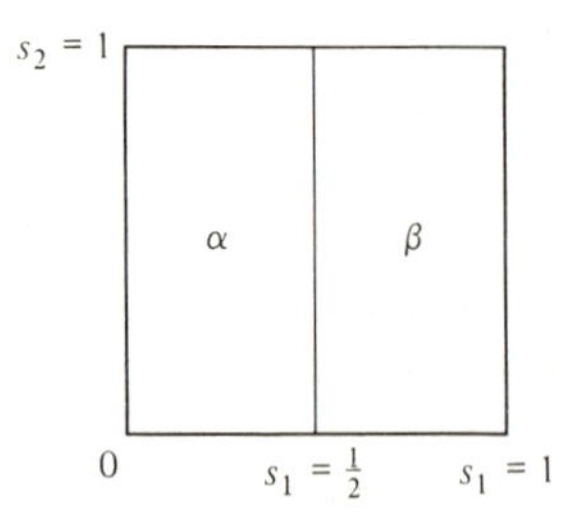

Representation of $\alpha\beta$

The meaning of the diagram is that the mapping α of I^2 is used to define $\alpha\beta$ on the left-hand rectangle of parameter space, by composition with the affine homeomorphism $s_1 \to 2s_1$ of the rectangle and the unit square, and β is used to define $\alpha\beta$ on the right-hand rectangle, by composition with the affine homeomorphism $s_1 \to 2s_1 - 1$.

1.1 Lemma: Suppose that α_0, α_1, β_0, and β_1 are n-cubes at b in X. If $\alpha_0 \simeq \alpha_1$ rel∂I^n and $\beta_0 \simeq \beta_1$ rel∂I^n, then $\alpha_0\beta_0 \simeq \alpha_1\beta_1$ rel∂I^n.

Proof: If α_t and β_t, $0 \le t \le 1$, are homotopies of α_0 to α_1 and β_0 to β_1, respectively, then $\alpha_t\beta_t$ is a homotopy from $\alpha_0\beta_0$ to $\alpha_1\beta_1$. □

Now we define the product of two homotopy classes $[\alpha]$ and $[\beta]$ to be the homotopy class of $[\alpha\beta]$,

$$[\alpha][\beta] = [\alpha\beta].$$

That the definition is independent of the particular representatives α and β follows from Lemma 1.1.

1.2 Theorem: With the product operation defined above, $\pi_n(X,b)$ is a group.

Proof: Note one important feature of the definition of the product of n-cubes. The variables $s_2,\ldots,s_n$ play an entirely passive role in the formula for the product, and the variable s_1 undergoes the same treatment as for the product of paths. Thus any reasonable argument involving products of paths carries over to products of n-cubes by simply adjoining the passive variables $s_2,\ldots,s_n$. This allows us to carry over the proof that $\pi_1(X,b)$ is a group, given in Section III.2, to the case at hand. Let us illustrate this point by carrying over the proof of Lemma III.2.8.

Let α be an n-cube at b in X. Define an n-cube α^{-1} by

$$\alpha^{-1}(s_1,s_2,\ldots,s_n) = \alpha(1 - s_1,s_2,\ldots,s_n).$$

We claim that $\alpha\alpha^{-1}$ is homotopic to the constant map $I^n \to b$. Indeed, following the proof of Lemma III.2.8, we define

$$\gamma_t(s_1,s_2,\ldots,s_n) = \begin{cases} \alpha(2s_1,s_2,\ldots,s_n), & 0 \le s_1 \le \dfrac{t}{2}, \\ \alpha(t,s_2,\ldots,s_n), & \dfrac{t}{2} \le s_1 \le 1 - \dfrac{t}{2}, \\ \alpha(2 - 2s_1,s_2,\ldots,s_n), & 1 - \dfrac{t}{2} \le s_1 \le 1. \end{cases}$$

The γ_t's form a continuous family of n-cubes at b in X, $\gamma_0 = b$, and $\gamma_1 = \alpha\alpha^{-1}$. Thus the adjunction of the passive variables converts the proof for the case of paths to that for n-cubes.

Similarly, by simply adjoining the passive variables, one proves, as in Section III.2, that $(\alpha\beta)\gamma \simeq \alpha(\beta\gamma)$ rel∂I^n, so that the associative law is valid in $\pi_n(X,b)$. If the constant n-cube at b is denoted by $\tilde{b}$, then $[\alpha\tilde{b}] = [\tilde{b}\alpha] = [\alpha]$ for any n-cube α, so that $[\tilde{b}]$ is an identity for $\pi_n(X,b)$. Finally, $[\alpha\alpha^{-1}] = [\alpha^{-1}\alpha] = [\tilde{b}]$, so that each homotopy class $[\alpha]$ has an inverse, namely, $[\alpha^{-1}]$. Thus $\pi_n(X,b)$ is a group. □

It is possible to place the remarks about generalizing the proofs from the case of paths ($n = 1$) to the general situation on a more precise basis. Let $\Omega_b(X)$ be the set of 1-cubes at b, that is, the set of closed paths in X based at b. If α is a 2-cube at b, then for each fixed s_1, the map $\alpha(s_1,\cdot)$ belongs to $\Omega_b(X)$. Thus the map $s_1 \to \alpha(s_1,\cdot)$ can be regarded as a path in $\Omega_b(X)$ which has initial and terminal point $\tilde{b} \in \Omega_b(X)$. The product operation on 2-cubes in X then becomes the product operation on paths in $\Omega_b(X)$. Once one introduces an appropriate topology on $\Omega_b(X)$, one can make an identification,

$$\pi_2(X,b) \cong \pi_1(\Omega_b(x),\tilde{b}).$$

Analogous identifications can be made for the homotopy groups $\pi_n(X,b)$, $n > 2$. We shall not pursue these ideas further.

The groups $\pi_n(X,b)$, $n \geq 2$, which are usually called the *higher homotopy groups* of X, have a remarkable property that is not in general shared by the fundamental group $\pi_1(X,b)$. In the proof of this special property, the variable s_2 ceases to play a passive role.

1.3 Theorem: The group $\pi_n(X,b)$ is abelian if $n \geq 2$.

Proof: Consider first the case where $n = 2$. We shall suggest the proof by means of diagrams representing 2-cubes, starting with the diagram representing $\alpha\beta$. By passing through a homotopy with intermediate states indicated in the diagram Stage 1,

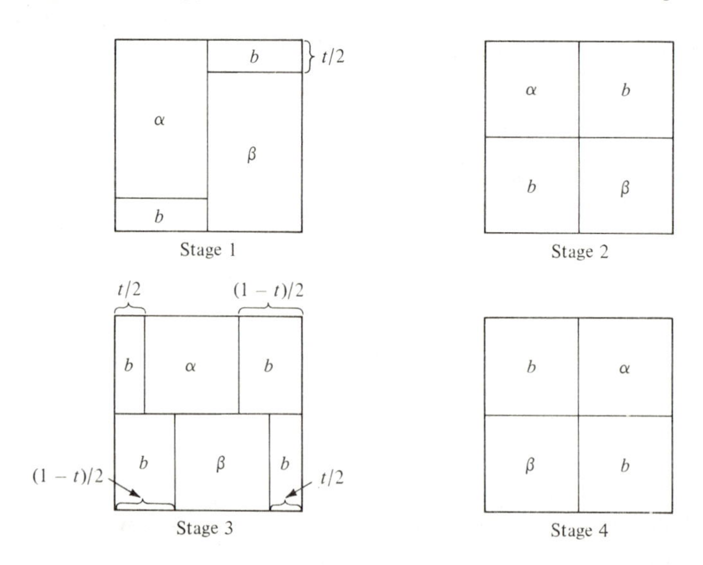

Stage 1
Stage 2
Stage 3
Stage 4

in which $t \in I$ is the homotopy parameter, we obtain a 2-cube homotopic to $\alpha\beta$, represented by Stage 2. This in turn may be transformed by the homotopy suggested by Stage 3 to obtain the homotopic 2-cube represented by Stage 4. Finally, the 2-cube of Stage 4 is homotopic to $\beta\alpha$, just as the 2-cube of Stage 2 is homotopic to $\alpha\beta$. We conclude that $\alpha\beta$ is homotopic to $\beta\alpha$ and $[\alpha][\beta] = [\beta][\alpha]$.

To treat the case where $n > 2$, one regards $s_3, \ldots, s_n$ as passive variables and defines all homotopies in terms of s_1 and s_2 only. □

We saw in Chapter III that $\pi_1(X,b)$ can be regarded as the set of homotopy classes of maps α of the circle S^1 into X that satisfy $\alpha(1) = b$. This is because the quotient space obtained from the unit interval I by identifying the two endpoints is homeomorphic to S^1.

Similarly, $\pi_n(X,b)$ can be regarded as the set of homotopy classes of maps α from the n-sphere S^n into X that satisfy $\alpha(p) = b$, where p is some fixed point (say the north pole) of the n-sphere. Indeed, the n-fold product of the open unit interval is homeomorphic to Euclidean space $\mathbb{R}^n$, so that the quotient space obtained from I^n by identifying ∂I^n to a point is homeomorphic to the one-point compactification of $\mathbb{R}^n$. In turn, the one-point compactification of $\mathbb{R}^n$ is homeomorphic to S^n, as can be seen by stereographic projection (Exercise II.7.7).

In order to produce an example of a space X for which $\pi_n(X,b) \neq 0$, it suffices to find an X and a map $\alpha : S^n \to X$ such that α is not homotopic to a constant. It turns out that the identity map $S^n \to S^n$ is not homotopic to a constant (this will be proved in the next section). Thus $\pi_n(S^n) \neq 0$ for $n \geq 1$.

What can be said about the higher homotopy groups of spheres? It is reasonably easy to show (Exercise 1) that

$$\pi_k(S^1) = 0, \qquad k > 1,$$

whereas (Exercise 4.8)

$$\pi_k(S^n) = 0, \qquad 1 \leq k < n.$$

It is somewhat more difficult to prove that

$$\pi_n(S^n) \cong \mathbb{Z}, \qquad n \geq 1.$$

That $\pi_n(S^n) \neq 0$ will be shown in Section 2, and in Section 5 the degree map from $\pi_n(S^n)$ to $\mathbb{Z}$ will be defined and some of its properties will be described. The proof that the degree map is an isomorphism of $\pi_n(S^n)$ and $\mathbb{Z}$ is somewhat beyond the scope of this book though references will be provided in Section 5. Meanwhile, the determination of the groups $\pi_k(S^n)$ for $k > n$ ($n \geq 2$) has turned out to be a very difficult problem which is not yet completely solved. It might come as a surprise that $\pi_3(S^2) \neq 0$. An example of a map from S^3 to S^2 that is not homotopic to a constant is the so-called "Hopf fibration," the fibration $\pi : S^3 \to CP^1$ introduced (in more generality) in Exercise II.13.9(c).

EXERCISES

1. Prove that $\pi_k(S^1,1) = 0$ if $k > 1$.
2. Prove that $\pi_k(X \times Y, (b,c)) \cong \pi_k(X,b) \times \pi_k(Y,c)$, $k \geq 1$.
3. Prove in detail that the multiplication defined on $\pi_k(X,b)$ is associative.

4. Let $a,b \in X$. Show that a path γ from a to b induces in a "natural" way an isomorphism of $\pi_k(X,b)$ and $\pi_k(X,a)$. Prove that if X is simply connected, this isomorphism is independent of the path γ. *Hint:* To the k-cube α at b in X, assign a k-cube $\tilde{\alpha}$ at a in X, as indicated by the accompanying figure, where the lines from the a-rim to the b-rim are to be each mapped to X by γ.

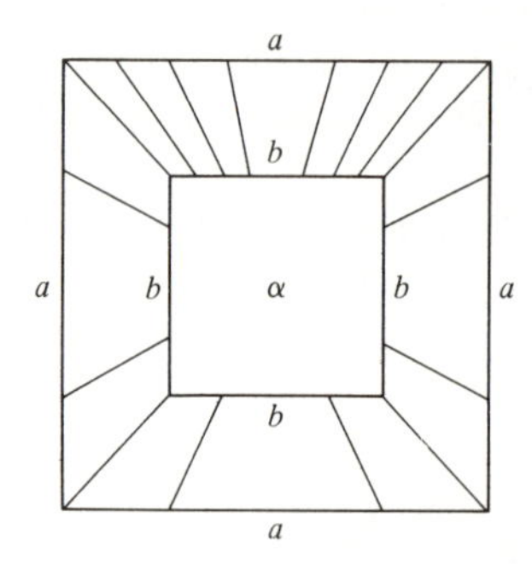

5. Let $f : S^k \rightarrow X$ satisfy $f(p) = b$ and suppose f is homotopic to a constant map. Show that f is homotopic to the constant map at b through a homotopy which fixes p, that is, that $[f]$ is the identity of $\pi_k(X,b)$.
6. Show that if X is contractible to a point b, then $\pi_k(X,b) = 0$ for all $k \geq 1$.
7. Show that a map $g : (Y,c) \rightarrow (X,b)$ induces in a "natural" way homomorphisms $g_* : \pi_k(Y,c) \rightarrow \pi_k(X,b)$, $k \geq 1$. Show that if $h : (W,a) \rightarrow (Y,c)$ is another map, then $(g \circ h)_* = g_* \circ h_*$.
8. Prove that if $g : (Y,c) \rightarrow (X,b)$ is a homotopy equivalence, then g_* is an isomorphism of $\pi_k(Y,c)$ and $\pi_k(X,b)$, $k \geq 1$.

2. NONCONTRACTIBILITY OF S^n

Many of the applications of the fundamental group discussed earlier depend only on knowing that $\pi_1(S^1) \neq 0$. The more detailed information that $\pi_1(S^1) \cong \mathbb{Z}$ played no additional role. Similarly, it is useful and important to know that $\pi_n(S^n) \neq 0$. To establish that $\pi_n(S^n) \neq \{0\}$, it is enough (in fact, equivalent) to show that the identity map of S^n is not homotopic to a constant. The main goal of this section is to establish this latter result. It is convenient to recall some terminology relevant to these considerations.

A topological space is *contractible* if the identity map of X is homotopic to a constant, that is, if there is a continuous function $F : X \times [0,1] \rightarrow X$ such that $F(x,1) = x$ for every $x \in X$ and $F(x,0) = F(y,0)$ for every pair of points $x,y \in X$. In this terminology, the main goal of this section is to prove the following theorem.

2.1 Theorem: The n-sphere S^n is not contractible.

In this section, two other important results that are closely related to Theorem 2.1 will be established. The first of these is most easily stated in the terminology of retracts. A subset Y of a topological space X is a *retract* of X if there is a continuous function $f : X \to X$ such that $f(X) \subseteq Y$ and $f(y) = y$ for all $y \in Y$. The following result generalizes Theorem III.6.2.

2.2 Theorem: The n-sphere S^n is not a retract of the $(n + 1)$-unit ball B^{n+1}.

We shall also prove the generalization of Theorem III.6.3 to all dimensions.

2.3 Theorem *(Brouwer Fixed Point Theorem):* Any (continuous) map from the ball B^{n+1} to B^{n+1} has a fixed point.

The three results Theorems 2.1, 2.2, and 2.3 are easily seen to be "equivalent" in the sense that, if one assumes the truth of any one of them, then the other two are easily deduced. These deductions are, in particular, much easier than the proof of any of the three (without assuming any other one of the three). Of course, since all three statements are true, it does not make genuine logical sense to prove their equivalence; in formal logic, any true statement implies any other. "Equivalence" in the present setup means only what was already noted, that the proof of any one of the three assuming any other is much easier than the real proof of any of the three. We discuss these easy inner implications first.

Consider first Theorems 2.1 and 2.2. If $F(x,r)$ is a contraction of S^n to a point, then

$$f(rx) = F(x,r), \qquad 0 \leq r \leq 1,\ x \in S^n, \tag{2.1}$$

defines a retraction f of B^{n+1} onto S^n. Conversely, if f is a retraction of B^{n+1} onto S^n, then (2.1) defines a contraction F of S^n to a point. Thus Theorems 2.1 and 2.2 are equivalent.

Next suppose there is a map $g : B^{n+1} \to B^{n+1}$ without a fixed point. For each $x \in B^{n+1}$, the ray issuing from $g(x)$ and passing through x hits the sphere S^n in a unique point $f(x)$. (See Exercise 1; this idea already appeared in the proof of Theorem III.6.3, which is the Brouwer Fixed Point Theorem for B^2.) The point $f(x)$ depends continuously on x, and $f(x) = x$ for $x \in S^n$, so that f is a retraction of B^{n+1} onto S^n. Conversely, if f is a retraction of B^{n+1} onto S^n, then we obtain a map $g : B^{n+1} \to B^{n+1}$ without a fixed point by setting $g(x) = -f(x)$, $x \in B^{n+1}$. Thus Theorems 2.2 and 2.3 are also equivalent.

We shall establish Theorem 2.2, and thus also, as noted, Theorems 2.1 and 2.3, by using the calculus of differential forms and Stokes' Theorem. In this context, the use of differential forms is really just a notational convenience. A proof that avoids differential forms, though it involves in effect the same calculations and depends upon Lemma 2.4, is sketched in Exercise 5.

The first step of the proof is to reduce considerations to the case wherein the maps considered are differentiable of all orders. Precisely, a function $f : U \to \mathbb{R}$ on an open set U in $\mathbb{R}^n$ is said to be C^∞, or *smooth*, if it is continuous and its partial derivatives of all orders exist and are continuous. A function $f : X \to \mathbb{R}$ defined on an arbitrary subset X of $\mathbb{R}^n$ is C^∞, or *smooth*, if there is an open subset V with $X \subseteq V$ and a smooth function $F : V \to \mathbb{R}$ with $F|_X = f$. Finally, a function with values in $\mathbb{R}^k$ is *smooth*, or C^∞, if its component functions are; here we are using the obvious idea that a function into $\mathbb{R}^k$ can be thought of as an ordered k-tuple of $\mathbb{R}$-valued functions, its *component functions*.

The following lemma shows that we can restrict our attention to smooth functions in proving Theorem 2.2.

2.4 Lemma: If there is a continuous retraction of B^{n+1} onto S^n, then there is a smooth retraction of B^{n+1} onto S^n.

Proof: Let $f = (f_1, \ldots, f_{n+1})$ be a retraction of B^{n+1} onto S^n, so that $f_j(x) = x_j$ for $x \in S^n$. Let $\varepsilon > 0$ be small. Choose $0 < \rho < 1$ so near to 1 that $|f_j(x) - x_j| < \varepsilon$ if $\rho \leq |x| \leq 1$. Let $\chi : [0,1] \to \mathbb{R}$ be a smooth function such that $0 \leq \chi(r) \leq 1$ for $0 \leq r \leq 1$, $\chi(r) = 0$ for $0 \leq r \leq \rho$, and $\chi(1) = 1$. (For an explicit construction of χ, see Exercise 2.) By the Weierstrass Approximation Theorem, there are polynomials $h_1, \ldots, h_{n+1}$ in the variables $x_1, \ldots, x_{n+1}$ such that $|h_j(x) - f_j(x)| < \varepsilon$ for $x \in B^{n+1}$, $1 \leq j \leq n+1$. Define

$$g_j(x) = \chi x_j + (1 - \chi)h_j, \qquad 1 \leq j \leq n+1.$$

From the expression

$$g_j - f_j = [\chi x_j - \chi f_j] + [(1 - \chi)h_j - (1 - \chi)f_j]$$

we obtain

$$|g_j - f_j| \leq \chi|x_j - f_j| + (1 - \chi)|h_j - f_j|.$$

The choices of χ and h_j show that each summand on the right is bounded by ε, so that $|g_j - f_j| < 2\varepsilon$. If ε is sufficiently small, the range of $g = (g_1, \ldots, g_{n+1})$ lies near the unit sphere S^n, so that in particular $\Sigma\, g_j^2 > 0$ on B^{n+1}. Hence we may define a smooth map G from B^{n+1} to S^n by

$$G_j(x) = g_j(x) \Big/ \left(\sum_{k=1}^{n+1} g_k^2 \right)^{1/2}, \qquad 1 \leq j \leq n+1.$$

Since $g_j(x) = x_j$ for $x \in S^n$, G is the identity on S^n. Hence G is a retraction of B^{n+1} onto S^n. □

Proof of Theorem 2.2: Suppose that $f = (f_1, \ldots, f_{n+1})$ is a smooth retraction of B^{n+1} onto S^n. This supposition will be shown to lead to a contradiction. The theorem then follows from Lemma 2.4.

Since $f(x) = x$ on S^n, we have

$$\int_{S^n} f_1 df_2 \wedge \cdots \wedge df_{n+1} = \int_{S^n} x_1 dx_2 \wedge \cdots \wedge dx_{n+1}. \tag{2.2}$$

By Stokes' Theorem, the left-hand side of (2.2) is equal to

$$(2.3) \quad \int_{B^{n+1}} df_1 \wedge \cdot \cdot \cdot \wedge df_{n+1} = \int_{B^{n+1}} \det\left(\frac{\partial f_j}{\partial x_k}\right) dx_1 \wedge \cdot \cdot \cdot \wedge dx_{n+1}.$$

We claim that this integral is zero. Indeed, differentiating the expression $\Sigma f_j^2 = 1$, we obtain

$$\sum_{j=1}^{n+1} f_j \frac{\partial f_j}{\partial x_k} = 0, \qquad 1 \leq k \leq n + 1.$$

Thus at each point the vector $(f_1, . . ., f_{n+1})$ is orthogonal to each of the vectors $\left(\frac{\partial f_1}{\partial x_k}, . . ., \frac{\partial f_{n+1}}{\partial x_k}\right)$, $1 \leq k \leq n + 1$. Hence the latter set of $n + 1$ vectors is linearly dependent, so that their determinant vanishes and the intergral in (2.3) is zero. One may also see that the integral vanishes from the fact that (2.3) can be interpreted as the $(n + 1)$-dimensional volume of the range of f, counting multiplicity. Since the range of f is contained in S^n, which has zero volume, the integral is zero.

In any event, the left-hand side of (2.2) vanishes. However, the right-hand side of (2.2) is equal by Stokes' Theorem to

$$\int_{B^{n+1}} dx_1 \wedge \cdot \cdot \cdot \wedge dx_{n+1},$$

which is strictly positive. This contradiction establishes the theorem. □

EXERCISES

1. Let $x,y \in B^n$, $x \neq y$. Show that the ray in $\mathbb{R}^n$ issuing from x and passing through y meets S^{n-1} at a unique point $h(x,y)$. Derive an explicit formula for $h(x,y)$. Deduce that the point $h(x,y)$ depends continuously on $x,y \in B^n$, $x \neq y$.
2. (a) Show that the function

$$\chi(t) = \begin{cases} \exp(-1/t^2), & t > 0, \\ 0, & t \leq 0, \end{cases}$$

is smooth (infinitely differentiable).

(b) Show that there is a smooth function Ψ on $\mathbb{R}$ such that $\Psi \geq 0$, $\Psi = 0$ outside the closed interval $[0,1]$, but Ψ is not identically zero. *Hint:* Consider $\chi(t)\,\chi(2 - t)$.

(c) Show that there is a smooth function φ on $\mathbb{R}$ such that $0 \leq \varphi \leq 1$, $\varphi(s) = 0$ for $s \leq 0$, and $\varphi(s) = 1$ for $s \geq 1$. *Hint:* Consider the indefinite integral of Ψ.

3. Let X be compact. The *cone over* X is the topological space Λ obtained from the product space $X \times [0,1]$ by identifying the slice $X \times \{1\}$ to a point. Prove that X is contractible if and only if there is a retraction of Λ onto the "slice" $X \times \{0\}$.
4. Show by elementary reasoning that if Theorems 2.1 to 2.3 are valid for an integer n, then they are valid for all integers less than n.
5. Prove Theorem 2.2 by using Lemma 2.4 and carrying out in detail the argument outlined as follows. For each smooth map $f = (f_1, \ldots, f_{n+1})$ from B^{n+1} to $\mathbb{R}^{n+1}$, define V_f to be the integral over B^{n+1} of the Jacobian determinant of f:

$$V_f = \int \cdots \int_{B^{n+1}} \det\left(\frac{\partial f_j}{\partial x_k}\right) dx_1 \cdots dx_{n+1}$$

Let g be another smooth map such that $g = f$ on S^n. Define $F(t,x) = tf(x) + (1 - t)g(x)$ and $V(t) = tV_f + (1 - t)V_g$, $0 \leq t \leq 1$. Then

$$\frac{dV}{dt} = \int \cdots \int_{B^{n+1}} \frac{\partial}{\partial t} \det\left(\frac{\partial F_j}{\partial x_k}\right) dx_1 \cdots dx_{n+1}.$$

If Q_l is the matrix obtained from the Jacobian matrix of F by replacing the lth column by the partial derivatives $\frac{\partial F_j}{\partial t}$, $1 \leq j \leq n + 1$, then

$$\frac{\partial}{\partial x_l} \det Q_l = \begin{vmatrix} \frac{\partial F_1}{\partial x_1 \partial x_l} & \frac{\partial F_1}{\partial x_2} & \cdots & \frac{\partial F_1}{\partial t} & \cdots & \frac{\partial F_1}{\partial x_{n+1}} \\ \vdots & \vdots & & \vdots & & \vdots \\ \frac{\partial F_{n+1}}{\partial x_1 \partial x_l} & \frac{\partial F_{n+1}}{\partial x_2} & \cdots & \frac{\partial F_{n+1}}{\partial t} & \cdots & \frac{\partial F_{n+1}}{\partial x_{n+1}} \end{vmatrix} + \cdots$$

$$+ \begin{vmatrix} \frac{\partial F_1}{\partial x_1} & \cdots & \frac{\partial F_1}{\partial x_l \partial t} & \cdots & \frac{\partial F_1}{\partial x_n} \\ \vdots & & \vdots & & \vdots \\ \frac{\partial F_{n+1}}{\partial x_1} & \cdots & \frac{\partial F_{n+1}}{\partial x_l \partial t} & \cdots & \frac{\partial F_{n+1}}{\partial x_{n+1}} \end{vmatrix} + \cdots$$

$$+ \begin{vmatrix} \frac{\partial F_1}{\partial x_1} & \cdots & \frac{\partial F_1}{\partial t} & \cdots & \frac{\partial^2 F_1}{\partial x_{n+1} \partial x_l} \\ \vdots & & \vdots & & \vdots \\ \frac{\partial F_{n+1}}{\partial x_1} & \cdots & \frac{\partial F_{n+1}}{\partial t} & \cdots & \frac{\partial^2 F_{n+1}}{\partial x_{n+1} \partial x_l} \end{vmatrix}$$

On account of cancellation by pairs,

$$\frac{\partial}{\partial t} \det \left(\frac{\partial F_j}{\partial x_k} \right) = \sum_{l=1}^{n+1} \frac{\partial}{\partial x_l} \det Q_l.$$

Since $\frac{\partial F_j}{\partial t}(t,x) = 0$ if $x \in S^n$, also $\det Q_l(t,x) = 0$ if $x \in S^n$, and consequently

$$\int \cdots \int_{B^{n+1}} \frac{\partial}{\partial x_l} \det Q_l \, dx_1 \cdots dx_{n+1} = 0.$$

It follows that $dV/dt = 0$ and $V_f = V_g$. Now if g is the identity map $g(x) = x$, $x \in B^{n+1}$, then $V_g > 0$, whereas if $f(B^{n+1}) \subset S^n$, then the Jacobian determinant of f vanishes and $V_f = 0$. Thus no such f can coincide with the identity on S^n.

6. (For readers familiar with differential forms on manifolds.) Using differential forms, show that if M is a compact oriented manifold with boundary ∂M, then there is no smooth retraction of M onto ∂M. *Hint:* Let α be an $(n-1)$-form on ∂M such that $\int_{\partial M} \alpha > 0$ and let $\beta = F^*(\alpha)$, an $(n-1)$-form on M that coincides with α on ∂M. Apply Stokes' Theorem to $\int_{\partial M} \beta$.

3. SIMPLEXES AND BARYCENTRIC SUBDIVISION

In this section, we introduce a class of geometric objects in $\mathbb{R}^n$, each called a simplex. The simplex is the natural extension to n dimensions of the triangle in $\mathbb{R}^2$ or the tetrahedron in $\mathbb{R}^3$. The use of simplexes serves to simplify the treatment of certain areas in homotopy theory that would be extremely complicated if we continued to use n-cubes. The principal useful features of the simplex idea involve the barycentric subdivision of the simplex and the related piecewise linear maps defined on the simplex. These concepts will be introduced in precise form in this section and the next. The machinery built up in this section will be used in Section 4 to prove a theorem on the approximation of continuous mappings by piecewise linear mappings and to use this approximation to obtain a second proof of the noncontractibility of S^n. This second proof introduces a method, the counting of preimages of a point for a given mapping, that is important in further developments in topology. The analogous method for C^∞ objects and maps is discussed in Section 5.

A set $\{v_0, \ldots, v_k\}$ of $k + 1$ points in $\mathbb{R}^n$ is *affinely independent* if whenever $t_0, \ldots, t_k \in \mathbb{R}$ satisfy $\sum_{j=0}^k t_j v_j = 0$ and $\sum_{j=0}^k t_j = 0$, then $t_0 = \cdots = t_k = 0$. Thus any linearly independent set is affinely independent. In fact, the following is true.

3.1 Lemma: The set $\{v_0, \ldots, v_k\}$ is affinely independent if and only if the k vectors $\{v_1 - v_0, v_2 - v_0, \ldots, v_k - v_0\}$ are linearly independent.

Proof: Suppose $\{v_0, \ldots, v_k\}$ are affinely independent. Let $t_1, \ldots, t_k$ satisfy $t_1(v_1 - v_0) + \cdots + t_k(v_k - v_0) = 0$. If $t_0 = -(t_1 + \cdots + t_k)$, then

$\Sigma_{j=0}^{k} t_j v_j = 0$ and $\Sigma_{j=0}^{k} t_j = 0$. By definition of affine independence, $t_0 = t_1 = \cdots = t_k = 0$, and $\{v_1 - v_0, \ldots, v_k - v_0\}$ are linearly independent.

Conversely, suppose $\{v_1 - v_0, \ldots, v_k - v_0\}$ are linearly independent. Suppose $t_0, t_1, \ldots, t_k$ satisfy $\Sigma\, t_j v_j = 0$ and $\Sigma\, t_j = 0$. Then $\Sigma_{j=1}^{k} t_j(v_k - v_0) = 0$, so that $t_1 = \cdots = t_k = 0$ and hence also $t_0 = 0$. Thus $\{v_0, \ldots, v_k\}$ are affinely independent. □

Suppose now that $\{v_0, \ldots, v_k\}$ is an affinely independent subset of $\mathbb{R}^n$. The *k-simplex generated by* $\{v_0, \ldots, v_k\}$, denoted by $\langle v_0, \ldots, v_k \rangle$, is defined to be the subset of $\mathbb{R}^n$ of convex combinations of the vectors $v_0, \ldots, v_k$. In other words, $\langle v_0, \ldots, v_k \rangle$ consists of precisely the vectors of the form $v = \Sigma_{j=0}^{k} t_j v_j$, where $t_j \geq 0$, $0 \leq j \leq k$, and $\Sigma\, t_j = 1$. Note that the representation of $v \in \langle v_0, \ldots, v_k \rangle$ as a convex combination $v = \Sigma\, t_j v_j$ is unique. Indeed, if $v = \Sigma\, s_j v_j$, where $\Sigma\, s_j = 1$, then $\Sigma\, (t_j - s_j)v_j = 0$ and $\Sigma\, (t_j - s_j) = 0$, which according to the definition of affine independence implies that $t_j = s_j$, $0 \leq j \leq k$.

The 0-simplex $\langle v_0 \rangle$ generated by the point v_0 is just the singleton $\{v_0\}$. The 1-simplex $\langle v_0, v_1 \rangle$, $v_0 \neq v_1$, is the closed-line segment joining v_0 to v_1. A 2-simplex is a triangle, and a 3-simplex is a tetrahedron.

Each subset of $l + 1$ distinct indices $\{i_0, \ldots, i_l\}$ determines an l-simplex $\langle v_{i_0}, \ldots, v_{i_l} \rangle$, which is called a (closed) *l-face* of $\langle v_0, \ldots, v_k \rangle$. It consists of all convex combinations $\Sigma\, t_j v_j$ for which $t_j = 0$ whenever $j \notin \{i_0, \ldots, i_l\}$. The 0-faces of $\langle v_0, \ldots, v_k \rangle$ are just the singletons $\langle v_j \rangle$, $0 \leq j \leq k$, and these are called the *vertices* of $\langle v_0, \ldots, v_k \rangle$.

Let $\{v_0, \ldots, v_k\}$ be an affinely independent set in $\mathbb{R}^n$. The *barycentric subdivision* of $\langle v_0, \ldots, v_k \rangle$ is the collection of k-simplexes generated by the vectors

$$\{v_{i(0)}, (v_{i(0)} + v_{i(1)})/2, \ldots, (v_{i(0)} + \cdots + v_{i(k)})/(k + 1)\}. \tag{3.1}$$

where $\{i(0), \ldots, i(k)\}$ is a permutation of the indices $\{0, \ldots, k\}$. Using the criterion of Lemma 3.1, one checks easily that the vectors in (3.1) are affinely independent.

Thus the barycentric subdivision of a 1-simplex $\langle v_0, v_1 \rangle$ corresponds to subdividing the interval in half into two subintervals. The barycentric subdivision of a triangle (2-simplex) consists of six triangles, as in the diagram.

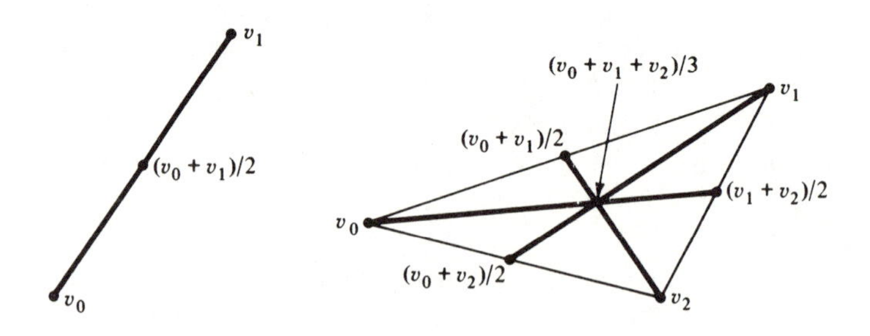

The family of simplexes in the barycentric subdivision of $\langle v_0, \ldots, v_k \rangle$ will be denoted by $\mathcal{B}^{(1)}$. Some elementary properties of $\mathcal{B}^{(1)}$ are given in the following lemma. While these properties are intuitively clear, at least in lower dimensions, formal proofs are required, if for no other reason than to confirm our intuition.

3.2 Lemma: Let $v_0,\ldots,v_k \in \mathbb{R}^n$ be affinely independent.

(i) $\langle v_0,\ldots,v_k\rangle$ is the union of the simplexes in $\mathcal{B}^{(1)}$.

(ii) The intersection of any two simplexes in $\mathcal{B}^{(1)}$ is the simplex generated by their common vertices.

(iii) The intersection of any two faces of simplexes in $\mathcal{B}^{(1)}$ is either empty or else the simplex generated by their common vertices.

(iv) The intersection of a face R of $\langle v_0,\ldots,v_k\rangle$ and a simplex $S \in \mathcal{B}^{(1)}$ is either empty or else a face of a simplex in the barycentric subdivision of R, generated by the vertices of S that lie in R.

Proof: Let $v = \Sigma\, t_j v_j \in \langle v_0,\ldots,v_k\rangle$, where $t_j \geq 0$, $1 \leq j \leq k$, and $\Sigma\, t_j = 1$. Assume that

$$t_{i(0)} \geq t_{i(1)} \geq \cdots \geq t_{i(k)} \geq 0, \tag{3.2}$$

where $\{i(0),\ldots,i(k)\}$ is a permutation of the indices. Consider the system of linear equations

$$\left\{\begin{aligned} s_0 + \frac{s_1}{2} + \cdots \quad + \frac{s_k}{k+1} &= t_{i(0)}, \\ &\;\;\vdots \\ \frac{s_{k-1}}{k} + \frac{s_k}{k+1} &= t_{i(k-1)}, \\ \frac{s_k}{k+1} &= t_{i(k)}. \end{aligned}\right. \tag{3.3}$$

These have a unique solution $s_0,\ldots,s_k$, which is obtained by solving them in reverse order. Moreover, condition (3.2) guarantees that $s_j \geq 0$ for $0 \leq j \leq k$. Summing the $k + 1$ equations, we obtain $\Sigma\, s_j = \Sigma\, t_j = 1$. Thus the system yields v as a convex combination of vectors in (3.1):

$$v = s_0 v_{i(0)} + s_1\left(\frac{v_{i(0)} + v_{i(1)}}{2}\right) + \cdots + s_k\left(\frac{v_{i(0)} + \cdots + v_{i(k)}}{k+1}\right). \tag{3.4}$$

This proves statement (i) of the lemma.

Note that the argument above is reversible. If $v = \Sigma\, t_j v_j$ can be expressed as the convex combination (3.4), then $s_0,\ldots,s_k$ satisfy the system (3.3). Since the s_j's are nonnegative, we see from (3.3) that (3.2) is valid. Thus the simplex with vertices (3.1) consists of precisely those convex combinations $v = \Sigma\, t_j v_j$ for which (3.2) is valid.

Now fix permutations $\{i(0),\ldots,i(k)\}$ and $\{j(0),\ldots,j(k)\}$ of the indices. We wish to describe the intersection of the corresponding simplexes in the barycentric subdivision. For this, we may assume that the j's are the identity permutation, that is, $\{j(0), j(1),\ldots,j(k)\} = \{0,1,\ldots,k\}$. We also partition the ordered set of i's into blocks

$$\{i(0),i(1),\ldots,i(k)\} = \{B_1,B_2,\ldots,B_l\},$$

where each

$$(3.5) \qquad B_m = \{i(\beta_{m-1} + 1),\ldots, i(\beta_m)\}$$

is a permutation of $\{\beta_{m-1} + 1,\ldots,\beta_m\}$. Since the intersection of any two such partitions is another such partition, we may assume that the blocks B_m have minimal size.

Suppose $v = \Sigma\, t_j v_j$ is such that (3.2) is valid and also

$$(3.6) \qquad t_p = t_q, \qquad \text{all } p,q \text{ in } B_m, \qquad 1 \le m \le l.$$

Then also

$$(3.7) \qquad t_0 \ge t_1 \ge \cdots \ge t_k \ge 0,$$

so that v belongs to the simplexes corresponding to both permutations. Conversely, suppose v belongs to the simplexes corresponding to both permutations. Then (3.2) and (3.7) are both valid. Write

$$t_{i(0)} = \cdots = t_{i(\gamma_0)} > t_{i(\gamma_0+1)} = \cdots = t_{i(\gamma_1)} > t_{i(\gamma_0+1)} = \cdots.$$

Condition (3.7) then implies that $\{i(0),\ldots,i(\gamma_0)\}$ is a permutation of $\{0,\ldots,\gamma_0\}$, and so on. By the minimality of the blocks B_m, we conclude that $\{i(0),\ldots,i(\gamma_0)\}$ is a union of certain of the B_m's, and so on. It follows that (3.6) is valid. Thus the intersection of the two simplexes consists of precisely those $v = \Sigma\, t_j v_j$ such that (3.2) and (3.6) are valid.

Now fix a block B_m as in (3.5). Referring to equations (3.3), we see that condition (3.6) is met if and only if

$$0 = s_{i(\beta_{m-1})} = \cdots = s_{i(\beta_m - 1)}$$

Thus v belongs to the intersection of the simplexes if and only if, in expression (3.4), we have $s_j = 0$ whenever j is an index such that $i(j)$ is not the last element of a block B_m. In other words, the intersection of the simplexes is precisely the simplex generated by the vectors

$$(v_{i(0)} + v_{i(1)} + \cdots + v_{i(\beta_m)})/(\beta_m + 1),\ 1 \le m \le l.$$

This proves (ii).

Let $S_1, S_2 \in \mathcal{B}^{(1)}$ and let R_1 and R_2 be faces of S_1 and S_2, respectively. By (ii), $S_1 \cap S_2 = \langle w_0,\ldots,w_q\rangle$, where $w_0,\ldots,w_q$ are the common vertices of S_1 and S_2. Now R_1 consists of the convex combinations of certain of the vertices of S_1, so that $R_1 \cap S_1 \cap S_2$ consists of convex combinations of those w_j's that belong to R_1. The corresponding statement for $R_2 \cap S_1 \cap S_2$ is valid. Hence $R_1 \cap R_2$, which coincides with $R_1 \cap R_2 \cap S_1 \cap S_2$, consists of the convex combinations of those w_j's that belong to R_1 and to R_2. This proves (iii).

To establish (iv), we may assume that

$$S = \langle v_0, (v_0 + v_1)/2,\ldots,(v_0 + \cdots + v_k)/(k + 1)\rangle$$

and that R includes $v_0, v_1,\ldots,v_q$ but not v_{q+1}. Suppose $v = \Sigma\, t_j v_j \in S \cap R$. Then $v = \Sigma\, s_j(v_0 + \cdots + v_j)/(j + 1)$, where the s_j's are solutions of system (3.3), with

$i(j) = j$. Since $v_{q+1} \notin R$, $t_{q+1} = 0$ and the corresponding equation of system (3.3) yields $s_j = 0$ for $q + 1 \leq j \leq k$. If $q = -1$, that is, if $v_0 \notin R$, this contradicts $\Sigma s_j = 1$, so that $S \cap R$ must be empty. On the other hand, if $q \geq 0$, then this shows that $S \cap R$ consists of precisely those $v = \Sigma_{j=0}^{q} t_j v_j$ that can be expressed in the form

$$v = \sum_{j=0}^{q} s_j(v_0 + \cdots + v_j)/(j + 1),$$

where $s_j \geq 0$ and $\Sigma s_j = 1$. Thus $S \cap R$ is the face, corresponding to the first $q + 1$ vertices, of the simplex $\langle v_0,(v_0 + v_1)/2,\ldots,(v_0 + \cdots + v_q)/(q + 1),\ldots\rangle$ in the barycentric subdivision of R. □

Next we wish to estimate the diameters of the simplexes in a barycentric subdivision. The *diameter* of a simplex $\langle v_0,\ldots,v_k\rangle$ is defined to be

$$d = \sup\{|u - v| : u,v \in \langle v_0,\ldots,v_k\rangle\}.$$

Suppose $u = \Sigma s_i v_i$ and $v = \Sigma t_j v_j$ belong to $\langle v_0,\ldots,v_k\rangle$. Using the relations $\Sigma s_i = 1 = \Sigma t_j$, we obtain

$$u - v = \sum_{i,j} s_i t_j (v_i - v_j). \tag{3.8}$$

From (3.8) and the relation $\Sigma s_i t_j = 1$, we obtain

$$|u - v| \leq \sum_{i,j} s_i t_j |v_i - v_j| \leq \max_{i,j} |v_i - v_j|.$$

It follows that the diameter of a simplex is the maximum of the distances separating its vertices:

$$d = \max_{0 \leq i,j \leq k} |v_i - v_j|.$$

3.3 Lemma: Let $\{v_0,\ldots,v_k\}$ be affinely independent vectors in $\mathbb{R}^n$, let d be the diameter of $\langle v_0,\ldots,v_k\rangle$, and let d' be the diameter of a simplex in the barycentric subdivision of $\langle v_0,\ldots,v_k\rangle$. Then

$$d' \leq \left(\frac{k}{k + 1}\right) d. \tag{3.9}$$

Proof: The lemma is valid when $k = 1$. In this case, a 1-simplex is an interval and barycentric subdivision amounts to cutting the interval in half. We make the induction hypothesis that the lemma is valid for simplexes with at most k vertices.

Let S be a simplex in the barycentric subdivision of $\langle v_0,\ldots,v_k\rangle$, say for simplicity that

$$S = \left\langle v_0, \frac{v_0 + v_1}{2},\ldots,\frac{v_0 + \cdots + v_k}{k + 1}\right\rangle.$$

Let w_1 and w_2 be the two vertices of S that are farthest apart, so that $d' = |w_1 - w_2|$. If neither w_1 nor w_2 coincides with $(v_0 + \cdots + v_k)/(k + 1)$, then w_1 and w_2 are vertices of a simplex in the barycentric subdivision of $\langle v_0, \ldots, v_{k-1} \rangle$. By the induction hypothesis,

$$|w_1 - w_2| \le \left(\frac{k-1}{k}\right) \max_{0 \le i,j < k-1} |v_i - v_j| \le \left(\frac{k}{k+1}\right) d,$$

so that (3.9) is valid.

On the other hand, suppose that $w_2 = (v_0 + \cdots + v_k)/(\mathrm{k} + 1)$ and $w_1 = (v_0 + \cdots + v_j)/(j + 1)$. Then

$$|w_2 - v_i| = \left| \frac{v_0 + \cdots + v_k}{k+1} - v_i \right| = \frac{1}{k+1} \left| \sum_{j=0}^{k} (v_j - v_i) \right|.$$

Estimating $|v_j - v_i|$ by d for $j \ne i$ and 0 for $j = i$, we obtain

$$|w_2 - v_i| \le \frac{k}{k+1} d, \qquad 0 \le i \le k.$$

Hence

$$|w_2 - w_1| = \left| w_2 - \frac{v_0 + \cdots + v_j}{j+1} \right| \le \frac{1}{j+1} \sum_{i=0}^{j} |w_2 - v_i|$$

$$\le \max |w_2 - v_i| \le \frac{k}{k+1} d$$

and again we obtain (3.9). □

We have defned $\mathcal{B}^{(1)}$ to be the family of k-simplexes obtained from $\langle v_0, \ldots, v_k \rangle$ by barycentric subdivision. Proceeding by induction, we define for $N \ge 2$ the *Nth barycentric subdivision* $\mathcal{B}^{(N)}$ of $\langle v_0, \ldots, v_k \rangle$ to be the family of k-simplexes obtained from the simplexes in $\mathcal{B}^{(N-1)}$ by barycentric subdivision.

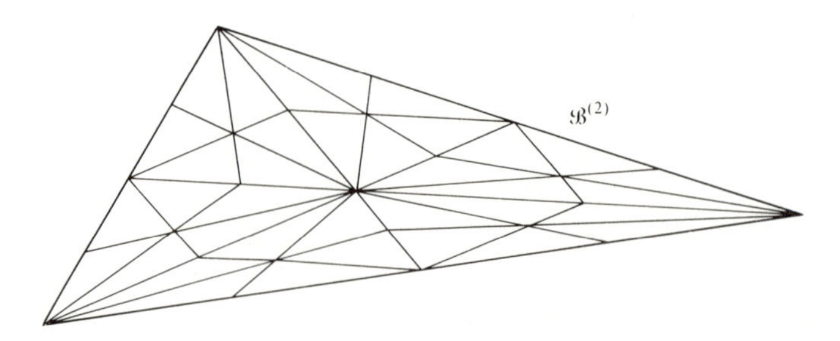

The properties of $\mathcal{B}^{(N)}$ that we require are summarized as follows.

3.4 Lemma: Let $v_0, \ldots, v_k \in \mathbb{R}^n$ be affinely independent and let $N \ge 1$.

(i) $\langle v_0, \ldots, v_k \rangle$ is the union of the simplexes in $\mathcal{B}^{(N)}$.

(ii) The intersection of any two simplexes in $\mathcal{B}^{(N)}$ is the simplex generated by their common vertices.

(iii) If $\langle v_0,\ldots,v_k\rangle$ has diameter d, then the simplexes in $\mathcal{B}^{(N)}$ have diameters at most $[k/(k+1)]^N d$.

Proof: (i) follows by induction from Lemma 3.2(i), and (iii) follows by induction from Lemma 3.3. To prove (ii), we may assume that the statement is true for $\mathcal{B}^{(N-1)}$.

Let $S,T \in \mathcal{B}^{(N)}$ and suppose S and T are obtained from the barycentric subdivision of S_0 and T_0, respectively, where $S_0,T_0 \in \mathcal{B}^{(N-1)}$. By the induction hypothesis, $F = S_0 \cap T_0$ is the simplex generated by the common vertices of S_0 and T_0. In particular, F is a face of both S_0 and T_0. By Lemma 3.2(iv), $F \cap S$ is a face of a simplex in the barycentric subdivision of F, generated by the vertices of S lying in F. A similar statement is valid for $F \cap T$. By Lemma 3.2(iii), $S \cap T = (F \cap S) \cap (F \cap T)$ is a simplex, generated by the common vertices of $F \cap S$ and $F \cap T$, that is, the common vertices of S and T. □

EXERCISES

1. Show that any k-simplex is homeomorphic to the unit ball B^k in $\mathbb{R}^k$ in such a way that the union of the $(k-1)$-faces corresponds to S^{k-1}.
2. Let S be a simplex in $\mathbb{R}^n$ and let f be a linear functional on $\mathbb{R}^n$ whose maximum value on S is c. Show that $f^{-1}(c)$ is a face of S and that furthermore any face of S can be obtained in such a way, as the maximum set of a linear functional on S.
3. How many simplexes are there in the Nth barycentric subdivision of a k-simplex?
4. Show that the intersection of any two faces of simplexes in $\mathcal{B}^{(N)}$, if not empty, is the simplex generated by their common vertices.
5. Show that the intersection of a face F of a simplex with a simplex T in its Nth barycentric subdivision, if not empty, is the face of T generated by the vertices of T that lie in F.
6. A subset S of a vector space V is *convex* if S includes the line segment between any two of its points, that is, if whenever $x,y \in S$ and $0 < t < 1$, then $tx + (1-t)y \in S$. A *convex combination* of vectors in a set S is a finite sum of the form $\Sigma\, t_jx_j$, where $x_j \in S$, $t_j \geq 0$, and $\Sigma\, t_j = 1$. Show that S is convex if and only if any convex combination of vectors in S again belongs to S.
7. Let S be a subset of a vector space V. Show that the set co(S) of convex combinations of elements of S is a convex subset of V, which is the smallest convex subset of V containing S. The set co(S) is called the *convex hull of* S.
8. Let S be a subset of $\mathbb{R}^n$ such that every element of co(S) has a unique representation as a convex combination of elements of S. Show that S is an affinely independent set and that co(S) is a simplex with vertices S.

4. APPROXIMATION BY PIECEWISE LINEAR MAPS

In this section, we use the machinery built up in the preceding section to prove that a continuous map from a simplex to $\mathbb{R}^m$ can be approximated uniformly by a map that is affine on each simplex in some barycentric subdivision of the simplex. Such a map is said to be "piecewise linear," though strictly speaking it is only a piecewise affine map. Intuitively, the proof is quite easy. One defines the approximator to coincide with the given map on the vertices of the simplexes in the Nth barycentric subdivision $\mathcal{B}^{(N)}$, where N is large, and one extends the approximator by linearity to each simplex in $\mathcal{B}^{(N)}$. As an application of the approximation theorem, we give another proof that the n-sphere is not contractible.

Recall that a transformation from one vector space to another is *affine* if it is the sum of a constant map and a linear transformation. Thus h is affine if and only if $h - h(0)$ is linear. This occurs if and only if h preserves convex combinations, that is, $h(\Sigma\, t_j x_j) = \Sigma\, t_j h(x_j)$ whenever $t_j \geq 0$ and $\Sigma\, t_j = 1$ (Exercise 1).

The following lemma contains a key property of simplexes.

4.1 Lemma: Let $\{v_0, \ldots, v_k\}$ be an affinely independent subset of $\mathbb{R}^n$. If $m \geq 1$ and if $w_0, \ldots, w_k \in \mathbb{R}^m$, then there is a unique affine map g from $\langle v_0, \ldots, v_k \rangle$ to $\mathbb{R}^m$ such that $g(v_j) = w_j$, $0 \leq j \leq k$.

Proof: The uniqueness of g is clear since any affine map preserves convex combinations. For the existence, let L be any linear transformation from $\mathbb{R}^n$ to $\mathbb{R}^m$ such that $L(v_j - v_0) = w_j - w_0$, $1 \leq j \leq k$, and set $g = L - L(v_0) + w_0$. The existence of L is guaranteed since $v_1 - v_0, \ldots, v_k - v_0$ are linearly independent. One checks that $g(v_j) = w_j$ for $0 \leq j \leq k$. □

The approximation theorem we have been aiming for is as follows.

4.2 Theorem: Let $\{v_0, \ldots, v_k\}$ be an affinely independent subset of $\mathbb{R}^n$ and let f be a continuous function from $\langle v_0, \ldots, v_k \rangle$ to $\mathbb{R}^m$. For each $\varepsilon > 0$, there exist $N > 1$ and a continuous map g from $\langle v_0, \ldots, v_k \rangle$ to $\mathbb{R}^m$ such that

$$|g(x) - f(x)| \leq \varepsilon, \qquad x \in \langle v_0, \ldots, v_k \rangle,$$

and g is affine on each k-simplex in the Nth barycentric subdivision $\mathcal{B}^{(N)}$ of $\langle v_0, \ldots, v_k \rangle$. Furthermore, g can be chosen so that $g = f$ on any face of $\langle v_0, \ldots, v_k \rangle$ on which f is affine.

Proof: Since f is uniformly continuous, there exists $\delta > 0$ such that $|f(x) - f(y)| < \varepsilon$ whenever $x,y \in \langle v_0, \ldots, v_k \rangle$ satisfy $|x - y| \leq \delta$. By Lemma 3.4(iii), we can choose N so large that each of the simplexes in $\mathcal{B}^{(N)}$ has diameter less than δ. Then $|f(x) - f(y)| < \varepsilon$ whenever x and y belong to the same simplex in $\mathcal{B}^{(N)}$.

Let T be a simplex in $\mathcal{B}^{(N)}$. By Lemma 4.1 there is a unique affine function g_T on T such that $g_T(v) = f(v)$ for each vertex v of T. If S is another simplex in $\mathcal{B}^{(N)}$, then g_S and g_T coincide on the common vertices of S and T. By Lemma 3.4(ii), $S \cap T$ is the simplex generated by the common vertices of S and T. Consequently g_S

coincides with g_T on $S \cap T$. Thus the various g_T's patch together to define a continuous function g on $\langle v_0, \ldots, v_k \rangle$, so that $g = g_T$ on T.

Let $w_0, \ldots, w_k$ be the vertices of T and let $w = \sum t_j w_j \in T$. Then $g(w) = \sum t_j g(w_j) = \sum t_j f(w_j)$, so that

$$|g(w) - f(w)| = \left| \sum t_j [f(w_j) - f(w)] \right| \leq \sum t_j \varepsilon = \varepsilon,$$

since $\sum t_j = 1$. Thus g is the desired approximant of f. □

The following theorem is an equivalent form of Theorem 2.2, that the n-sphere is not contractible (see Exercise 3.1). This proof is independent of the calculus.

4.3 Theorem: Let $T = \langle v_0, \ldots, v_n \rangle$ be the simplex generated by the affinely independent set $\{v_0, \ldots, v_n\}$ in $\mathbb{R}^n$. Then there is no continuous map from T to ∂T that is the identity on ∂T.

Proof: Suppose that f is such a map, so that $f : T \rightarrow \partial T$ is the identity on ∂T. We shall eventually derive a contradiction.

By translating T if necessary, we may assume that 0 belongs to the interior of T. Let ε be half the distance from 0 to ∂T and choose N and g as in Theorem 4.2. By the final assertion of Theorem 4.2, we may arrange that

$$g(x) = x, \qquad x \in \partial T.$$

Since $f(y) \in \partial T$ and $|f(y) - g(y)| \leq \varepsilon$, we have $|g(y)| \geq \varepsilon$ by the choice of ε. In particular,

$$0 \notin g(T).$$

For each $y \in T \backslash \{0\}$, let R_y denote the ray issuing from 0 and passing through y:

$$R_y = \{ty : t > 0\}.$$

Then R_y cuts ∂T in precisely one point, call it $P(y)$. This defines a continuous function P on $T \backslash \{0\}$ that projects $T \backslash \{0\}$ radially onto ∂T. We are interested in the map $h = P \circ g$ from T to ∂T. Note that h is the identity on ∂T. Though not piecewise linear, the map h is more tractable than the original map f.

Let E be an $(n - 2)$-face of some simplex in $\mathcal{B}^{(N)}$, say $E = \langle w_0, \ldots, w_{n-2} \rangle$. Since g is affine on E, $g(E)$ consists of convex combinations of the $n - 1$ vectors $g(w_0), \ldots, g(w_{n-2})$ and $g(E)$ lies in a subspace V of $\mathbb{R}^n$ of dimension at most $n - 1$. The radial projection of any element of V lies on $V \cap \partial T$. Hence $h(E) \subseteq V \cap \partial T$. Since V does not contain any $(n - 1)$-face of T, V meets each $(n - 1)$-face of T in a closed nowhere-dense subset, in fact, in a subset of dimension at most $n - 2$ (Exercise 6). Hence $h(E)$ is a compact nowhere-dense subset of ∂T.

From the Baire Category Theorem (Section I.2), applied to a finite union, we conclude that the image under h of the union of all $(n - 2)$-faces of simplexes in $\mathcal{B}^{(N)}$ is a nowhere-dense subset of ∂T. In particular, there exists $u \in T$ such that $h^{-1}(u)$ does not meet any $(n - 2)$-face of any simplex in $\mathcal{B}^{(N)}$. For this fixed u, we shall analyze carefully the compact set Q, defined by

$$Q = h^{-1}(u) = g^{-1}(R_u).$$

First observe that since g is the identity on ∂T, h is also the identity on ∂T and

$$Q \cap \partial T = \{u\}.$$

Fix a simplex $S \in \mathcal{B}^{(N)}$ such that $Q \cap S \neq \emptyset$. Suppose that $g = v_0 + \Lambda$ on S, where Λ is linear.

We claim that $Q \cap S$ contains at least two points. Indeed, let $x_0 \in Q \cap S$. Suppose first that Λ is not invertible. Choose $y \neq 0$ such that $\Lambda(y) = 0$ and choose $t \in \mathbb{R}$, $t \neq 0$, such that $x_0 + ty \in S$. Since $g(x_0 + ty) = g(x_0) \in R_u$, we obtain $x_0 + ty \in Q \cap S$. Suppose on the other hand that Λ is invertible. Choose z such that $\Lambda(z) = u$ and choose t small, $t \neq 0$, such that $x_0 + tz \in S$. Then $g(x_0 + tz) = g(x_0) + tu \in R_u$ and $x_0 + tz \in Q \cap S$. This establishes the claim.

Next we claim that $Q \cap S$ cannot contain three noncolinear points. Indeed, suppose that x_0, x_1, and x_2 are three noncolinear points in $Q \cap S$. Let

$$E = \{x_0 + s(x_1 - x_0) + t(x_2 - x_0) : s,t \in \mathbb{R}\} \cap S,$$

the intersection of S and the translate by x_0 of a two-dimensional subspace. It is easy to see (Exercise 5) that E meets an $(n - 2)$-face of S. One checks that

$$g(x_0 + s(x_1 - x_0) + t(x_2 - x_0)) = (1 - s - t)g(x_0) + sg(x_1) + tg(x_2),$$

from which it follows that the range of g on E consists of multiples of u. Furthermore, since the range of g on E is connected and does not include 0, $g(E)$ consists only of positive multiples of u, that is, $g(E) \subseteq R_u$. Hence $E \subseteq Q$. However, Q does not meet any $(n - 2)$-face of S. This contradiction establishes the claim.

Let L be the line passing through two points of $Q \cap S$. We have shown that $Q \cap S \subseteq L$. As before, we see that $g(x) \in R_u$ for all $x \in L \cap S$, so that $L \cap S \subseteq Q$ and $L \cap S = Q \cap S$. Since $Q \cap S$ meets no $(n - 2)$-face of S, we find that $Q \cap S$ is a closed interval, passing through the interior of S, with endpoints lying on two distinct $(n - 1)$-faces of S.

Now we are in a position to complete the proof, as follows.

The point u lies on an $(n - 1)$-face of some simplex S_1 in $\mathcal{B}^{(N)}$, and $Q \cap S_1$ is a straight-line segment from u to a point u_1 on another $(n - 1)$-face of S_1. Since $u_1 \notin \partial T$, u_1 lies on the $(n - 1)$-face of precisely one other simplex $S_2 \in \mathcal{B}^{(N)}$, and

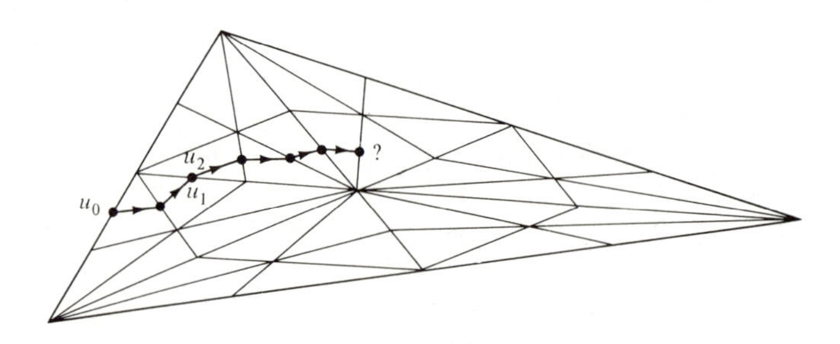

$Q \cap S_2$ is a straight-line segment from u_1 to a point u_2 on another $(n - 2)$-face of S_2. Again $u_2 \notin \partial T$, and we may continue the process to obtain an infinite chain of distinct simplexes $S_1,S_2,S_3,\ldots$ in $\mathcal{B}^{(N)}$. This is an absurdity, and the theorem is thereby established. □

In closing, we mention that piecewise linear topological proofs often have differential topological analogues. For instance, the proof we have just given, of the noncontractability of spheres, has a counterpart that depends upon notions of differential topology that run along the following lines. Suppose that there exists a map f from B^n to S^{n-1} that is the identity on S^{n-1}. We may assume that f is smooth. Let E be the set of $x \in B^n$ at which the rank of the Jacobian matrix of f is strictly less than $n - 1$ (its maximum possible value). According to Sard's Theorem, a fundamental result of differential topology, the set $f(E)$ is a closed nowhere-dense subset of S^{n-1}. In particular, there exists $u \in S^{n-1}$ such that the compact set $f^{-1}(u)$ is disjoint from E. From the Implicit Function Theorem, it follows that each point $y \in f^{-1}(u)$ has a neighborhood $B(y;\varepsilon)$ such that $f^{-1}(u) \cap B(y;\varepsilon)$ is a smooth curve joining two points of $\partial B(y;\varepsilon)$. By starting at u and proceeding along $f^{-1}(u)$ into B^n, one finds that one can follow the compact curve $f^{-1}(u)$ as long as one wishes without doubling back on oneself, yet one cannot exit from B^n. This is again an absurdity, which establishes the theorem.

EXERCISES

1. Prove that a map h from $\mathbb{R}^n$ to $\mathbb{R}^m$ is affine if and only if h preserves convex combinations.
2. Prove that any two k-simplexes are affinely homeomorphic.
3. Let V be a vector space. A *flat of dimension k,* or a *k-flat,* is a subset of V of the form $x_0 + W$, where $x_0 \in V$ is fixed and W is a k-dimensional subspace of V.
   (a) Show that the maximum number of affinely independent vectors in a k-flat is $k + 1$.
   (b) Show that each set of $k + 1$ affinely independent vectors $w_0,\ldots,w_k$ is contained in a unique k-flat, consisting of vectors of the form $w_0 + \sum_{j=1}^{k} s_j(w_j - w_0)$, $s_1,\ldots,s_k \in \mathbb{R}$.
4. Let $g : \mathbb{R}^n \to \mathbb{R}^m$ be affine. Prove the following assertions:
   (a) If E is a flat in $\mathbb{R}^n$, then $g(E)$ is a flat in $\mathbb{R}^m$.
   (b) If F is a flat in $\mathbb{R}^m$, then $g^{-1}(F)$ is a flat in $\mathbb{R}^n$.
5. Let S be a k-simplex in $\mathbb{R}^n$ and let $w_0,\ldots,w_q \in S$ be affinely independent. Show that the q-flat generated by $w_0,\ldots,w_q$ meets some $(k - q)$-face of S.
6. Let S be an n-simplex in $\mathbb{R}^n$ and let E be a k-flat in $\mathbb{R}^n$ that passes through the interior of S. Show that the intersection of E and any k-face of S is a nowhere-dense subset of the k-face.

The following sequence of exercises establishes the fact that the dimension corresponding to a nonempty open subset of Euclidean space is a topological property.

7. Prove that if f and g are maps from S^n to S^n such that $|f(x) - g(x)| < 2$ for all $x \in S^n$, then f and g are homotopic.

Hint: Observe first that $tf(x) + (1 - t)g(x) \neq 0$ for $0 \leq t \leq 1$, $x \in S^n$.

8. Prove that $\pi_k(S^n) = 0$ if $1 < k \leq n$. *Hint:* Using the technique of the proof of Theorem 4.2, show that any map $f : S^k \to S^n$ can be approximated by a map $g : S^k \to S^n$ such that the range of g is a proper subset of S^n.
9. Prove that S^m is not homeomorphic to S^n if $m \neq n$.
10. Prove that $\mathbb{R}^n$ is not homeomorphic to $\mathbb{R}^m$ if $n \neq m$. *Hint:* If X and Y are homeomorphic, then the one-point compactifications of X and Y are also homeomorphic.
11. Prove that $\pi_n(\mathbb{R}^{n+1}\backslash\{0\}) \neq 0$ for $n \geq 1$.
12. Prove that if U is an open subset of $\mathbb{R}^{n+1}$ and if $p \in U$, then $\pi_n(U\backslash\{p\}) \neq 0$.
13. Prove that if B is a ball in $\mathbb{R}^{n+1}$ and if $p \in B$, then $\pi_k(B\backslash\{p\}) = 0$ for $1 \leq k < n$.
14. Prove that if $n \neq m$, then no nonempty open subset of $\mathbb{R}^n$ is homeomorphic to an open subset of $\mathbb{R}^m$.

Remark: This result is called, for historical reasons, the Invariance of Domain Theorem. For the proof, use Exercises 12 and 13.

5. DEGREES OF MAPS

The purpose of this section is to provide intuitive insight into some of the concepts that are important in the further development of (algebraic) topology. The precise development of these ideas requires lengthy considerations of a technical nature, but it is possible to achieve a useful informal understanding without too much difficulty.

The starting point for our considerations is the idea, which arises from the techniques of the previous sections, that there is some homotopic significance to the number of times that points are covered by a map $f : S^n \to S^n$ or, more precisely, to the number of points in $f^{-1}(\{p\})$ for $p \in S^n$. Thinking about maps $f : S^1 \to S^1$ shows that to make real sense of this idea one has to count some points in $f^{-1}(\{p\})$ negatively. Namely, one would hope that if $f : S^1 \to S^1$ is homotopic to a constant, then the number of points in $f^{-1}(\{p\})$ would be (for most p) zero, since it is zero for a constant map (for all but one p). However, the map that first winds once around counterclockwise and then unwinds once around clockwise shows that the literal number of points in $f^{-1}(\{p\})$ can be always nonzero even if f is homotopic to a constant (Exercise 1). Nevertheless, if one takes the difference between the number of points in $f^{-1}\{p\}$ where f is winding and subtracts the number where f is unwinding, then one does obtain zero (for almost every choice of $p \in S^1$) if f is homotopic to a constant. (See Exercise 2 for a precise version of this statement that applies to a wide class of maps.)

There are some serious problems in attempting to formulate a precise version of this idea for general continuous maps $f : S^1 \to S^1$. For one thing, it may happen that $f^{-1}(\{p\})$ is an infinite set for every $p \in S^1$ (Exercise 3). It turns out, however, that these difficulties evaporate almost completely if one restricts attention to C^∞ maps.

This restriction involves no real loss of homotopic information since every map is homotopic to a C^∞ map. For C^∞ maps $f : S^1 \to S^1$, the following theorem holds and is even quite easy to prove, using the method of Exercise 4.

5.1 Theorem: If $f : S^1 \to S^1$ is a C^∞ map, then there is a closed nowhere-dense subset C of S^1 such that each $p \in S^1 \backslash C$ has the following properties:

(a) $f^{-1}(\{p\})$ is finite,

(b) $\dfrac{d}{d\theta} \arg f(e^{i\theta})$ does not vanish at any point of $f^{-1}(\{p\})$,

(c) the number of points of $f^{-1}(\{p\})$ at which $\dfrac{d}{d\theta} \arg f(e^{i\theta})$ is positive, minus the number at which it is negative, equals the index of f (as defined in Chapter III).

The theorem just stated is interesting in its own right, and it formalizes an intuition that is quite convincing. However, its greatest interest derives from the following observation. Suppose that we did not already know what the group $\pi_1(S^1)$ is, but that we could show that the number of "derivative positive" points in $f^{-1}(\{p\})$ minus the number of "derivative negative" points in $f^{-1}(\{p\})$ depended only on the homotopy class of f. Then we could infer, for instance, that the maps $e^{i\theta} \to e^{ik\theta}$ and $e^{i\theta} \to e^{il\theta}$ are not homotopic if $k \neq l$.

Now we seek to adapt this approach in order to describe the groups $\pi_n(S^n)$ for $n \geq 2$. At this point we know that $\pi_n(S^n)$ has at least two elements since S^n is not contractable, but we do not know what the group is nor even whether it is finite or infinite.

Again the basic strategy will be to count the number of points in $f^{-1}(p)$, with appropriate signs, and to show that this number depends only on the homotopy class of f. As before, attention is restricted to maps $f : S^n \to S^n$ which are C^∞. In order to find a replacement for the condition on the derivative of the argument $\arg f(e^{i\theta})$, we proceed as follows.

Extend f to the punctured $(n + 1)$-space $\mathbb{R}^{n+1}\backslash\{0,\ldots,0)\}$ by declaring the extension to be linear on rays issuing from the origin, that is, by setting

$$\hat{f}(v) = \|v\| f(v/\|v\|), \qquad v \in \mathbb{R}^{n+1},\ v \neq (0,\ldots,0).$$

Then $\hat{f}$ is a C^∞ map of $\mathbb{R}^{n+1}\backslash\{(0,\ldots,0)\}$ that maps each sphere of radius r, centered at the origin, to itself. Thus $\hat{f}$ preserves direction and distances in the "radial direction." A point $p \in S^n$ is said to be a *nonsingular point for* f if the Jacobian of $\hat{f}$ is nonzero at p; otherwise p is a *singular point for* f. If the Jacobian of f is positive at a nonsingular point p, then p is said to be an *orientation-preserving point* for f; if it is negative, p is an *orientation-reversing point* for f. A point $q \in S^n$ is a *critical point of* f if $q = f(p)$ for some singular point p of f. The point $q \in S^n$ is a *noncritical point of* f if every point in $f^{-1}(\{q\})$ is a nonsingular point of f. Finally, if $q \in S^n$ is a noncritical point for f, then the *degree of* f *at* q, denoted by $\deg(f,q)$, is the number of orientation-preserving points in $f^{-1}(\{q\})$ minus the number of orientation-reversing points in $f^{-1}(\{q\})$. With this terminology, we now state the following (true!) theorem, which generalizes Theorem 5.1.

5.2 Theorem: If $f : S^n \to S^n$ is a C^∞ map, then the set of noncritical points of f is an open dense subset of S^n and the degree function $q \to \deg(f,q)$ is constant on the set of noncritical points of f. Moreover, if $g : S^n \to S^n$ is another C^∞ map homotopic to f, then the constant value of $\deg(f,q)$, q noncritical for f, coincides with the constant value of $\deg(g,q')$, q' noncritical for g.

Of course, the noncritical sets of f and q may be different, though by the Baire Category Theorem they have dense intersection. Theorem 5.2 allows us to define unambiguously the *degree* of a C^∞ map $f : S^n \to S^n$ to be the value of $\deg(f,q)$ on the set of noncritical points of f; the degree of f is denoted by $\deg(f)$. The second assertion of Theorem 5.2 then states that $\deg(f) = \deg(g)$ whenever f and g are homotopic. Thus "deg" determines a function from $\pi_n(S^n)$ to the integers $\mathbb{Z}$, assigning a degree to each homotopy class. It is quite easy to see that this degree map is a group homeomorphism from $\pi_n(S^n)$ to $\mathbb{Z}$. It is also easy to see that all integers occur as degrees. It is true, though far from obvious (and not to be proved here), that any two maps with the same degree are homotopic. Thus, the following is true.

5.3 Theorem: The degree map defined above is an isomorphism from $\pi_n(S^n)$ to the group $\mathbb{Z}$ of integers.

Another approach to defining degrees, which fortunately yields the same answers, is to pursue the line of thought used in Section 2, based on certain integrals over S^n. Specifically, one would then define, for a C^∞ map $f : S^n \to S^n$, the degree of f to be

$$\frac{1}{\mathrm{Vol}(S^n)} \int_{S^n} J_{\hat{f}}\, dV,$$

where $J_{\hat{f}}$ is the Jacobian determinant of $\hat{f}$, dV is the Euclidean n-dimensional volume (area) element on S^n, and

$$\mathrm{Vol}(S^n) = \int_{S^n} dV$$

is a normalizing factor chosen so that the identity map has degree 1. It is true though not obvious that the degree thus defined in terms of integrals is the same as that defined earlier for every C^∞ map $f : S^n \to S^n$. A more detailed consideration of this idea would lead us into the subject area of the cohomology of differential forms, or deRham cohomology. The reader is invited to consider how to modify the argument in Section 2 to show that the integral degree is the same for two homotopic maps. (The argument as actually given there shows directly only that the integral degree of any map homotopic to a constant is zero.) With this information in hand, the verification that the two ideas of degree coincide would need to be carried out only for one map in each homotopy class of maps from S^n to S^n, since both concepts are homotopy invariants. It is easy to check this coincidence of the two concepts for certain standard maps which realize each integer as degree (in the original sense). Thus, if one assumes that maps of equal degree (in the first sense) are homotopic, the coincidence of the two ideas of degree follows in all cases.

Before turning to the idea of the proof of Theorem 5.2, we present an application of the theorem, which generalizes Theorem III.9.2.

Recall that a (tangent) *vector field* V on S^n is a continuous function $V : S^n \to \mathbb{R}^{n+1}$ with the property that $\langle p, V(p)\rangle = 0$ for every $p \in S^n \subset \mathbb{R}^{n+1}$, where $\langle \cdot,\cdot \rangle$ denotes the inner product on $\mathbb{R}^{n+1}$. On a sphere of odd dimension, there is always a tangent vector field that is nowhere zero. For example, the vector field V defined when n is odd by

$$V(x_1,\ldots,x_{n+1}) = (-x_2,x_1,-x_4,x_3,\ldots,-x_{n+1},x_n)$$

has this property. As we have seen, the situation is different for S^2, and in fact the next theorem asserts that every tangent vector field on any even-dimensional sphere has a zero.

5.4 Theorem: If n is an even positive integer and if $V : S^n \to \mathbb{R}^{n+1}$ is a tangent vector field on S^n, then there is a point $p \in S^n$ such that $V(p) = (0,\ldots,0)$.

Proof: Suppose that there were a nowhere-vanishing tangent vector field $V : S^n \to \mathbb{R}^{n+1}\backslash\{(0,\ldots,0)\}$. Then there would be a homotopy from the identity map of S^n to the antipodal map $A : S^n \to S^n$ defined by $A(x_1,\ldots,x_{n+1}) = (-x_1,\ldots,-x_{n+1})$. The homotopy could be obtained by moving each point $p \in S^n$ along the great circle with tangent direction $V(p)$ until $-p$ is reached (Exercise 6). Thus to prove Theorem 5.4 it is enough to show that A is not homotopic to the identity. To find the degree of A, note that $\hat{A}$ is just the map $\mathbb{R}^{n+1}\backslash\{(0,\ldots,0)\} \to \mathbb{R}^{n+1}\backslash\{(0,\ldots,0)\}$ that takes $(x_1,\ldots,x_{n+1})$ to $(-x_1,\ldots,-x_{n+1})$. The Jacobian determinant of this map is everywhere $(-1)^{n+1}$. Thus every point of S^n is noncritical for A, and for each $q \in S^n$, $A^{-1}(\{q\})$ is a single point at which A is orientation-reversing if n is even. Hence $\deg(A) = -1$ if n is even. Since the identity map has degree 1, Theorem 5.2 shows that A is not homotopic to the identity map. Thus Theorem 5.4 is proved. □

Note that when n is odd, $\deg(A) = 1$ and A is homotopic to the identity, as required for consistency with the already observed existence of nowhere-zero tangent vector fields on S^n, n odd.

The precise proof of Theorem 5.2, the vector field application of which has just been discussed, requires disposing of some rather difficult technical points. Some of these points we shall simply omit (and refer the reader to other sources); others we shall treat in exercises. If these technical points are assumed handled, however, then the proof of the theorem is quite straightforward in concept. The first thing to note is that the set C of critical points of a C^∞ map $f : S^n \to S^n$ is closed; this is easy (Exercise 7). To see that the set C is nowhere dense, it is enough to see that it has zero n-dimensional volume. To show this on a precise level, it is convenient to use the concepts of Lebesgue measure theory, but the intuition of the situation is almost obvious. Since the Jacobian is zero at a singular point, the Jacobian is arbitrarily small (less than ε, for $\varepsilon > 0$ given) on a neighborhood of the singular point, so that the image of such a neighborhood has a volume that is a small multiple of the volume of the neighborhood itself. Since the total volume of such neighborhoods can be taken not to exceed the volume of S^n, it follows that for any $\varepsilon > 0$, the set C is contained in a set of volume $\varepsilon \mathrm{Vol}(S^n)$. Thus C can have no interior! Here we have used the fact that $\hat{f}$ is isometric in the radial direction, so that the size of the Jacobian determinant of $\hat{f}$ controls the n-dimensional volume behavior of f on S^n (Exercise 8).

To investigate the situation when $f : S^n \to S^n$ and $g : S^n \to S^n$ are homotopic, suppose $F : S^n \times [0,1] \to S^n$ is a C^∞ homotopy,

$$F(v,0) = f(v), \quad F(v,1) = g(v), \qquad v \in S^n.$$

(The existence of a continuous homotopy from f to g implies the existence of a C^∞ one; see Exercise 9.) Then volume reasoning again shows that there is an open dense set of points in S^n such that, if q is in this set, then the Jacobian matrix of $\hat{F}$ has rank $n + 1$ at every point of $F^{-1}(\{q\})$, where

$$\hat{F}(v,t) = \|v\|F(v/\|v\|,t), \qquad v \in \mathbb{R}^{n+1}\backslash\{(0,\ldots,0)\},\ t \in [0,1].$$

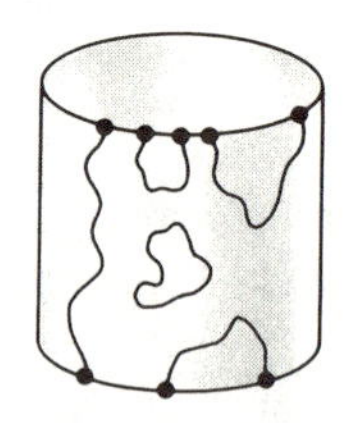

It follows from the Implicit Function Theorem that then $F^{-1}(\{q\})$ is a finite union of C^∞ curves in $S^n \times [0,1]$ which either are simple closed curves or else have endpoints in $(S^n \times \{0\}) \cup (S^n \times \{1\})$. The situation for S^1 is illustrated in the figure.

The counting of points in $f^{-1}(\{q\})$ can now be related to the counting for $g^{-1}(\{q\})$ as follows. Every point p_1 in $f^{-1}(\{q\})$ is the endpoint of a curve in $F^{-1}(\{q\})$. If this curve terminates in $F^{-1}(\{q\})$, in say p_2, then it is easy to see that the orientations of f at p_1 and p_2 are opposite, so that p_1 and p_2 cancel in the counting that computes $\deg(f,q)$. On the other hand, if the curve from p_1 terminates in $S^n \times \{1\}$, in say p_2, then p_2 occurs in the counting for $\deg(g,q)$ with the same orientation as p_1 in $\deg(f,q)$. It follows that $\deg(f,q) = \deg(g,q)$. It is straightforward, by the same type of reasoning, to see that $\deg(f,q)$ is independent of the choice of q in the noncritical set of f. Thus Theorem 5.2 is proved at least in outline.

The proof sketch just given is of course closely related to the actual proof in Section 4 of the noncontractibility of S^n. Essentially only two changes have been made. First, piecewise linear maps have been replaced by C^∞ maps. This change is only for convenience and intuitive clarity; the same reasoning can be carried out on the piecewise linear level. Second, orientations have been taken into account. This could also be done on the piecewise linear level. What was actually done in Section 4 was to count "mod 2," that is, to keep track of only evenness or oddness of the number of points in $f^{-1}(\{p\})$. In the mod 2 context, $+1$ and -1 are of course the same, so that for mod 2 degree, orientation can be ignored!

The concept of degree has already been seen to yield much information about maps of S^n to S^n. It turns out that, in addition, this concept can be used to analyze maps of a much wider class of topological spaces. The topological spaces to which it applies are called (compact) *manifolds*. Informally speaking, a manifold is a topological space which looks locally like a Euclidean space. More precisely, an *n-dimensional manifold* is a topological space such that for each point of the space there is an open set containing the point that is homeomorphic to an open set in $\mathbb{R}^n$. The "mod 2" degree idea applies to maps from any compact n-dimensional manifold to any other. To define the integer-valued degree, where orientations of points of $f^{-1}(\{q\})$ are counted, the manifolds themselves must be "orientable." An *oriented manifold* is one in which the homeomorphisms in the definition of manifold can be and have been chosen to be coherent with respect to the orientation of $\mathbb{R}^n$ in the sense that if $\varphi : U \to \varphi(U)$ and $\psi : V \to \varphi(V)$ are two such coordinatizing homeomorphisms of open subsets of M onto open subsets of $\mathbb{R}^n$, then $\psi \circ \varphi^{-1} : \varphi(U \cap V) \to \psi(U \cap V)$ is orientation-preserving in a suitable sense. The sense that is appropriate in this general setting is explained in Exercise 10. A manifold is *orientable* if it can be given an oriented manifold structure. For an example of a manifold that is *not* orientable, see Exercise 11.

An n-dimensional manifold M is said to be a *differentiable manifold* if the coordinatizing homeomorphisms in the definition of a manifold can be chosen so that whenever φ and ψ are such homeomorphisms, then the "overlap map" $\psi \circ \varphi^{-1}$ is a C^∞ map where defined (i.e., on $\varphi(U \cap V)$). This concept is a natural generalization of the idea of a smooth surface in $\mathbb{R}^3$. For a differentiable manifold, the maps $\psi \circ \varphi^{-1}$ are orientation-preserving whenever $\psi \circ \varphi^{-1}$ has a positive Jacobian everywhere on its domain of definition. Thus a collection of coordinatizing homeomorphisms for a manifold M with C^∞ "overlap maps" makes M an oriented manifold if all the overlap maps have everywhere-positive Jacobian. (The following converse statement is true, but much less obvious: If M is a differentiable manifold and if M is orientable in the topological sense, then M is orientable in the differentiable sense that M can be covered by coordinatizing homeomorphisms with C^∞ overlaps and with everywhere-positive Jacobian.) As an example, it is easy to see that S^n is an orientable differentiable manifold (Exercise 12), as is any product of orientable differentiable manifolds, such as the torus $S^1 \times S^1$.

For compact differentiable manifolds, there are vector field results which generalize the results about tangent vector fields on S^n. First, one notes that there is a natural idea of a C^∞ vector field on a differentiable manifold (Exercise 13). Then one considers vector fields with isolated zeros, i.e., vector fields V on a compact manifold M such that $\{p \in M : V(p) = 0\}$ is a discrete and hence finite set. Suppose p is a point such that $V(p) = 0$. Choose a C^∞ coordinatizing homeomorphism $\varphi : U \to \mathbb{R}^n$ for an open set U containing p. Then V determines a C^∞ vector field $\tilde{V}$ on $\varphi(U) \subseteq \mathbb{R}^n$ with $\tilde{V}(\varphi(p)) = 0$. Choose $\varepsilon > 0$ so small that $\varphi(U)$ includes the closed ball in $\mathbb{R}^n$ of radius ε centered at $\varphi(p)$ and set

$$\sigma_\varepsilon(v) = \tilde{V}(\varphi(p) + \varepsilon v)/\|\tilde{V}(\varphi(p) + \varepsilon v)\|, \qquad v \in S^n.$$

Then σ_ε maps S^n to S^n, so that we may define the index of V at p to be the degree of σ_ε, provided this degree is independent of the choice of ε. For all sufficiently small

$\varepsilon > 0$, the various maps σ_ε are homotopic, so that the index of V at p indeed does not depend on ε. The index of V at p is also independent of the homeomorphism coordinatizing a neighborhood of p, providing the homeomorphism is C^∞.

It is a remarkable result proved by H. Poincaré and H. Hopf that the sum of the indices of the vector field V over all the zeros of V is independent of which vector field V (with isolated zeros) is considered. This number depends only on the topology of the manifold; it is called the *Euler-Poincaré characteristic* of the manifold. The Euler-Poincaré characteristic of S^n is $1 + (-1)^n$, i.e., 0 if n is odd and 2 if n is even (Exercise 14). This shows the relationship to Theorem 5.4. A vector field without zeros has sum of indices equal to 0 (since an empty sum is zero), but for n even, every vector field on S^n has index sum 2. Hence every vector field on S^n, n even, has a zero.

The Poincaré-Hopf Theorem relates local behavior of some differentiable object, namely C^∞ vector fields, to the global topological structure of the manifold. Numerous theorems of this general type, relating local invariants to global structure, have been discovered. These form a large but unified subject, called *global differential geometry* or *differential topology*, depending on the particular emphasis of the theorems. It is one of the most interesting and important parts of modern mathematics.

EXERCISES

1. Fix an integer k and define a function $f : S^1 \to S^1$ by $f(e^{i\theta}) = e^{2ki\theta}$ if $0 \le \theta < \pi$ and $f(e^{i\theta}) = e^{-2ki\theta}$ if $\pi \le \theta < 2\pi$. Show that f is continuous and homotopic to a constant map. For each $e^{i\theta} \in S^1$, how many points are there in $f^{-1}(\{e^{i\theta}\})$?
2. Suppose $f : S^1 \to S^1$ is a C^∞ mapping such that (for convenience) $f((1,0)) = (1,0)$. Suppose also that $\frac{d}{d\theta} \arg f(e^{i\theta})$ vanishes at only finitely many points of S^1. Suppose finally that f is homotopic to a constant mapping. Prove that, for any $p \in S^1$ that is not the image under f of one of the (finitely many) points at which $\frac{d}{d\theta} \arg f(e^{i\theta}) = 0$, the number of points in $f^{-1}(\{p\})$ is finite. Prove also that for such p the number of points in $f^{-1}(\{p\})$ at which $\frac{d}{d\theta} \arg f(e^{i\theta}) > 0$ equals the number of points of $f^{-1}(\{p\})$ at which $\frac{d}{d\theta} \arg f(e^{i\theta}) < 0$. *Hint:* Think of f as a map from $[0,1]$ to S^1 with $f(0) = f(1)$. Then consider the "lift" of f to a map $h : [0,1] \to \mathbb{R}$, where $e^{ih(t)} = f(t)$, as in the discussion of covering spaces in Section III.5. Then $h(0) = h(1) = 0$ because h is homotopic to a constant. Now look at the graph of h and count the number of points where the graph hits a fixed horizontal line (or a fixed finite collection of horizontal lines that are 2π apart).

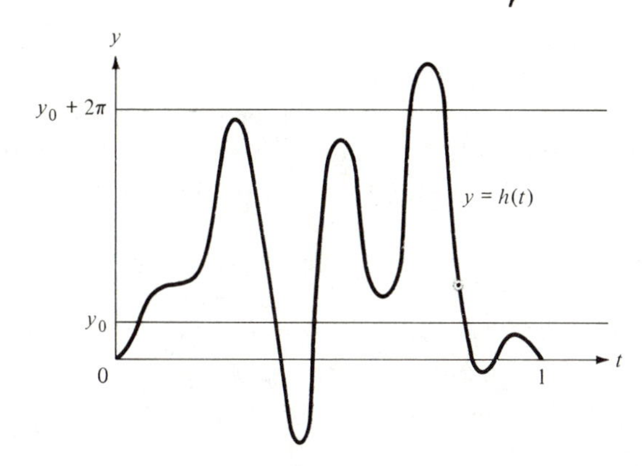

3. Construct a map f of S^1 to S^1 such that $f^{-1}(\{q\})$ is an infinite set for every $q \in S^1$. *Hint:* Recall the Peano curve $h : [0,1] \to [0,1] \times [0,1]$ described in Exercise II.6.11. Consider maps of the form $f(e^{i\theta}) = e^{2\pi i g}$, where g is the x-component of $h(\theta/2\pi)$.
4. Establish the following for a C^∞ map from S^1 to S^1:
   (a) If $f^{-1}(\{p\})$ is infinite for some $p \in S^1$, then there is a point of $f^{-1}(\{p\})$ at which $\dfrac{d}{d\theta}$ arg $f(e^{i\theta})$ vanishes. *Hint*: Look at a limit point of a sequence of distinct points of $f^{-1}(\{p\})$.
   (b) If A is the set of points in S^1 at which $\dfrac{d}{d\theta}$ arg $f(e^{i\theta})$ vanishes, then $f(A)$ is a closed set with empty interior. *Hint:* Cover A with a finite number of open arcs of total length at most 2π such that $\left|\dfrac{d}{d\theta} \arg f(e^{i\theta})\right| < \varepsilon/2\pi$ at every point of each of the arcs. Then obtain a cover of $f(A)$ by a finite number of arcs of total length less than ε.
5. Show that $f^{-1}(\{q\})$ is finite if q is a noncritical point of a C^∞ map $f : S^n \to S^n$. *Hint:* Show that $f^{-1}(\{q\})$ has no limit points by applying the Inverse Function Theorem to the radial extension $\hat{f}$ of f.
6. Show that if there is a tangent vector field $V : S^n \to \mathbb{R}^{n+1}$ on S^n, then the antipodal map is homotopic to the identity map of S^n to S^n. *Hint:* In the case $|V| = 1$ on S^n, consider

$$H(p,t) = (\cos(\pi t))p + (\sin(\pi t))V(p), \qquad p \in S^n,\ 0 \le t \le 1.$$

7. Prove that the set of critical points of a C^∞ map $f : S^n \to S^n$ is closed.
8. Suppose that f is a C^∞ map from S^n to S^n and that $\hat{f}$ is the extension of f, defined in this section. Prove that the Jacobian determinant J of $\hat{f}$ satisfies $J(\lambda v) = J(v)$ for all positive $\lambda \in \mathbb{R}$ and all $v \in \mathbb{R}^{n+1} \backslash \{(0, \ldots, 0)\}$. Deduce that the set of critical points of f on S^n has empty interior in S^n if and only if the set of critical points of $\hat{f}$ (i.e., the image under $\hat{f}$ of the zero set of J) has empty interior in $\mathbb{R}^{n+1}$.

9. Prove that if $f,g : S^n \to S^n$ are C^∞ maps that are homotopic (as continuous maps), then they are C^∞-homotopic, i.e., there is a C^∞ map $H : S^n \times [0,1] \to S^n$ such that $H(x,0) = f(x)$ and $H(x,1) = g(x)$ for all $x \in S^n$. *Hint*: Using Exercise 2.2, first construct an auxiliary C^∞ function $\rho_\varepsilon : [0,1] \to [0,1]$, for $\varepsilon > 0$ small, such that $\rho_\varepsilon(t) = 0$ if $\varepsilon \leq t \leq 1 - \varepsilon$ and $\rho_\varepsilon(t) = 1$ if $0 \leq t \leq \varepsilon/2$ or if $1 - \varepsilon/2 \leq t \leq 1$. Then let F be a continuous homotopy of f and g, approximate F by a function $G(x,t)$ that is a polynomial in $x_1, \ldots, x_n$ and t, and set $H_0(x,t) = \rho_\varepsilon(t)f(x) + [1 - \rho_\varepsilon(t)]G(x,t)$, $H(x,t) = H_0(x,t)/|H_0(x,t)|$, for $x \in S^n$ and $0 \leq t \leq 1/2$, with a similar formula for $1/2 \leq t \leq 1$.

10. Let $U,V \subseteq \mathbb{R}^n$ be connected open sets and let $f : U \to V$ be a homeomorphism. For each p in U, choose an $\varepsilon > 0$ such that $B(p;2\varepsilon) \subset U$. Consider the open set $f(B(p;2\varepsilon)) \subset V$ and choose $\delta > 0$ such that $B(f(p),2\delta) \subset f(B(p;2\varepsilon))$. Define a map $\sigma : S^n \to B(p;2\varepsilon)\backslash\{p\}$ by $\sigma(x) = f^{-1}(f(p) + \delta x)$.
    (a) Show that $\sigma : S^n \to B(p;2\varepsilon)\backslash\{p\}$ is not homotopic to a constant map of S^n into $B(p;2\varepsilon)\backslash\{p\}$. Deduce that the degree of the map $x \to (\sigma(x) - p)/\|\sigma(x) - p\|$ is not zero.
    (b) Show that this degree is independent of the choices of δ and ε.

Remark: The map f is *orientation-preserving* at p if this degree is positive and *orientation-reversing* at p if this degree is negative.

   (c) Show that either f is orientation-preserving at every point of U or f is orientation-reversing at every point of U.
   (d) Show that the composition of two orientation-preserving homeomorphisms is orientation-preserving and that the inverse of an orientation-preserving homeomorphism is orientation-preserving.
   (e) Show that if f is C^∞ with a C^∞ inverse, then f is orientation-preserving at p if and only if the Jacobian determinant of f at p is positive.

11. Let X be the quotient space obtained from the semiopen square $\{(x,y) : 0 \leq x \leq 1, 0 < y < 1\}$ by identifying $(0,y)$ with $(1,1 - y)$, $0 < y < 1$. Prove that X is a differentiable manifold. Prove that X is not orientable.

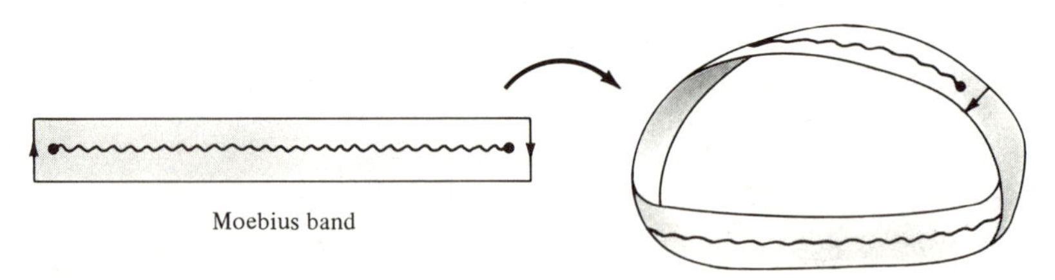

Moebius band

Remark: The manifold X is known as a *Moebius band*.

12. Prove that S^n is a differentiable manifold and that it is orientable. *Hint*: To coordinatize open subsets of S^n, consider maps from subsets of S^n to $\mathbb{R}^n$ of the form $(x_1, \ldots, x_{n+1}) \to (x_1, \ldots, x_{k-1}, x_{k+1}, \ldots, x_{n+1})$.

13. Suppose that M is a differentiable manifold, $\{U_\alpha\}_{\alpha \in A}$ is an open cover of M, and $\psi_\alpha : U_\alpha \to \mathbb{R}^n$ are homeomorphisms such that $\psi_\alpha \circ \psi_\beta^{-1}$ is a C^∞ map

where defined (on $\psi_\alpha(U_\alpha \cap U_\beta)$). A *vector field on M* is a collection of vector fields $V_\alpha : \psi_\alpha(U_\alpha) \to \mathbb{R}^n$ on the open sets $\psi_\alpha(U_\alpha)$ with the property that for every $\alpha,\beta \in A$ and every $p \in \psi_\alpha(U_\alpha \cap U_\beta)$, the image of $V_\alpha(p)$ under the Jacobian linear mapping of $\psi_\beta \circ \psi_\alpha^{-1}$ is $V_\beta(\psi_\beta(\psi_\alpha^{-1}(p)))$. Show that a tangent vector field on S^n in the sense of the text is associated to a tangent vector field on S^n considered as a differentiable manifold as in Exercise 12 by the following association. If $V : S^n \to \mathbb{R}^{n+1}$ is a tangent vector field (in the sense of the text) and if $\psi : U \to \mathbb{R}^{n+1}$, $U \subset S^n$, is a coordinatizing homeomorphism, then set $V_\psi(p)$, $p \in \psi(U)$, equal to the tangent vector at $t = 0$ to the curve $\gamma(t)$ defined by $\gamma(t) = \psi(\gamma_1(t))$, where $\gamma_1(t) = (p + tV(p))/\|p + tV(p)\|$.

14. Show that the Euler characteristic of S^n is zero if n is odd and 2 if n is even. *Hint:* For n odd, there is a tangent vector field on S^n without zeros. For n even, consider the tangent vector field V defined by

$$V(x_1,\ldots,x_{n+1}) = (-x_1x_{n+1},\ldots,-x_nx_{n+1},1 - x_{n+1}^2), \qquad x \in S^n.$$

Show that it is a tangent vector field and find the indices of its zeros.

Bibliography

N. Bourbaki, *Topologie Générale*. Paris: Hermann, 1953.

W. Chinn and N. Steenrod, *First Concepts of Topology*. New York: Random House, 1966.

S. Eilenberg and N. Steenrod, *Foundations of Algebraic Topology*. Princeton, N.J.: Princeton University Press, 1952.

M. Greenberg, *Lectures on Algebraic Topology*. New York: Benjamin, 1967.

V. Guillemin and A. Pollack, *Differential Topology*. Englewood Cliffs, N.J.: Prentice-Hall, 1974.

J. Hocking and G. Young, *Topology*. Reading, Mass,: Addison-Wesley, 1961.

S. Hu, *Homotopy Theory*. New York: Academic Press, 1959.

J. Kelley, *General Topology*. New York: Van Nostrand, 1955; New York: Springer-Verlag, 1975.

J. Milnor, *Topology from a Differential Viewpoint*. Charlottesville: University Press of Virginia, 1965.

J. Munkres, *Topology: A First Course*. Englewood Cliffs, N.J.: Prentice-Hall, 1975.

I. Singer and J. Thorpe, *Lecture Notes on Elementary Topology and Geometry*. New York: Springer-Verlag, 1976.

E. Spanier, *Algebraic Topology*. New York: McGraw-Hill, 1966.

J. Vick, *Homology Theory*. New York: Academic Press, 1973.

A. Wallace, *An Introduction to Algebraic Topology*. Elmsford, N.Y.: Pergamon Press, 1957.

List of Notations

$\cup$	set theoretic union
$\cap$	set theoretic intersection
$S \backslash T$	set of points in S that do not belong to T (I.1)
$\varnothing$	empty set
$\Pi_{\alpha \in A} X_\alpha$, ΠX_α	product of spaces X_α (II.12)
π_α	projection map onto coordinate space X_α (II.12)
$X/\sim$	quotient space of X (II.13)
$\mathbb{R}$	real numbers
$\mathbb{R}^n$	n-dimensional Euclidean space
$\mathbb{C}$	complex numbers
$\mathbb{Z}$	integers
$\mathbb{Z}_m$	congruence classes of integers modulo m (III.1)
$\mathbb{R}_0$	rational numbers
B^n	closed unit ball in $\mathbb{R}^n$
S^n	n-sphere, boundary of B^{n+1}
I^n	n-cube (IV.1)
T^n	n-torus, product of n circles
P^n	real projective space (II.13)
CP^n	complex projective space (II.13)
$B(x;r)$	open ball with center x and radius r (I.1)
$\text{int}(Y)$	interior of Y (I.1, II.1)
$\overline{Y}$	closure of Y (I.1, II.1)
∂E	boundary of E (I.1, II.1)
$\\|\cdot\\|$	norm (I.7)
l^p	p-summable sequences (I.7)
$\\|\cdot\\|_p$	p-norm (I.7)
$\mathcal{B}(\mathfrak{X},\mathfrak{Y})$	bounded operators from $\mathfrak{X}$ to $\mathfrak{Y}$ (I.7)

$G'(x)$	Frechet derivative (I.9)
C^k	k times continuously differentiable (I.9)
C^∞	continuous derivatives of all orders (IV.2)
$G \times H$	product of groups (III.1)
$[\gamma]$	homotopy class of γ (III.2)
$\alpha\beta$	path product (III.2)
α^{-1}	path inverse (III.2)
(X,b)	pointed space (III.3)
$\pi_1(X,b)$, $\pi_1(X)$	fundamental group (III.3)
$\pi_n(X,b)$	homotopy groups (IV.1)
$\gamma_0 \cong \gamma_1 \text{ rel}\{0,1\}$	homotopic paths, endpoints fixed (III.2)
$f \simeq g \text{ rel} A$	homotopic maps relative to A (III.7)
α_*	isomorphism induced by α (III.3)
f_*	homomorphism induced by f (III.4, IV.1)
$\text{ind}(\gamma)$	index of path γ (III.5)
$\text{ind}(f)$	index of f (III.8)
$\deg(f)$	degree of f (IV.5)
$\hat{f}$	extension of f used to define degree (IV.5)
$\langle v_0, \ldots, v_k \rangle$	k-simplex (IV.3)
$\mathcal{B}^{(1)}$, $\mathcal{B}^{(N)}$	barycentric subdivisions (IV.3)

SOLUTIONS TO SELECTED EXERCISES

Chapter I. Metric Spaces

Section I.1. Open and Closed Sets

1. The typical proof that two sets A and B coincide breaks into two steps: first show that if $x \in A$ then $x \in B$; then show that if $x \in B$ then $x \in A$. For (a), suppose $x \in (U \cup V)\backslash W$. Then either $x \in U$ or $x \in V$, while $x \notin W$. Thus either $x \in U\backslash W$ or $x \in V\backslash W$. Consequently $x \in (U\backslash W) \cup (V\backslash W)$. This shows that $(U \cup V)\backslash W \subseteq (U\backslash W) \cup (V\backslash W)$. For the reverse inclusion, suppose $x \in (U\backslash W) \cup (V\backslash W)$. Then either $x \in (U\backslash W)$ or $x \in (V\backslash W)$. In either case, $x \in (U\backslash W) \cup (V\backslash W)$. Thus $(U\backslash W) \cup (V\backslash W) \subseteq (U \cup V)\backslash W$. Since each is contained in the other, the two sets coincide. Parts (b) and (c) are similar. The two set-theoretic identities referred to in (d) are called *de Morgan's formulas.* To prove the first, note that $x \in X\backslash \cup E_\alpha$ if and only if $x \in X$ while $x \notin E_\alpha$ for all α. This occurs if and only if $x \in X\backslash E_\alpha$ for all α, which occurs if and only if $x \in \cap(X\backslash E_a)$. This establishes the first identity. The proof of the second identity is similar.

2. The metric properties are trivial to verify. Since each ball $B(x; 1/2)$ reduces to the singleton set $\{x\}$, every subset is a union of open balls, hence every subset is open.

3. Follow the hint.

4. For any $\varepsilon > 0$, the ball $B(1; \varepsilon)$ meets the complement of $(0, 1]$, so the set is not open. By the same token, each ball $B(0; \varepsilon)$ meets $(0, 1]$, so the complement of $(0, 1]$ is not open, and $(0, 1]$ is not closed.

5. For each $s \in S$, $|f(s) - h(s)| \leq |f(s) - g(s)| + |g(s) - h(s)| \leq d(f, g) + d(g, h)$. Taking the supremum over $s \in S$, we obtain the triangle inequality for d. The remaining metric properties are trivial to verify. For a generalization, see Exercise 6.10(a).

6. Let U be the union of all open sets contained in Y. Then U is open and $U \subset Y$. If $x \in U$, there is a ball centered at x and contained in U, hence in Y. Hence each $x \in U$ is an interior point of Y, and U is contained in the interior of Y. On the other hand, the interior of Y is an open set contained in Y, so U contains the interior of Y, consequently U coincides with the interior of Y.

7. Proceed in analogy with Exercise 6.

8. Denote $E = \{x : d(x, x_0) \leq r\}$. Suppose $z \in X\backslash E$, and set $\delta = d(z, x_0) - r > 0$. If $x \in B(z, \delta)$, then $d(x, x_0) \geq d(z, x_0) - d(x, z) > d(z, x_0) - \delta = r$, so $x \notin E$ and $B(z, \delta) \subset X\backslash E$. Thus $X\backslash E$ contains a ball about each of its points, hence $X\backslash E$ is open, and E is closed. The statement about open balls being dense in closed balls holds in $\mathbf{R}^n$, but it does not hold in metric spaces in general. For example, if d is the discrete metric given by $d(x, y) = 1$ for all $x \neq y$, then any open ball of radius one contains only one point, its center, while any closed ball of radius one coincides with the entire space.

9. Use Theorem 1.11, that an element belongs to the closure of S if and only if it is the limit of a sequence in S.

10. For the first statement, if x is a limit point of S, let x_k be any point in $S \cap B(x; 1/k)$ such that $x_k \neq x$. For the second statement, denote by E the set of limit points of S. If $y \notin E$, then there is a open ball B containing y such that B contains only finitely many points of S. Then no point of B can be a limit point of S, that is, B is contained in the

complement of E. Hence the complement of E is open, and E is closed.

11. Supppose x is not a limit point of S. Then there is $\varepsilon > 0$ such that $B(x;\varepsilon)$ contains only finitely points of S. By shrinking ε, we can assume that $B(x;\varepsilon)$ contains no points of S other than possibly x itself. If x is in the closure of S, then this ball meets S, and consequently x itself belongs to S, in which case x is an isolated point of S.

12. A sequence $\{x_n\}$ converges in a metric space to x if and only if given any open set U containing x, there exists $N \geq 1$ such that $x_n \in U$ for $n \geq N$. It follows that two metric spaces with the same open sets have the same convergent sequences. On the other hand, from Theorem 1.11 we see that a set in a metric space is closed if and only if it contains the limit of any sequence in the set that converges. Thus if two metrics have the same convergent sequences, then they have the same closed sets, consequently they have the same open sets.

13. It is trivial to check that $\rho(x_n, x) \to 0$ if and only if $d(x_n, x) \to 0$. Hence the metrics have the same convergent sequences. By Exercise 12, they are equivalent.

14. Suppose E is open. Then $X \backslash E$ is closed, and no point of E is adherent to $X \backslash E$. Hence $E \cap \overline{X \backslash E} = \emptyset$, and $E \cap \partial E = \emptyset$. Conversely, suppose $E \cap \partial E = \emptyset$. Then no point of E is adherent to $X \backslash E$, consequently $X \backslash E$ is closed, and E is open. The proof of the second statement is similar.

Section 2. Completeness

1. By the triangle inequality, if $n > m$ then $d(x_m, x_n) \leq d(x_m, x_{m+1}) + \cdots + d(x_{n-1}, x_n)$, and this tends to 0 as $m, n \to \infty$.

2. Choose $n_k > n_{k-1}$ so that $d(x_{n_k}, x_m) < 1/2^k$ for all $m > n_k$.

3. For each point $x \in X$ that is *not* an isolated point of X, define $U_x = X \backslash \{x\}$. Each such U_x is open and dense in X, and the intersection of the U_x's consists precisely of the isolated points of X. By the Baire Category Theorem, the intersection of the U_x's is dense in X.

4. If $F = \cup F_n$, where each F_n is nowhere dense, then $E = \cup(E \cap F_n)$, and each $E \cap F_n$ is nowhere dense. For the example, note the $\mathbf{R}$ is of first category in $\mathbf{R}^2$, but $\mathbf{R}$ is not of first category in itself.

5. A countable union of countable unions is a countable union.

6. The metric properties are straightforward to verify. For the triangle inequality, first show that $\min(1, d(f(s), h(s))) \leq \min(1, d(f(s), g(s))) + \min(1, d(g(s), h(s)))$ for each fixed s, then take the supremum over S. The characterization of convergence follows immediately from the observation that if $\varepsilon < 1$, then $\rho(f_n, f) \leq \varepsilon$ if and only if $d(f_n(s), f(s)) \leq \varepsilon$ for all $s \in S$. Suppose X is complete. Let $\{f_n\}$ be a Cauchy sequence in $\mathcal{F}$. The definition of ρ shows that $\{f_n(s)\}$ is a Cauchy sequence in X for each fixed $s \in S$. Define $f(s)$ to be the limit of this sequence. To see that f_n converges to f in $\mathcal{F}$, suppose $0 < \varepsilon < 1$. Choose N so that $\rho(f_n, f_m) < \varepsilon$ for $n, m > N$. Then $d(f_n(s), f_m(s)) < \varepsilon$ for each $s \in S$. Passing to the limit as $m \to \infty$ and using an easy estimate (see Exercise 6.1), we obtain $d(f_n(s), f(s)) \leq \varepsilon$ for all $s \in S$. Hence $\rho(f_n, f) \leq \varepsilon$ for $n > N$, and $f_n \to f$ in $\mathcal{F}$. Thus $\mathcal{F}$ is complete. On the other hand, if X is not complete, and $\{x_n\}$ is a Cauchy sequence in X that does not converge, then the constant functions $f_n(s) = x_n$ form a Cauchy sequence in $\mathcal{F}$ that does not converge.

7. (a) The reflexiveness of the relation follows from the defining property (1.2) of a

metric, the symmetry of the relation follows from the corresponding property (1.3) of the metric, and the transivity follows from the triangle inequality (1.4). (b) From the triangle inequality we obtain $|d(s_k, t_k) - d(s_j, t_j)| \leq d(s_k, s_j) + d(t_j, t_k)$, which tends to 0 as $j.k \to \infty$. Thus $\{d(s_k, t_k)\}$ is a Cauchy sequence of real numbers, and it converges to some limit. If $\{r_k\} \sim \{s_k\}$, then the estimate $|d(r_k, t_k) - d(s_k, t_k)| \leq d(r_k, s_k)$ shows that the limit does not depend on the choice of sequence in the equivalence class, and thus determines a function $\rho(\tilde{s}, \tilde{t})$ on $\tilde{X}$. The verification that ρ has the metric properties is straightforward. (c) Suppose that $\{\tilde{s}^{(n)}\}$ is a Cauchy sequence in $\tilde{X}$. For each m, choose N_m such that $|s_j^{(m)} - s_k^{(m)}| < 1/m$ for $j, k \geq N_m$. Define $t_m = s_{N_m}^{(m)}$. It is straightforward to verify that $\{t_m\}$ is a Cauchy sequence, and that $\rho(\tilde{s}^{(m)}, \tilde{t}) \to 0$. (d) That the correspondence is an isometry follows immediately from the definition of ρ. If $\tilde{s}$ is the equivalence class of a sequence $\{s_k\}$, and $\tilde{s}_k$ is the equivalence class of the constant sequence at s_k, then $\tilde{s}_k$ converges to $\tilde{s}$ with respect to the metric ρ. Thus X is identified with a dense subset of $\tilde{X}$. (e) Let Y be any complete metric space containing X as a dense subset. Define a map of $\tilde{X}$ into Y by sending the equivalence class of $\{s_k\}$ to the limit of $\{s_k\}$. It is straightforward to check that this is an isometry of $\tilde{X}$ onto Y.

8. Suppose x and y both belong to $\cap E_k$. Then for each k, x and y are both in E_k, so $d(x, y) \leq \operatorname{diam} E_k$. Since this tends to 0, we obtain $d(x, y) = 0$, and $x = y$. To see that the intersection is nonempty, select for each k a point $x_k \in E_k$. If $n > k$, then $x_n \in E_k$, so that $d(x_k, x_n) \to 0$ as $k, n \to \infty$. Thus $\{x_k\}$ is a Cauchy sequence. Since X is complete, the sequence converges to a point $x \in X$. Since $x_n \in E_k$ for $n > k$, and since E_k is closed, the limit x belongs to E_k, this for all k, so $x \in \cap E_k$. This proves the first statement. If X is not complete, we can still conclude that $\cap E_k$ has at most one point. However, if there is a Cauchy sequence $\{x_k\}$ that does not converge, we can define E_k to be the closure of the set $\{x_n : n \geq k\}$, and these E_k's form a decreasing sequence of closed sets whose diameters tend to 0 and whose intersection is empty. Thus the assertion of the exercise holds if and only of X is complete.

Section 3. The Real Line

1. If the integers are bounded above, then by the Least Upper Bound Axiom they have a least upper bound M. Since $M - 1$ is not an upper bound, there is an interger m such that $m > M - 1$. But then $m + 1 > M$, contradicting the fact that M is an upper bound. For (b), apply (a) to find an integer m such that $m > 1/\varepsilon$, and then $0 < 1/m < \varepsilon$. For (c), choose a large integer N such that $N > -a$. Then $0 < a + N < b + N$. It suffices to find a rational number s between $a + N$ and $b + N$, since then $s - N$ is a rational number between a and b. We can thus assume that $0 < a < b$. Choose an integer m such that $0 < 1/m < b - a$. Let k be the largest integer such that $k < bm$, and set $s = k/m$. Then $s < b$, and since $k + 1 \geq bm$, we have $s = (k+1)/m - 1/m \geq b - 1/m > b - (b - a) = a$. Part (d) follows immediately from (c).

2. If t is any irrational number, and if r is rational, then $r + t/n$ is irrational, and $r + t/n \to r$.

3. The rationals have no isolated points. This does not contradict Exercise 2.3, because the rationals are not complete.

4. For each $x \in U$, let I_x be the union of all open intervals containing x that are contained in U. Show that each I_x is an open interval (possibly infinite or semi-infinite), any two I_x's either coincide or are disjoint, and the union of the I_x's is U.

5. Suppose the irrationals can be expressed as the union of a sequence of closed sets E_n. Then the reals can be expressed as the union of a sequence of closed sets, the E_n's and the singleton sets $\{r\}$ for r rational. By the Baire Category Theorem, one of these closed sets has nonempty interior. It cannot be one of the singletons, and since any nonempty interval contains rational numbers, it cannot be one of the E_n's. This contradiction establishes the assertion.

6. Take complements in the preceding exercise.

7. Since it is the intersection of closed sets, the Cantor set is closed. Since each interval of E_n has length $1/3^n$, each point of the Cantor set is within $1/3^n$ of a point in the complement of the Cantor set, and the Cantor set has empty interior. Since the endpoints of the middle-third intervals form a dense subset of the Cantor set, and none of these are isolated, the Cantor set has no isolated points. By Exercise 2.3, the Cantor set is uncountable.

8. Define $f(x) = 1/(2 \cdot 3^n) - |x - (m + 1/2)/3^n|$ for x in the middle-third interval between $m/3^n$ and $(m + 1)/3^n$, and define $f(x) = 0$ elsewhere. Then $f(x) = 0$ if and only if x is in the Cantor set. The Lipschitz estimate holds for s and t in the same middle-third interval, and it also hold when s and t are both in the Cantor set. Suppose s is in a middle-third and t is in the Cantor set. Let r be the endpoint of the middle-third interval containing s that is between s and t. Then $|f(s) - f(t)| = |f(s)| \leq |s - r| \leq |s - t|$. The case when s and t belong to different middle-thirds is similar.

9. For (a) the interior is $(0, 1)$ and the closure is $[0, 1]$. Every point of the closure is a limit point, so there are no isolated points. For (b) and (c), the interior is empty and the closure is $\mathbf{R}$. Every point is a limit point, and there are no isolated points. For (d), suppose $\{(1/m_j) + (1/n_j)\}$ is a sequence in the set that converges to a point x. Passing to a subsequence, we can assume either that $m_j \to \infty$, or else that there is an integer m such that $m_j = m$ for all j. Passing to a further subsequence, we can assume either that $n_j \to \infty$, or else that there is an integer n such that $n_j = n$ for all j. This gives a total of four cases. If $m_j \to \infty$ and $n_j \to \infty$, then the sequence tends to $x = 0$. If $m_j \to \infty$ and $n_j = n$, then the sequence tends to $x = 1/n$, which is already in the set, since $1/n = (1/2n) + (1/2n)$. Something similar occurs if we switch the m's and n's. In the fourth case, the sequence is eventually constant, and $x = (1/m) + (1/n)$. Thus the closure of the set is the set with 0 adjoined, while the limit points of the set are the rationals of the form $1/n$, $n \geq 1$, together with 0. The remaining points of the set are isolated points. Since the set consists of only rational numbers, it has empty interior.

10. We apply the Baire Category Theorem. For each $m \geq 1$, let E_m be the set of points $s \in \mathbf{R}$ for which there is a sequence $s_n \to s$ with $|f(s_n)| \leq m$. The E_m's increase to $\mathbf{R}$. Indeed, $s \in E_m$ just as soon as $m \geq |f(s)|$, since we can then take $s_n = s$ in the definition of E_m. We claim that E_m is closed. Indeed, let $t_k \in E_m$ converge to t. Choose s_k such that $|t_k - s_k| < 1/k$ and $|f(s_k)| \leq m$. Then $s_k \to t$, so that $t \in E_m$. By the Baire Category Theorem, one of the sets E_M has nonempty interior U.

Section 4. Products of Metric Spaces

1. The proof that (4.1) satisfies the triangle inequality is covered in Exercise 1.3. The verifications of the metric properties for (4.2) and (4.3) are straightforward.

2. Denote the respective unit balls by B_1, B_2, and B_3. Then B_1 is the unit disk, B_2 is a square with vertices at $(\pm 1, \pm 1)$, and B_3 is a diamond with vertices at $(\pm 1, 0)$ and $(0, \pm 1)$. Evidently $B_3 \subset B_1 \subset B_2$. The sketches are similar to those given in the text, with

the centers placed at the origin.

3. The estimate (4.5) holds for these metrics, so that convergence in the product space implies convergence of the components. These metrics are dominated by n times the maximum of $d_k(x_k, y_k)$, so that convergence of the components implies convergence in the product space.

4. Part (a): interior $= \{(x, y) : 0 < x^2 + y^2 < 1\}$; closure $=$ limit points $= \{(x, y) : x^2 + y^2 \leq 1\}$; no isolated points. Part (b): empty interior; closure $=$ limit points $= \mathbf{R} \times \{0\}$; no isolated points. Part (c): empty interior; closed; no limit points; each point is isolated. Part (d): limit points $= \{(0,0)\} \cup \{(1/n, 0) : n \geq 1\} \cup \{(0, 1/m) : m \geq 1\}$; closure $=$ union of set and its limit points; empty interior; each point is isolated.

5. We can take the metric on the product to be $\max d_j(x_j, y_j)$. By Theorem 4.2, $\tilde{X}_1 \times \cdots \times \tilde{X}_n$ is complete. The inclusion maps $X_j \to \tilde{X}_j$ are isometries, so that the inclusion map $X_1 \times \cdots \times X_n \to \tilde{X}_1 \times \cdots \times \tilde{X}_n$ is also an isometry. Since each X_j is dense in $\tilde{X}_j$, also $X_1 \times \cdots \times X_n$ is dense in $\tilde{X}_1 \times \cdots \times \tilde{X}_n$. Hence $\tilde{X}_1 \times \cdots \times \tilde{X}_n$ is the completion of $X_1 \times \cdots \times X_n$.

6. Let $\{x^{(j)} = (x_1^{(j)}, \ldots, x_n^{(j)})\}$ be a sequence in X. Suppose first that $x^{(j)}$ converges to $x = (x_1, \ldots, x_n)$ in X. Let U_k be an open subset of X_k containing x_k. Then $U = X_1 \times \cdots \times X_{k-1} \times U_k \times X_{k+1} \times \cdots \times X_n$ is an open subset of X containing x. Hence $x^{(j)} \in U$ for j large, and consequently $x_k^{(j)} \in U_k$ for j large. Thus $x_k^{(j)} \to x_k$. Conversely, suppose that each component sequence $\{x_k^{(j)}\}$ converges to x_k in X_k. Let U be an open subset of X containing x. By hypothesis, there are open subsets U_k of X_k such that $x_k \in U_k$ and $U_1 \times \cdots \times U_n \subset U$. For each k, $x_k^{(j)} \in U_k$ for j large. Hence $x^{(j)} \in U$ for j large, and $x^{(j)} \to x$ in X.

7. It is straightforward to check that d is a metric. For (b), suppose $\{x^{(k)}\}$ converges to x in X. Since $d_j(x_j^{(k)}, x_j) \leq 2^j d(x^{(k)}, x)$, for each fixed j we have $x_j^{(k)} \to x_j$ as $k \to \infty$. For the converse, suppose for each fixed j that $x_j^{(k)} \to x_j$ as $k \to \infty$. Let $\varepsilon > 0$. Choose N so that $\sum_{N+1}^{\infty} 1/2^j < \varepsilon/2$. Then choose n so that $d_j(x_j^{(k)}, x_j) \leq \varepsilon/2N$ for $j \geq n$ and $1 \leq k \leq N$. Then if $k \geq n$ we have

$$d(x^{(k)}, x) = \sum_{j=1}^{\infty} \frac{1}{2^j} \min(1, d_j(x_j^{(k)}, x_j)) \leq \sum_{j=1}^{N} \frac{1}{2^j} d_j(x_j^{(k)}, x_j) + \sum_{j=N+1}^{\infty} \frac{1}{2^j},$$

which is estimated easily by ε. Hence $x^{(k)} \to x$, and (b) holds. Suppose U_k is open in X_k for $1 \leq k \leq m$. From (b) we see that $X_1 \times \cdots \times X_{k-1} \times (X_k \backslash U_k) \times X_{k+1} \times \cdots$ is closed under sequential convergence, hence its complement $X_1 \times \cdots \times X_{k-1} \times U_k \times X_{k+1} \times \cdots$ is open. Since a finite intersection of open sets is open, $U_1 \times \cdots \times U_m \times X_{m+1} \times \cdots$ is open. Let U be an arbitrary open subset of X, and let $x \in U$. Choose $\varepsilon > 0$ so that $B(x; \varepsilon) \subset U$. Choose m so that $1/2^m < \varepsilon/2$. For $1 \leq k \leq m$, let U_k be the open ball in X_k centered at x_k of radius $\varepsilon/(2m)$. Then a simple estimate shows that $U_1 \times \cdots \times U_m \times X_{m+1} \times \cdots \subset B(x; \varepsilon)$. It follows that U is a union of product sets of the required form.

Section 5. Compactness

1. Any bounded non-closed subset of $\mathbf{R}^n$ is totally bounded but not compact.

2. Any closed unbounded subset of $\mathbf{R}^n$ is complete but not compact.

3. Let $\varepsilon > 0$. For each $x \in X$, let U_x be the open ball of radius ε centered at x.

The collection of open sets U_x, $x \in X$, forms an open cover of X. By compactness, there is a finite subcover, that is, there are a finite number of points $x_1, \cdots, x_n$ such that X is the union of the open balls of radius ε centered at the x_j's. Thus X is totally bounded.

4. Suppose the statement in (a) fails. Then taking $\varepsilon = 1/k$, we find $x_k \in X$ such that the open ball $B(x_k, 1/k)$ is not contained in any of the U_α's. Since X is sequentially compact, we may pass to a subsequence and assume that $x_k \to x$. Choose α such that $x \in U_\alpha$. Since U_α is open, there is $\delta > 0$ such that $B(x, \delta) \subset U_\alpha$. If k is so large that $d(x_k, x) < \delta/2$ and $(1/k) < \delta/2$, then $B(x_k, 1/k)$ is contained in $B(x, \delta)$, hence in U_α, a contradiction. For part (b), if one of the open sets in the cover is proper, the Lebesgue numbers are bounded. Since the cover is finite, it is easy to see that the least upper bound of the Lebesgue numbers is again a Lebesgue number.

5. Suppose $X_1, \ldots, X_n$ are compact. Let $\{x^{(k)} = (x_1^{(k)}, \ldots, x_n^{(k)})\}$ be a sequence in $X_1 \times \cdots \times X_n$. There is a subsequence $j_1 < j_2 < \cdots$ such that the first component sequence $x_1^{(j_i)}$ converges in X_1. This subsequence has a further subsequence for which the second component converges in X_2. Repeating this procedure n times, we arrive at a subsequence $k_1 < k_2 < \cdots$ for which $x_m^{(k_i)}$ converges in X_m for all m, $1 \le m \le n$. Then $x^{(k_i)}$ converges in X. Thus X is sequentially compact, hence compact.

6. Use Theorem 5.7, that a subspace of a separable metric space is separable.

7. Follow the hint.

8. Let E, F, and G be nonempty closed sets. Let $x \in E$. Given $\varepsilon > 0$, choose $y \in F$ such that $d(x,y) \le d(x,F) + \varepsilon$, and choose $z \in G$ such that $d(y,z) \le d(y,G) + \varepsilon$. Then $d(x,z) \le d(x,y) + d(y,z) \le \rho(E,F) + \rho(F,G) + 2\varepsilon$. Take the infimum over $z \in G$ and let ε tend to 0, to obtain $d(x,G) \le \rho(E,F) + \rho(F,G)$ for all $x \in E$. The same estimate holds for $d(z,E)$ for all $z \in G$, and these estimates yield the triangle inequality for ρ. The remaining metric properties for ρ are easy to verify. For the compactness assertion, let $\{E_n\}$ be a sequence of closed subsets of X. For each $k \ge 1$, let A_k be a finite subset of X such that the open balls of radius $1/k$ centered at points of A_k cover X. Set $E_{0j} = E_j$, and for each $k \ge 1$ define a subsequence $\{E_{kj}\}_{j=1}^\infty$ of $\{E_n\}$ so that $\{E_{kj}\}$ is a subsequence of $\{E_{k-1,j}\}$, and such that for each $x \in A_k$, either $B(x, 1/k) \cap E_{k,j} = \emptyset$ for all j, or else $B(x,1/k) \cap E_{k,j} \ne \emptyset$ for all j. Set $F_m = E_{mm}$, the diagonal sequence. Then $\{F_m\}$ is a subsequence of $\{E_n\}$, and the F_m's for $m \ge k$ are a subsequence of $\{E_{k,j}\}_{j=1}^\infty$. Let F be the set of all $x \in X$ such that each open neighborhood of x contains points of F_m for infinitely many m. If y is in the closure of F, and U is any open neighborhood of y, then U is also an open neighborhood of some point of F, so that U meets infinitely many of the F_m's. Consequently $y \in F$, and F is closed. Let $x \in F_k$. Choose $z \in A_k$ such that $d(z,x) < 1/k$. Then $B(z,1/k) \cap F_k \ne \emptyset$, so by construction, $B(z,1/k) \cap F_m \ne \emptyset$ for all $m \ge k$. Let $y_m \in B(z,1/k) \cap F_m$, and let y be any limit point of the y_m's as $m \to \infty$. Since $d(z,y) \le 1/k$, we have $d(x,y) \le 2/k$. Since $y \in F$, we obtain $d(x,F) \le 2/k$, and sup $d(x,F)$ over $x \in F_k$ is at most $2/k$. Now let $y \in F$. Choose $z \in A_k$ such that $d(z,y) < 1/k$. Then $B(z,1/k)$ is an open neighborhood of y, so it meets infinitely many of the F_m's. By construction, $B(z,1/k)$ meets F_m for all $m \ge k$, so that in particular $B(z,1/k) \cap F_k \ne \emptyset$. Thus $d(y,F_k) \le 2/k$, and sup $d(y,F_k)$ over $y \in F$ is at most $2/k$. It follows that $\rho(F,F_k) \le 2/k$, and F_k converges to F.

Section 6. Continuous Functions

1. The triangle inequality gives $d(x,z) - d(y,z) \le d(x,y)$ and also $d(y,z) - d(x,z) \le d(x,y)$, so that $|d(x,z) - d(y,z)| \le d(x,y)$. This estimate shows immediately that the function $f(x) = d(x,z)$ is uniformly continuous from X to $\mathbf{R}$; indeed, if $d(x,y) < \varepsilon$, then

$|f(x)-f(y)|<\varepsilon$. The continuity of $d(x,z)$ as a function of the two variables x and z follows from the estimate $|d(x,z)-d(y,w)| \leq |d(x,z)-d(y,z)|+|d(y,z)-d(y,w)| \leq d(x,y)+d(z,w)$.

2. This follows immediately from the characterization given in Exercise 1.12, that two metrics are equivalent if and only if they have the same convergent sequences.

3. The line through $(s,0)$ and $(0,1)$ has the parametric representation $t \rightarrow (0,1)+t[(s,0)-(0,1)] = (ts, 1-t)$. This meets the circle when $t^2s^2+(t-1/2)^2 = 1/4$, or $t=0$ and $t = 1/(1+s^2)$. The solution $t=0$ corresponds to $(0,1)$, and the other solution yields up the formula $h(s) = (s/(1+s^2), s^2/(1+s^2))$. Thus h is continuous. If (x,y) lies on the circle, and $s = y/x$, then $h(s) = (x,y)$. Thus $(x,y) \rightarrow y/x$ provides a continuous inverse for h. Since h is bicontinuous, the metrics have the same open sets, hence are equivalent (cf. Exercise 1.12). The sequence $s_n = n$ is a Cauchy sequence with respect to the metric ρ, though it does not converge. Since $(\mathbf{R},\rho)$ is isometric to a dense subset of the circle, the completion of $(\mathbf{R},\rho)$ can be identified with the circle. Since the circle is totally bounded, $(\mathbf{R},\rho)$ is also totally bounded. (The point $(0,1)$ on the circle can be thought of as the point at infinity of $\mathbf{R}$, where we identify $+\infty$ and $-\infty$.)

4. This follows immediately from the definition of continuity in terms of mapping convergent sequences to convergent sequences.

5. Suppose $x_n \in X_0$ converges to x_0. Since f is continuous at x_0, $f(x_n) \rightarrow f(x_0)$. Since g is continuous at $f(x_0)$, $(g \circ f)(x_n) = g(f(x_n)) \rightarrow g(f(x)) = (g \circ f)(x)$. Hence $g \circ f$ is continuous at x_0.

6. Let $E = \{s \in [0,1] : f(s) \geq s\}$. Then E is a closed subset of $[0,1]$. Since $0 \in E$, E is nonempty. Let t be the least upper bound of E. Since E is closed, t is in E, that is, $f(t) \geq t$. If $t<1$, then $f(t+\varepsilon) < t+\varepsilon$ for $\varepsilon > 0$ small. Passing to the limit as $\varepsilon \rightarrow 0$, we obtain $f(t) \leq t$, hence $f(t) = t$. If $t = 1$, we have $f(t) \leq t$ by default, so again $f(t) = t$.

7. A proof in a more general context will be given in the next chapter. A direct proof based on sequential compactness is as follows. Let $\{y_n\}$ be any sequence in $f(X)$, say $y_n = f(x_n)$. Since X is compact, the sequence $\{x_n\}$ has a convergent subsequence, say $x_{n_k} \rightarrow x$. Since f is continuous, the subsequence $\{y_{n_k} = f(x_{n_k})\}$ converges to $f(x)$. Thus $f(X)$ is sequentially compact, hence compact.

8. Let X be a nonempty compact metric space. If $f: X \rightarrow \mathbf{R}$ is continuous, then $f(X)$ is a compact subset of $\mathbf{R}$, by the preceding exercise, and it is nonempty. Hence the least upper bound M of $f(X)$ exists and belongs to $f(X)$. If $x \in X$ satisfies $f(x) = M$, then f attains its maximum at x.

9. An outline for constructing such an f is as follows. If X is not complete, let $\{x_n\}$ be a Cauchy sequence in X that does not converge. Then $\lim d(x,x_n)$ exists and defines a continuous function $f(x)$ on X that is strictly positive. Further, $f(x_n) \rightarrow 0$ as $n \rightarrow \infty$, so $1/f$ is a continuous unbounded function on X. If X is complete but not compact, then X is not totally bounded, and for some $c>0$ there is a sequence $\{x_n\}$ in X such that $d(x_n,x_m) \geq 4c$ whenever $m \neq n$. Define $f(x) = n(1-d(x,x_n)/c)$ if $d(x,x_n) < c$, and set $f(x) = 0$ elsewhere. This defines a continuous function f with $f(x_n) \rightarrow \infty$.

10. The metric properties for d immediately imply the metric properties for ρ. Suppose X is complete. Let $\{f_n\}$ be a Cauchy sequence in $B(S,X)$. The definition of ρ shows that $\{f_n(s)\}$ is a Cauchy sequence in X for each fixed $s \in S$. Define $f(s)$ to be the limit of this sequence. To see that f_n converges to f in $B(S,X)$, let $\varepsilon > 0$. Choose N so that $\rho(f_n,f_m) < \varepsilon$ for $n,m > N$. Then $d(f_n(s),f_m(s)) < \varepsilon$ for each $s \in S$. Passing to the limit

as $m \to \infty$ and using the continuity of d (Exercise 1), we obtain $d(f_n(s), f(s)) \leq \varepsilon$ for all $s \in S$. Hence $\rho(f_n, f) \leq \varepsilon$ for $n > N$, and $f_n \to f$ in $B(S, X)$. Thus $B(S, X)$ is complete. On the other hand, if X is not complete, and $\{x_n\}$ is a Cauchy sequence in X that does not converge, then the constant functions $f_n(s) = x_n$ form a Cauchy sequence in $B(S, X)$ that does not converge. Part (c) boils down to the standard three-epsilon proof that a uniform limit of continuous functions is continuous. Suppose $\{f_n\}$ is a sequence in $BC(S, X)$ that converges to f uniformly on S, that is, with respect to the metric ρ. We must show that f is continuous. Let $s \in S$ and let $\varepsilon > 0$. Choose n such that $\rho(f_n, f) < \varepsilon$. Choose $\delta > 0$ such that $d(f_n(t), f_n(s)) < \varepsilon$ if t belongs to the δ-ball in S centered at s. Then for t in this δ-ball we have $d(f(t), f(s)) \leq d(f(t), f_n(t)) + d(f_n(t), f_n(s)) + d(f_n(s), f(s)) < 3\varepsilon$. Hence f is continuous at s.

11. Follow the hint.

12. Fix a point $y_0 \in Y$. For each $n \geq 1$, define E_n to be the set of points $x \in X$ such that $d(f_\alpha(x), y_0) \leq n$ for all α. Since the f_α's are continuous, E_n is closed. By hypothesis, X is the union of the E_n's. By the Baire Category Theorem, some E_n has nonempty interior, which we take to be U.

13. Let $\varepsilon > 0$. For each $N \geq 1$, define E_N to be the set of points $x \in X$ such that $\rho(g_n(x), g_m(x)) \leq \varepsilon/2$ for all $m, n \geq N$. Since the g_j's are continuous, E_N is closed. On account of the hypothesis, X is the union of the E_N's. By the Baire Category Theorem, some E_N has nonempty interior, which we take to be U.

14. Follow the hint. For part (d), apply (c) to the complements of one-point sets. This shows that the complement of any countable subset of a Polish space is Polish.

15. Follow the hint.

Section 7. Normed Linear Spaces

1. For the 1-norm, 2-norm, and ∞-norm, see Exercise 4.2. The unit ball in the 4-norm is a circle with bulges towards the vertices of the unit ball in the ∞-norm. The balls centered at $(2, 1)$ are translates of the unit balls.

2. For part (a), note that if $x \in \mathbf{R^n}$, then $||x||_\infty = |x_k|$ for some k, and consequently $||x||_\infty = (|x_k|^p)^{1/p} \leq (\sum |x_j|^p)^{1/p} = ||x||_p$. For (b), use the definition of $||x||_p$ and the estimate $|x_j| \leq ||x||_\infty$. Part (c) follows from (a) and (b).

3. The estimates in part (b) assert that the identity map of the space is bounded from each norm to the other. By Theorem 7.5, this is equivalent to bicontinuity. That the estimates in (a) are equivalent to those in (b) follows from Lemma 7.3. Part (c) is a straightforward verification.

4. Let $|| \cdot ||$ be a norm on $\mathbf{R^n}$. For $1 \leq j \leq n$, let e_j be the vector in $\mathbf{R^n}$ with entries 1 in the jth position and 0 elsewhere, and let $C = \max\{||e_j|| : 1 \leq j \leq n\}$. From $x = (x_1, \ldots, x_n) = \sum x_j e_j$, we obtain $||x|| \leq C \sum |x_j| = C||x||_1$. Thus the identity map is continuous from the 1-norm to the given norm. Since the unit sphere in the 1-norm is compact, and continuous functions map compact sets to compact sets, it is compact in the metric determined by the norm $|| \cdot ||$. By Lemma 7.2, the norm function is continuous, so it attains its minimum on any compact set. We conclude that $\inf\{||x|| : ||x||_1 = 1\}$ coincides with the norm of some nonzero vector, consequently the infimum is strictly positive, and this implies by Exercise 3(c) that the norm $|| \cdot ||$ is equivalent to the 1-norm.

5. Let L be a linear functional on $\mathbf{R^n}$. Let the e_j's be the standard basis vectors (as above), and take C to be the maximum of the $|L(e_j)|$'s. From $L(x) = \sum x_j L(e_j)$ we estimate $|L(x)| \leq C \sum |x_j|$. Thus L is continuous with respect to the 1-norm, hence by Exercise 4 with respect to any norm. Any linear transformation $T : \mathbf{R^n} \to \mathcal{X}$ can be expressed in the form $T(x) = \sum L_j(x) w_j$, where the L_j's are the coordinate functionals $L_j(x) = x_j$ on $\mathbf{R^n}$, and each $w_j = T(e_j)$ is a fixed element of $\mathcal{X}$. Since each L_j is continuous, T is continuous.

6. Parts (a) and (b) are straightforward, and (c) follows easily from the homogeneity of T with respect to scalar multiplication.

7. For part (a), show by calculus that for fixed $t > 0$ the function $f(s) = st - s^p/p$ attains its maximum at $s = t^{1/(p-1)} = t^{1/q}$, and the maximum value is t^q/q. For part (b), apply (a) with $s = |x_j|$ and $t = |y_j|$ and sum. This works if $||x||_p = ||y||_q = 1$. The general case follows by homogeneity; apply the special case to $x/||x||_p$ and $y/||y||_q$.

8. That $||\cdot||_p$ is a norm on ℓ^p follows easily from the finite-dimensional case in Exercise 7. The crucial point to verify is that ℓ^p is complete. We assume for convenience that $p = 1$. Suppose $\{x^{(n)}\}_{n=1}^{\infty}$ is a Cauchy sequence in ℓ^1. For each fixed j we have $|x_j^{(n)} - x_j^{(m)}| \leq ||x^{(n)} - x^{(m)}||$, so that $\{x_j^{(n)}\}_{n=1}^{\infty}$ is a Cauchy sequence of scalars. Set $x_j = \lim x_j^{(n)}$. Since $\sum_{j=1}^{N} |x_j| = \lim_{n\to\infty} \sum_{j=1}^{N} |x_j^{(n)}| \leq \lim_{n\to\infty} ||x^{(n)}||$, we see that $\sum_1^\infty |x_j|$ is finite, and $x \in \ell^1$. Let $\varepsilon > 0$. Choose n so large that $||x^{(n)} - x^{(m)}|| < \varepsilon$ for $m > n$. Choose N so large that $\sum_{j=N+1}^{\infty} |x_j^{(n)} - x_j| < \varepsilon$. Then

$$||x^{(m)} - x|| \leq \sum_{j=1}^{N} |x_j^{(m)} - x_j| + \sum_{j=N+1}^{\infty} |x_j^{(m)} - x_j^{(n)}| + \sum_{j=N+1}^{\infty} |x_j^{(n)} - x_j|.$$

The first sum tends to 0 as $m \to \infty$, while the second and third are each bounded by ε. It follows that $\limsup_{m\to\infty} ||x^{(m)} - x|| \leq 2\varepsilon$. Since this holds for all $\varepsilon > 0$, $x^{(m)}$ converges to x in norm, and ℓ^1 is complete. Now for part (b), let $x \in \ell^p$. Then $x_j \to 0$, so $|x_j| < 1$ for j large. Hence $|x_j|^r \leq |x_j|^p$ for j large, and $x \in \ell^r$. If $x \in \ell^r$, and $x^{(m)}$ is the sequence whose entries are x_j for $1 \leq j \leq m$ and 0 for $j > m$, then $x^{(m)} \in \ell^p$ and $x^{(m)}$ converges to x in ℓ^r. Consequently ℓ^p is dense in ℓ^r. The sequence defined by $x_j = j^{-1/p}$ belongs to ℓ^r but not to ℓ^p, so that ℓ^p is a proper dense subset of ℓ^r. In (c), the closure of ℓ^1 in ℓ^∞ coincides with the space of sequences that converge to 0.

9. Let $s \in \ell^\infty$. If $x \in \ell^1$, then $|L_s(x)| \leq \sum |s_j||x_j| \leq ||s||_\infty ||x||_1$. Hence the linear functional L_s is bounded on ℓ^1, with norm $||L_s|| \leq ||s||_\infty$. Let $e_k \in \ell^1$ be the sequence with all zeros, except for a 1 in the kth place. Then $||e_k|| = 1$, and $L_s(e_k) = s_k$. Hence $||L_s|| \geq |s_k|$, this for all k, and $||L_s|| = ||s||_\infty$. Suppose L is an arbitrary continuous linear functional on ℓ^1. Define a sequence s by $s_k = L(e_k)$. Then $|s_k| \leq ||L||\,||e_k|| = ||L||$, so $s \in \ell^\infty$. Evidently $L(x) = L_s(x)$ for any finite linear combination x of the e_k's. Since these are dense in ℓ^1, L coincides with L_s.

10. If $f \in C[0,1]$ is not identically zero, then $|f|$ is a nonnegative continuous function on $[0,1]$ that is not identically zero. Hence $|f| \geq \varepsilon$ on some subinterval of positive length L, and $\int |f| \geq \varepsilon L$, so $||f||_1 > 0$. The other norm properties follow similarly from elementary properties of the integral.

11. The sequence $\{f_n\}$ defined in the figure is a Cauchy sequence. Suppose that it converges in the norm $||\cdot||_1$ to a continuous function $g(s)$, that is, $\int_0^1 |f_n(s) - g(s)|\,ds \to 0$. Since $f_n(s) = 0$ for $0 \leq s \leq 1/2$, and $f_n(s)$ converges uniformly to 1 on any closed subinterval of $(1/2, 1]$, we obtain $\int_0^{1/2} |g(s)|\,ds + \int_{1/2}^1 |1 - g(s)|\,ds = 0$. Since g is continuous, we obtain

$g = 0$ on $[0, 1/2]$ and $g = 1$ on $[1/2, 1]$, a contradiction. Thus the space is not complete.

12. For (a) and (b), use the hint and proceed as in Exercise 1.3. The complex version is similar to the real version, except that one begins by considering the integral of $f - \lambda g$ times its complex conjugate.

13. Given d, set $||x|| = d(x, 0)$. Conversely, given $||\cdot||$, define $d(x, y) = ||x - y||$.

14. First show that $|r-t|/(1+|r-t|) \leq |r-s|/(1+|r-s|) + |s-t|/(1+|s-t|)$ for real numbers r, s, and t. By translation invariance, we can assume that $s = 0$, and then the inequality reduces to $(x+y)/(1+x+y) \leq x/(1+x) + y/(1+y)$ for $x, y > 0$, which can be established by studying the function $x/(1+x)$. From the scalar inequality, it follows easily that d is a metric, which is clearly invariant under translations. Since the metric is not homogeneous (in fact, it is bounded), it cannot arise from a norm.

15. The estimate $\sup\{||(T_n - T)x|| : ||x|| \leq M\} \leq M||T_n - T||$ shows that if T_n converges to T in the operator norm, then T_n converges to T uniformly on any bounded subset of $\mathcal{X}$. Conversely, if T_n converges to T uniformly on the unit ball of $\mathcal{X}$, then from the definition of the norm we have $||T_n - T|| \to 0$.

16. This follows from repeated application of Lemma 7.3. If $||x|| \leq 1$, then $||TSx|| \leq ||T||\,||Sx|| \leq ||T||\,||S||\,||x||$. Hence TS is bounded, and $||TS|| \leq ||T||\,||S||$.

17. The norm properties are easy to verify. The completeness follows from the observation that a sequence in the direct sum is a Cauchy sequence if and only if the component sequences are Cauchy sequences in $\mathcal{X}$ and $\mathcal{Y}$ respectively.

18. The Principle of Uniform Boundedness shows that the norms of the T_n's are bounded, hence Theorem 7.6 applies.

19. Since $|f(t_j)| \leq ||f||$, we have $|L(f)| \leq \sum |a_j|\,||f||$, and $||L|| \leq \sum |a_j|$. To see that equality holds, choose $f \in C[0,1]$ such that $f(t_j) = +1$ if $a_j \geq 0$, $f(t_j) = -1$ if $a_j < 0$, and $|f| \leq 1$. Then $L(f) = \sum |a_j|$, and consequently $||L|| = \sum |a_j|$.

20. Define a linear functional L_n on $C[0,1]$ by $L_n(f) = \sum_j a_j^{(n)} f(t_j^{(n)})$. By the preceding exercise, $||L_n|| = \sum_j a_j^{(n)} = \int_0^1 1\,dt = 1$. Further, L_n converges to the continuous linear functional $L(f) = \int_0^1 f(t)\,dt$ on polynomials, which form a dense subset of $C[0,1]$. By Theorem 7.6, L_n converges on all $f \in C[0,1]$ and the limit defines a continuous linear functional, so the limit must coincide with L.

21. Since $|L(f)| \leq \int |f(t)h(t)|dt \leq ||f|| \int |h(t)|dt$, we have $||L|| \leq \int |h(t)|dt = ||h||_1$. We claim that $||L|| = ||h||_1$. Our strategy is to consider the function F that is 1 where $h > 0$ and -1 where $h < 0$, since then $|F| \leq 1$ and $\int Fhdt = ||h||_1$. Unfortunately F need not be continuous, so we approximate F by continuous functions, as follows. By compactness, we can cover the set where $h = 0$ by a finite number of open intervals on which $|h| < \varepsilon$. Let I be the union of these intervals, and let $J = [0,1] \backslash I$. Since h is continuous, it does not change sign on any of the intervals in J. Define f to be 1 on the intervals in J where $h > 0$ and to be -1 on the intervals of J where $h < 0$. We extend f continuously to each interval of I so that $|f| \leq 1$ there. Then f is continuous on $[0,1]$, and $||f|| \leq 1$. Now $\int_J f(t)h(t)dt = \int_J |h(t)|dt$, while $|\int_I f(t)h(t)dt| \leq \int_I |h(t)|dt \leq \varepsilon$. It follows that $|L(f)| \geq \int |h(t)|dt - 2\varepsilon$, from which we obtain $||L|| \geq ||h||_1 - 2\varepsilon$. Now let ε tend to 0.

22. From $|(Tf)(s)| \leq ||f|| \int_0^1 |k(s,t)|\,dt$, we obtain

$$||T|| \leq \sup_{0 \leq s \leq 1} \int_0^1 |k(s,t)|\,dt.$$

We claim that equality holds here. To see this, it suffices to find for fixed s_0 a function satisfying $|f| \leq 1$ and $(Tf)(s_0) \approx \int |k(s_0,t)|\,dt$. We proceed as in Exercise 21. Let $\varepsilon > 0$. Let f be a continuous function such that $f(t) = +1$ on the set where $k(s_0,t) \geq \varepsilon$, $f(t) = -1$ on the set where $k(s_0,t) \leq -\varepsilon$, and $|f| \leq 1$. We readily estimate that $(Tf)(s_0)$ differs from $\int_0^1 |k(s_0,t)|\,dt$ by at most 2ε, which establishes the assertion.

Section 8. The Contraction Principle

1. First observe that for any $\varepsilon > 0$, there is a constant $c = c(\varepsilon) < 1$ such that $d(\Phi(x),\Phi(y)) \leq c\,d(x,y)$ for all $x,y \in X$ satisfying $d(x,y) \geq \varepsilon$. This follows from the compactness of X and the continuity of the distance function. Now let $x_0 \in X$ be arbitrary. The sequence $d(\Phi^n(x_0),\Phi^{n+1}(x_0))$ is monotone decreasing, hence converges to a limit r. If $r > 0$, then when $d(\Phi^n(x_0),\Phi^{n+1}(x_0))$ approaches close to r, we would obtain $d(\Phi^{n+1}(x_0),\Phi^{n+2}(x_0)) \leq c(r)\,d(\Phi^n(x_0),\Phi^{n+1}(x_0)) < r$, a contradiction. Thus $r = 0$. Let z be adherent to the sequence $\Phi^n(x_0)$, say $z = \lim \Phi^{n_j}(x_0)$. Then $d(z,\Phi(z)) = \lim d(\Phi^{n_j}(x_0),\Phi^{n_j+1}(x_0)) = 0$, so $z = \Phi(z)$, and z is a fixed point. The fixed point is evidently unique. This proof shows in fact that $\Phi^n(x_0)$ converges to the fixed point for any starting point x_0.

2. A rotation of a circle is an isometry with no fixed points, unless it is the identity, in which case every point is fixed.

3. Let x_0 be the unique fixed point for Φ^m. Since $\Phi^m(\Phi(x_0)) = \Phi(\Phi^m(x_0)) = \Phi(x_0)$, $\Phi(x_0)$ is also a fixed point of Φ, and consequently $\Phi(x_0) = x_0$. The uniqueness is clear.

4. For (a) note first that $\Phi^m(v) = u + T(u) + \cdots + T^{m-1}(u) + T^m(v)$, and consequently $d(\Phi^m(v),\Phi^m(w)) = ||\Phi^m(v) - \Phi^m(w)|| = ||T^m(v) - T^m(w)|| \leq ||T^m||\,||v-w|| = ||T^m|| d(v,w)$. For (b) we observe that the solutions of the equation are precisely the fixed points of Φ, so that Exercise 3 gives the result.

5. For (a), set $\Phi(v) = (u + T(v))/\lambda$ and proceed as in Exercise 4. For (d), choose $r < |\lambda|$ such that $||T^n||^{1/n} \leq r$ for n large. Then $||T^n||/|\lambda^{n+1} \leq r^n/|\lambda|$ for n large, and the sequence of partial sums of the series (8.15) is a Cauchy sequence, in fact, a fast Cauchy sequence (cf. Exercise 2.1). Thus the series converges in norm. That the sum of the series is $(\lambda I - T)^{-1}$ can be seen by pre- and post-multiplying the series by $\lambda I - T$. It can also be seen by noting that the v_m's converge to the solution v to (8.14), by the proof of Theorem 8.1, and that (8.14) is equivalent to $v = (\lambda I - T)^{-1}u$.

6. Define the Volterra integral operator T by

$$(Tv)(s) = \int_a^s K(s,t)v(t)\,dt, \qquad a \leq s \leq b.$$

We prove by induction that $|(T^m v)(s)| \leq M^m(s-a)^m/m!$. This is true for $m = 0$. If it holds for m, then $|(T^{m+1}v)(s)| \leq \int_a^s |K(s,t)(T^m v)(t)|\,dt| \leq \int_a^s M \cdot M^m(s-a)^m\,ds/m! \leq M^{m+1}(s-a)^{m+1}/(m+1)!$, as required. It follows that $||T^m|| \leq M^m(b-a)^m/m!$. Consequently $||T^m|| < 1$ for m large, and Exercise 4 applies.

7. Suppose $v(t)$ is a solution on $[a,\beta]$ that is different from u. Let α be the infimum of $s > a$ such that $u(s) \neq v(s)$. Then $u(t) = v(t)$ for $a \leq t \leq \alpha$. The estimates in the

proof show that $|u(t) - \xi| < r$ for $a \le t < \beta$, so that $|v(\alpha) - \xi| < r$. Then $|v(t) - \xi| < r$ on some interval $a \le t \le \alpha + \varepsilon$. Thus v lies in the space E corresponding to the interval $[a, \alpha+\varepsilon]$, so by the uniqueness assertion of the Contraction Mapping Theorem, $v = u$ there, contradicting the choice of α.

8. One solution is $u(t) = t^2/4$, another is $u(t) \equiv 0$. This does not contradict the Cauchy-Picard theorem, since $F(u,t) = \sqrt{u}$ does not satisfy the Lipschitz condition (8.8).

9. If (8.7) and (8.8) hold for ξ in some set S, with constants r, c, and M independent of ξ, then the solution u_ξ depends continuously on ξ, in the norm of uniform convergence. This follows from Theorem 8.5, applied to the appropriate Φ_ξ.

10. Assume u is not defined on all of $[a,b]$. Let β be the supremum of $s > a$ such that the solution $u(t)$ exists and is in U for $a \le t < s$. On account of the uniformity of the constants on compact subsets, for each compact subset K of U there is an $\varepsilon = \varepsilon(K)$ such that if $u(t) \in K$, then the solution exists and remains in U for $t \le s \le t + \varepsilon$. Thus if $s < \beta$ and $u(s) \in K$, then we must have $s + \varepsilon \le \beta$. It follows that $u(s) \notin K$ for $\beta - \varepsilon < s < \beta$. Hence $u(s) \to \partial U$ as $s \to \beta$.

11. Since $d(x_0, \Phi_s(x_0))$ is a continuous function of s on the compact set S, it is bounded, say by M. Theorem 8.2 then yields the estimate $d(x_m(s), x^*(s)) \le Mc^m/(1-c)$, which shows that $x_m(s)$ converges to $x^*(s)$ uniformly on S.

Section 9. The Frechet Derivative

1. If G were Frechet differentiable at 0, then $G'(0)$ would be 0, by Theorem 9.4. However, $G(t,t)/||(t,t)|| = 1/2$ does not tend to 0 with t.

2. This is proved in any reputable multivariable calculus text. To display the idea of the proof, we treat the special case $n = 2$, $m = 1$, so that $G(x,y)$ is a real-valued function of two variables whose partial derivatives G_1 and G_2 are continuous in a neighborhood of $(0,0)$. By subtracting an appropriate linear combination $Ax + By + C$ from G, we can assume that $G(0,0) = G_1(0,0) = G_2(0,0) = 0$. We aim to show that G has Frechet derivative 0 at the origin, and for this we must show that $|G(x,y)|/||(x,y)|| \to 0$ as $||(x,y)|| \to 0$. Let $\varepsilon > 0$. Choose $\delta > 0$ such that $|G_1| < \varepsilon$ and $|G_2| < \varepsilon$ on the δ-ball B_δ centered at the origin. Let $(x,y) \in B_\delta$. By the Mean Value Theorem, there is a point $\xi = \xi(y)$ between 0 and x such that $G(x,y) = G(0,y) + G_1(\xi,y)x$. By the Mean Value Theorem again there is a point η between 0 and y such that $G(0,y) = G_2(0,\eta)y$. Then $G(x,y) = G_1(\xi,y)x + G_2(0,\eta)y$ is estimated by $|G(x,y)| \le 2\varepsilon||(x,y)||$. Hence the limsup of $|G(x,y)|/||(x,y)||$ as $(x,y) \to (0,0)$ is at most 2ε. This holds for all $\varepsilon > 0$, so the limit is 0.

3. Suppose $F(x_0 + v) = F(x_0) + F'(x_0)v + R(v)$ and $G(F(x_0) + w) = G(F(x_0)) + G'(F(x_0))w + S(w)$, where $R(v)/||v||$ and $S(w)/||w||$ tend to 0 as $v, w \to 0$. Let $\varepsilon > 0$. Choose $\eta > 0$ such that $||S(w)|| \le \varepsilon||w||$ for $||w|| \le \eta$. Choose $\delta > 0$ such that $||F'(x_0)v + R(v)|| \le \eta$ and $||R(v)|| \le \varepsilon||v||$ for $||v|| \le \delta$. Now $G(F(x_0 + v)) = G(F(x_0)) + G'(F(x_0))F'(x_0)v + T(v)$, where $T(v) = G'(F(x_0))R(v) + S(F'(x_0)v + R(v))$. For $||v|| \le \delta$ we obtain by $||T(v)|| \le ||G'(F(x_0))||\varepsilon||v|| + \varepsilon||F'(x_0)v + R(v)|| \le \varepsilon(||G'(F(x_0))|| + ||F'(x_0)|| + \varepsilon)||v||$. Hence $||T(v)||/||v|| \to 0$ as $v \to 0$, and $G \circ F$ is Frechet differentiable at x_0 with derivative $G'(F(x_0))F'(x_0)$.

4. Define a linear operator A on $\mathbf{R}^2$ by $A(x,y) = (x, y/2)$. One readily checks that $||A|| = 1$ while $||A^{-1}|| = 2$.

5. If A is invertible, and $A_n \to A$, then $A_n^{-1} \to A^{-1}$, and $||A_n^{-1}TA_n^{-1} - A^{-1}TA^{-1}|| \le$

$||A_n^{-1}T(A_n^{-1} - A^{-1})|| + ||(A_n^{-1} - A^{-1})TA^{-1}|| \leq (||A_n^{-1}|| + ||A^{-1}||)\,||A_n^{-1} - A^{-1}||\,||T||$. This shows that the operators $T \to A_n^{-1}TA_n^{-1}$ converge to the operator $T \to A^{-1}TA^{-1}$, in the operator norm.

6. By the Mean Value Theorem, for each $x_0, x \in \mathbf{R}$ and $s \in [a,b]$, there is a point x_1 between x_0 and x such that $F(x,s) - F(x_0,s) = F_1(x_1,s)(x - x_0)$. Let $\varepsilon > 0$. Fix M large, $M > ||u_0||$. Choose $\delta > 0$ such that if $|x_0| \leq M$ and $|x_1 - x_0| \leq \delta$, then $|F_1(x_1,s) - F_1(x_0,s)| < \varepsilon$ for all $a \leq s \leq b$. Then if $|x - x_0| \leq \delta$ we have $F(x,s) = F(x_0,s) + F_1(x_0,s)(x - x_0) + R(x,s)$, where $R(x,s) = (F_1(x_1,s) - F_1(x_0,s))(x - x_0)$ satisfies $|R(x,s)| \leq \varepsilon|x - x_0|$. Applying this estimate at $x_0 = u_0(s)$ and $x = u(s)$, we obtain for any $u \in C[a,b]$ satisfying $||u - u_0|| < \delta$,

$$(\Psi u)(t) = (\Psi u_0)(t) + \int_a^t F_1(u_0(s),s)[u(s) - u_0(s)]\,ds + S(u,t),$$

where $|S(u,t)| \leq \int_a^t |u(s) - u_0(s)|\,ds \leq \varepsilon(b-a)||u - u_0||$. It follows that Ψ is differentiable at u_0, with derivative T.

7. From $\Phi(\xi,u_0) - \Phi(\xi_0,u_0) = -(\xi - \xi_0)$, we see that $\Phi_1(\xi_0,u_0)$ is the continuous linear functional on $\mathbf{R}$ given by $\Phi_1(\xi_0,u_0)(t) = -t$. The Frechet derivative $\Phi_2(\xi_0,u_0)$ is calculated as in the preceding exercise to be the linear operator $I - T$ on $C[a,b]$. For (b) and (c), follow the hint, and apply the Implicit Function Theorem.

8. The Inverse Mapping Theorem shows that $f(U)$ contains a neighborhood of each of its points, so that $f(U)$ is open. Applying the same observation to the restriction of f to W, we see that $f(W)$ is open for any open subset W of U.

9. Differentiate the formula $f'(x) = -G_2(x,f(x))^{-1}G_1(x,f(x))$ with respect to the x_j's, using the usual rules for differentiation.

10. Follow the hint.

Chapter II. Topological Spaces

Section 1. Topological Spaces

1. This follows from the definition of a topology.

2. Finite intersections of sets with finite complements have finite complements, since the complement of an intersection is the union of the complements and a finite union of finite sets is finite. The remaining properties are immediate.

3. If U has finite complement, and if N is the largest integer in the complement of U, then $n \in U$ for $n > N$, so the sequence $\{1,2,3,\dots\}$ eventually lies in U. The definition of the cofinite topology shows that the sequence converges to each point. A sequence $\{n_1, n_2, n_3 \dots\}$ converges to an integer m if and only if each integer $j \neq m$ occurs in the sequence only a finite number of times. Thus the convergent sequences are exactly those for which at most one value occurs infinitely often.

4. The closure of S in the discrete topology is S itself, since all sets, and in particular S itself, are closed. The closure of S in the indiscrete topology is $\emptyset$ if $S = \emptyset$, otherwise is X. The closure of S in the cofinite topology is S itself if S is finite (or empty) and is all of X if S is infinite; this follows from the fact that the closed sets in the cofinite topology are $\emptyset$, X, and the finite subsets of X.

5. Suppose x is a point in the closure of the set of limit points of S. Then every open

set U containing x also contains a limit point y of S. Since $U\backslash\{x\}$ is open, it must contain a point of S. Thus x is a limit point of S, and the set of limit points is closed. Clearly a point cannot be both a limit point of S and an isolated point. Isolated points of S belong to S, hence to $\bar{S}$. Also limit points of S belong to $\bar{S}$. Conversely, if $s \in S$, then s is either an isolated point of S or a limit point of S, while if $s \in \bar{S}$ but $s \notin S$, then s must be a limit point of S.

6. Let d be the metric for X, and set $d(p,q) = r > 0$. Then the open balls U and V centered at p and q respectively of radius $r/2$ are disjoint. If X is infinite and if U and V are nonempty open sets in the cofinite topology so that $X\backslash U$ and $X\backslash V$ are finite, then $U \cap V$, which has complement the finite set $(X\backslash U) \cup (X\backslash V)$, cannot be empty.

7. Follows from the definition of interior.

8. Let T be the intersection of all closed sets containing S. Then T is closed, and T contains S, so $T = \bar{T}$ contains $\bar{S}$. For the reverse inclusion, suppose $p \notin \bar{S}$. Then there exists an open set U containing p that does not meet S. Then $X\backslash U$ is a closed set containing S, so $X\backslash U$ contains T, and $p \notin T$. It follows that $T \subset \bar{S}$, and the sets coincide.

9. Follows by the same type of reasoning as used in Problem 8.

10. Let U be open. Note successively that U^- is the set of points x such that every neighborhood of x meets U; $U^{-\prime}$ is the set of points x such that some neighborhood of x is disjoint from U; $U^{-\prime-}$ is the set of points x such that every neighborhood of x contains a point y with some neighborhood of y disjoint from U; and $U^{-\prime-\prime}$ is the set of points x such that some neighborhood of x contains no point y with a neighborhood of y disjoint from U. This latter set coincides with the set of points x such that some neighborhood of x consists entirely of points y, every neighborhood of which intersects U, which in turn is the set of points x such that some neighborhood of x consists entirely of points of U^-. Thus $U^{-\prime-\prime-}$ is the set of points x such that every neighborhood of x contains a point y that has a neighborhood consisting entirely of points of U^-. Now we must show that this last set is equal to U^-. Indeed, if $x \in U^-$, then every open neighborhood V of x intersects U and if $y \in U \cap V$, then y has a neighborhood consisting entirely of points of U^-, namely $U \cap V$. (This is a neighborhood of y because U is open.) Conversely, if $x \in U^{-\prime-\prime-}$ as described, and if V is an open neighborhood of x, then there is a $y \in V$ with y having a neighborhood W consisting entirely of points of U^-. In particular, y itself is in U^- so that, because V is a neighborhood of y, it must be that $V \cap U \neq \emptyset$. So $x \in U^-$.

11. Solution follows from hint given and Problem 10.

12. (a) Use the fact that a finite union of countable sets is countable (essentially the same reasoning as the proof that the cofinite topology is a topology). (b) The sequence $\{x_n\}$ converges to x if and only if for some N, $x = x_n$ for all $n \geq N$. (c) The closed sets are the countable sets, the empty set, and X itself. So for an arbitrary $p \in X$, the smallest closed set containing $X\backslash\{p\}$ is X. Consequently the closure of the set contains p, though p is not the limit of a sequence in the set.

Section 2. Subspaces

1. A set F in E is open in E in the topology inherited from X if and only if $F = E \cap U$ for some open subset U of X. In that case $F = E \cap (S \cap U)$, and $S \cap U$ is relatively open in S, so F is open in E in the topology inherited from S. Conversely if F is open in E in the topology inherited from S then $F = E \cap V$ where $V \subset S$ and V is relatively open in S. Then $V = S \cap U$ for some open set U in X. Hence $F = E \cap V = E \cap (S \cap U) = E \cap U$, and

F is relatively open in the topology E inherits from X.

2. It suffices to show that if $x \in S$ is adherent to $A \cap S$ in the relative topology of S, then x is adherent to A in X. For this, suppose U is open in X and $x \in U$. Then $U \cap S$ is open in S so $U \cap S$ contains a point of $A \cap S$, by the adherence of x to $A \cap S$. Hence U contains a point of A. So x is adherent to A in X. For the example, let X be the closed interval $[-1, 1]$, let $A = [-1, 0)$, and let $S = [0, 1]$. Then $A \cap S$ is empty, as is the closure of $A \cap S$ in S. But $\bar{A} \cap S = \{0\} \neq \emptyset$.

3. This is straightforward from the definition of relative topology.

4. If S is open in X, then relatively open subsets of S are open in X, since they have the form $S \cap V$, where V is open in X. Conversely, S is always open in itself in the relative topology, so if every relatively open set in S is open in X, then S is open in X. The statement is also true for closed sets. Suppose every set in S that is closed in S is closed in X. Then, since S is closed in itself, S is closed in X. Conversely, suppose S is closed in X and C is a relatively closed set in S. Since $S \backslash C$ is open in S, it has the form $S \cap V$, where V is open in X. Then $X \backslash C = (X \backslash S) \cup (V \cap S) = (X \backslash S) \cup V$. Since $X \backslash S$ and V are open, $X \backslash C$ is open and C is closed in X.

5. Such a set A has the form $S \cap U$ where U is an open subset of X. Then $A \cap T = (S \cap U) \cap T = (S \cap T) \cap U$ so $A \cap T$ is relatively open in $S \cap T$.

6. If $A \subset S \cap T$ and if $A = S \cap U$ and $A = T \cap V$, then $A = (S \cup T) \cap (U \cap V)$. If U and V are open in X, then so is $U \cap V$, so A is open in $S \cup T$.

Section 3. Continuous Functions

1. Use the reasoning of the proof of Theorem I.6.2.

2. If f is continuous and E is closed in Y, then $Y \backslash E$ is open in Y, so $f^{-1}(Y \backslash E) = X \backslash f^{-1}(E)$, is open in X, and $f^{-1}(E)$ is closed in X. Conversely, suppose $f^{-1}(E)$ is closed in X when E is closed in Y. If V is open in Y, then $f^{-1}(Y \backslash V) = X \backslash f^{-1}(V)$ is closed in X, and $f^{-1}(V)$ is open in X.

3. (a) Note that $f^{-1}(S)$ is open in X for every subset S of Y, in particular for every open subset of Y. (b) Follow the hint. (c) Since $\emptyset$ and Y are the only open subsets of Y, and $f^{-1}(\phi) = \emptyset$ and $f^{-1}(Y) = X$ are open in X, f is continuous. (d) Consider the identity map from Y, with the indiscrete topology, to Y, with the given topology.

4. Any finite open interval is homeomorphic to $(0, 1)$ via a linear map $x \to ax + b$. The function $x \to x/(1 + x)$ maps $(0, +\infty)$ homeomorphically onto $(0, 1)$. The inverse tangent function maps the real line homeomorphically onto a finite open interval, which is homeomorphic to $(0, 1)$.

5. This is similar to Exercise 4.

6. The map sending (x, y) to $(2x - 1, 2y - 1)$ implements a homeomorphism of the given square and the square $\{|x| < 1, |y| < 1\}$. In turn, this latter square can be mapped homeomorphically to the open unit disk by contracting along radii. An explicit homeomorphism is given by sending $(x, y) = (r \cos \theta, r \sin \theta)$ to $|\cos \theta|(x, y)$ if $|x| \geq |y|$ and to $|\sin \theta|(x, y)$ if $|y| \geq |x|$.

7. In polar coordinates, map (r, θ) to $(r + 1, \theta)$.

8. First construct a homeomorphism of a closed disk minus the closed disk with

same center and half the radius, onto the same (full size) closed disk minus the center, with the additional property that it is constant on the boundary (e.g., for the closed unit disk, $(r, \theta) \to (2r - 1, \theta)$). Then apply this to appropriate disks centered at the integral lattice points in the plane.

9. One such homeomorphism is obtained by considering the right half plane as a subset of the complex z-plane, $z = x + iy$. The fractional linear transformation $f(z) = (z-1)/(z+1)$ gives the homeomorphism requested.

11. This is "metamathematically" obvious, since the definition of metrizability uses only the concept of open sets and the property of being open or not is preserved under homeomorphism.

12. Apply Theorem 3.2 plus the (easy) facts that the maps (x, y) to $x + y$, xy, and x/y are all continuous (where defined).

13. This follows from the definition of the relative topology and the fact that $(f|_S)^{-1}(V) = f^{-1}(V) \cap S$ for V in Y.

14. The metric space properties are easy to check. For completeness, suppose $\{f_n\}$ is a Cauchy sequence in $BC(X)$. For each $x \in X$, $\{f_n(x)\}$ is a Cauchy sequence or real numbers, so $g(x) = \lim f_n(x)$ exists and defines g. Clearly $f_n \to g$ in the metric d, and from this it is easy to see that g is bounded. We check that g is continuous. Let $\varepsilon > 0$ be given. Choose n such that $d(g, f_n) < \varepsilon/3$. For given $p \in X$, choose a neighborhood U of p such that $|f_n(p) - f_n(q)| < \varepsilon/3$ for $q \in U$. Then for $q \in U$ we have $|g(q) - g(p)| \leq |f_n(q) - g(q)| + |f_n(q) - f_n(p)| + |f_n(p) - g(p)| < \varepsilon/3 + \varepsilon/3 + \varepsilon/3 = \varepsilon$. (For virtually the same proof, see Exercise I.6.10.)

Section 4. Base for a Topology

1. If X is an infinite set, then any countably infinite subset S is dense in X. Indeed, if U is open, then $X \backslash U$ is finite, so $S \cap U \neq \emptyset$. To answer the question, note that if X is countable, then the collection of finite subsets of X is countable, so the family of sets with finite complement is also countable, and the family of all open sets forms a countable basis. On the other hand, suppose X is uncountable. Let $\{U_n\}$ be any countable collection of open sets. Since each $X \backslash \{x\}$ is open, $\cup(X \backslash U_n)$ is countable, and there is a point $x \notin \cup(X \backslash U_n)$. Then $X \backslash \{x\}$ is open but contains no U_n. So the cofinite topology of an uncountable set is not second countable.

2. Since the singleton sets of the form $\{x\}$, $x \in X$, are open, and since every open set is a union of base elements, every base contains all the singleton sets. On the other hand, every nonempty open set is a union of singletons, so the family of singleton sets forms a base.

3. If f is continuous, then since each $U \in \mathcal{B}$ is open in Y, $f^{-1}(U)$ is open in X. Conversely, suppose $f^{-1}(U)$ is open in X for each $U \in \mathcal{B}$. Any open set V in Y can be expressed as the union of sets U_λ in $\mathcal{B}$, where λ is in some index set. Then $f^{-1}(V) = f^{-1}(\cup U_\lambda) = \cup f^{-1}(U_\lambda)$ is open. Hence f is continuous.

4. In (a), take $U_n = B(x, 1/n)$. In (b), take the U_n's to be the sets in a countable base that contain x. For (c), if x is adherent to S, then each set $S \cap U_n$ is nonempty. If s_n is any point in $S \cap U_n$, then the sequence $\{s_n\}$ converges to x.

5. For (a), note that the family of sets consisting of $\emptyset$, X, and all unions of finite

intersections of sets in $\mathcal{S}$ forms a topology, and clearly any topology containing $\mathcal{S}$ contains these sets. Part (b) is clear from part (a). For (c), use the fact that inverse functions preserve unions and intersections.

6. (a) The intersection of two half-open intervals of the form $[a,b)$ is either empty or a half-open interval. Thus the family of all unions of half-open intervals together with the empty set is closed under finite intersections, hence forms a topology, which has the half-open intervals as a base. (b) Any finite open interval (a,b) is the union of half-open intervals of the form $[a+\varepsilon, b)$, and any open set of real numbers is a union of finite open intervals. (c) The complement of $[a,b)$ is the union of the $\mathcal{T}$-open sets $(-\infty, a)$, $[b, b+2)$, and $(b+1, +\infty)$, hence is $\mathcal{T}$-open, and $[a,b)$ is $\mathcal{T}$-closed. (d) If t is in the $\mathcal{T}$-closure of S, then the $\mathcal{T}$-open neighborhood $[t, t+1/n)$ contains a point t_n of S. The sequence $\{t_n\}$ has the desired properties. (e) Suppose f is $\mathcal{T}$-continuous. Let $t \in \mathbf{R}$ and $\varepsilon > 0$, and let U be the open interval $(f(t)-\varepsilon, f(t)+\varepsilon)$. Then $f^{-1}(U)$ is $\mathcal{T}$-open, so there is $\delta > 0$ such that $f^{-1}(U)$ contains $[t, t+\delta)$. Thus $|f(s)-f(t)| < \varepsilon$ for $t < s < t+\delta$, and f is continuous from the right at t. Conversely, suppose f is continuous from the right at t. Let U be an open subset of $\mathbf{R}$, in the usual topology, and let $t \in f^{-1}(U)$. Then $f(t) \in U$. Since f is continuous from the right, there is $\varepsilon > 0$ such that $f(s) \in U$ for $t < s < t+\varepsilon$, that is, for s in the $\mathcal{T}$-open set $[t, t+\varepsilon)$. Thus f is $\mathcal{T}$-continuous at t, and by Theorem 3.1, f is continuous.

7. If $\{U_n\}$ is a countable base for the topology of X, then the sets $S \cap U_n$ form a countable base for the relative topology of any subset S of X.

8. The definition of $\mathcal{T}$ as it stands is incorrect; we take $\mathcal{T}$ to consist of all sets of the form $U = V\backslash A$, where V is an open subset of the $\mathbf{R}^2$ (in the usual topology), and A is an arbitrary subset of $\mathbf{R}$. This is clearly a topology. Let $z = (x, 0) \in \mathbf{R}$. For each $r > 0$, the set $(B(z,r)\backslash\mathbf{R}) \cup \{\mathbf{z}\}$ is a $\mathcal{T}$-open set containing z, and every $\mathcal{T}$-open set containing z contains a set of this form. From this it follows easily that the points with rational coordinates are $\mathcal{T}$-dense, so that the space is separable. The relative $\mathcal{T}$-topology for $\mathbf{R}$ is the discrete topology, which is neither separable nor second-countable. In view of Exercise 7, the topology $\mathcal{T}$ is not second-countable.

Section 5. Separation Axioms

1. Suppose $p_n \to p$ and $p_n \to q$ with $p \neq q$. By the Hausdorff property, there are disjoint open sets U and V such that $p \in U$ and $q \in V$. By definition of convergence, there is an N such that $p_n \in U$ for $n > N$. Similarly, there is an M such that $p_n \in V$ for $n > M$. For n larger than both N and M, we have $p_n \in U \cap V$, contradicting $U \cap V = \emptyset$.

2. If X is a T_1-space, then one-point sets in X are closed and hence finite sets in X are closed. So the complements of finite sets in X are open in X. But then the identity map $I : X \to X_0$ is continuous, since I^{-1} of each open set in X_0 is open in X. Conversely, if the identity map I is continuous, then since inverse images of closed sets under a continuous map are closed (Exercise 3.2), each set S that is closed in X_0 is closed in X. Since one-point sets are closed in X_0, they are also closed in X, and hence X is a T_1-space.

3. Suppose Y is a subspace of X. If X is a T_1 space and $y \in Y$, then $X\backslash\{y\}$ is open in X, so by the definition of subspace, $Y\backslash\{y\} = (X\backslash\{y\}) \cap Y$ is open in Y. Thus $\{y\}$ is closed in Y, and Y is a T_1-space. If X is a Hausdorff space and $y, z \in Y$ are distinct, then there are disjoint open subsets U, V of X with $y \in U$, $z \in V$. Then $Y \cap U$ and $Y \cap V$ are disjoint relatively open sets in Y containing y and z respectively, and Y is Hausdorff. Finally, suppose X is regular. Let $y \in Y$, and suppose E is a relatively closed subset of Y such that $y \notin E$. Then $Y\backslash E$ is open in Y, so there is an open set W in X with $Y\backslash E = Y \cap W$.

Then $X\backslash W$ is closed in X, and $y \notin X\backslash W$. Since X is regular, there are disjoint open sets U, V with $y \in U$ and $X\backslash W \subset V$. Then $Y \cap U$ and $Y \cap V$ are disjoint open sets in Y with $y \in Y \cap U$ and $E \subset Y \cap V$. Thus Y is regular.

4. The forward implications are trivial, in the sense that they are just the results proved in the text. Suppose that Urysohn's Lemma is valid for X, and suppose E and F are disjoint closed sets in X. If $f : X \to [0,1]$ is a continuous function such that $f = 0$ on E and $f = 1$ on F, then $f^{-1}([0,1/2))$ and $f^{-1}((1/2,1])$ are disjoint open sets containing E and F respectively. Thus X is normal. Suppose next that Tietze's Theorem is valid. Suppose E and F are disjoint closed sets in X. Then $E \cup F$ is closed. Define $h : E \cup F \to [0,1]$ by $h = 0$ on E and $h = 1$ on F. Evidently h is continuous on $E \cup F$. Let $f : X \to [0,1]$ be a continuous extension of h to X. Again $f^{-1}([0,1/2))$ and $f^{-1}((1/2,1])$ are disjoint open sets containing E and F respectively, and X is normal.

5. Clearly E and F are disjoint. Since $X\backslash E$ is the intersection of X with the union of the open upper and left half-planes, $X\backslash E$ is relatively open in X. Hence E is a closed subset of X, as is F. For the continuous function f, pass to polar coordinates and set $f(r\cos\theta, r\sin\theta) = 1 - \theta/\pi$, where $0 \le \theta \le \pi$.

6. Following the hint, let h be an extension to X of $g \circ f$. Then $h^{-1}(\{-1,1\})$ and E are disjoint closed sets in X. Choose $F : X \to [0,1]$ such that F is continuous, $F = 0$ on $h^{-1}(\{-1,1\})$, and $F = 1$ on E. Then the range of $(g \circ f)F$ is contained in $(-1,+1)$, and the function $g^{-1} \circ ((g \circ f)F)$ gives the required extension of f.

7. Since one-point sets are closed in the cofinite topology, the cofinite topology is always a T_1-topology. If X is infinite, then any two nonempty open sets in the cofinite topology have nonempty intersection (in fact, their intersection is cofinite). It follows that the cofinite topology on X is a Hausdorff topology if and only if X is finite.

8. Since the complements of one-point sets are open in the usual topology, they are $\mathcal{T}$-open. Hence one-point sets are $\mathcal{T}$-closed, and $(\mathbf{R}, \mathcal{T})$ is a T_1-space. For normality, suppose E and F are disjoint $\mathcal{T}$-closed sets. Let $t \in \mathbf{R}\backslash\mathbf{E}$. Let a be the supremum of r for $r \in E$, $r < t$, and let b be the infimum of r for $r \in E$, $r > t$. We set $a = -\infty$ if $E \cap (-\infty, t) = \emptyset$, and similarly for b. Then (a,b) is disjoint from E, and since b is a decreasing limit of points in E, we have $b \in E$. While a is an increasing limit of points in E, it might occur that $a \notin E$, and in fact it might occur that $a \in F$ or that $a = t$. In any event, since $t \in \mathbf{R}\backslash\mathbf{E}$ is arbitrary, this shows that the complement of E is the disjoint union of sets of one of the forms (a,b) or $[a,b)$, where (ignoring the cases where $a = -\infty$ or $b = +\infty$) we have $b \in E$. For each such interval such that $[a,b) \cap F$ is nonempty, define c to be the infimum of $s \in [a,b) \cap F$. Then $c \in F$. Let W be the union of all intervals $[c.b)$ obtained in this way. Then W is $\mathcal{T}$-open, W contains F, and W is disjoint from E. We claim that W is also $\mathcal{T}$-closed. Indeed, suppose s is adherent to W. Then there is a decreasing sequence s_n in W that converges to s. Suppose that $[c_n, b_n)$ is the interval constructed above that contains s_n. If the intervals are different, then also c_n and b_n decrease to s, so that $s \in E \cap F$, contradicting the hypothesis that the sets are disjoint. Thus for large n the intervals are the same, say $[c,b)$, in which case the limit s also belongs to $[c,b)$, and W is closed. Thus E and F are separated by the $\mathcal{T}$-open sets W and $\mathbf{R}\backslash\mathbf{W}$.

9. The intersection of any two sets in $\mathcal{B}$ either is empty or belongs to $\mathcal{B}$, so $\mathcal{B}$ is a base for a topology $\mathcal{T}$. Since the usual topology for $\mathbf{R}$ is Hausdorff, and $\mathcal{T}$ includes the usual topology, $\mathcal{T}$ is Hausdorff. Since $\mathbf{R_0}$ is the union of sets of the form $(a,b) \cap \mathbf{R_0}$, $\mathbf{R_0}$ is $\mathcal{T}$-open, and $\mathbf{R}\backslash\mathbf{R_0}$ is $\mathcal{T}$-closed. Suppose f is a continuous function as in (d), and suppose $f(p) \neq 0$

for some $p \in \mathbf{R}_0$, say $f(p) = 1$. Then by continuity there is an open interval (a, b) containing p such that $f(s) > 3/4$ for $s \in (a,b) \cap \mathbf{R}_0$. Let q be any irrational number in (a, b). We are assuming that $f(q) = 0$. By continuity again there is an interval (c, d) containing q such that $f(s) < 1/4$ for all $s \in (c, d)$. Since there are rational numbers in $(c,d) \cap (a,b)$, this is a contradiction. We conclude that $f = 0$, and (d) is established. For (e), consider the disjoint $\mathcal{T}$-closed sets $\mathbf{R} \backslash \mathbf{R}_0$ and $\{0\}$. Any $\mathcal{T}$-open set containing 0 must contain an interval of rational numbers $J = (-a, a) \cap \mathbf{R}_0$. In turn, any open set containing an irrational number in $(-a, a)$ must contain an open interval, hence must meet J. It follows that the closed set and point are not contained in disjoint $\mathcal{T}$-open sets, and the topology is not regular.

Section 6. Compactness

1. Suppose Y is compact and $h : X \to Y$ is a homeomorphism. Let $\{U_\alpha : \alpha \in A\}$ be an open cover of X. Then $\{h(U_\alpha) : \alpha \in A\}$ is an open cover of Y. By compactness of Y, there are $\alpha_1, \ldots, \alpha_n$ such that $h(U_{\alpha_1}) \cup \cdots \cup h(U_{\alpha_n}) = Y$. Then $U_{\alpha_1} \cup \cdots \cup U_{\alpha_n} = X$, and X is compact.

2. Suppose X has the indicated property, which we call the *finite intersection property.* If $\{U_\alpha : \alpha \in A\}$ is an open cover of X, then the closed sets $\{X \backslash U_\alpha : \alpha \in A\}$ have empty intersection. Hence the intersection of some finite subfamily of these closed sets is empty, and the corresponding U_α's form a finite cover of X. Conversely, suppose X is compact. Let $\{E_\alpha : \alpha \in A\}$ be a family of closed sets with empty intersection. Then the sets $X \backslash E_\alpha$ form an open cover of X. We extract a finite subcover of the $X \backslash E_\alpha$'s, and the corresponding intersection of E_α's is empty. Thus X has the finite intersection property.

3. Suppose $\{U_\alpha\}$ is an open cover of X. Choose any index α_0 such that $U_{\alpha_0} \neq \emptyset$. Then $X \backslash U_{\alpha_0} = \{x_1, \ldots, x_n\}$ is finite. Choose U_{α_j} with $x_j \in U_{\alpha_j}$. Then $\{U_{\alpha_0}, U_{\alpha_1}, \ldots, U_{\alpha_n}\}$ is a finite subcover.

4. The family of one-point sets $\{x\}$, for $x \in X$, forms an open cover of X, which has a finite subcover only if X is finite. Thus a compact discrete space is finite. The converse is trivial.

5. If $f : X \to \mathbf{R}$ is continuous, and X is compact, then by Theorem 6.6, $f(X)$ is a compact subset of $\mathbf{R}$. By the Heine-Borel Theorem (Theorem I.5.5), $f(X)$ is a closed bounded subset of $\mathbf{R}$. Hence $f(X)$ contains its supremum and infimum, and consequently f attains its maximum and minimum values.

6. By Theorem 6.5, a compact Hausdorff space is normal. Since one-point sets in a Hausdorff space are closed, we can apply Urysohn's Lemma (Theorem 5.3) to the sets $\{x\}$ and $\{y\}$.

7. The construction is explained in the remark following the problem.

8. Follow the hint. Note that the final step in the proof is a 3ε-argument, which establishes that the subsequence is a Cauchy sequence with respect to the metric of uniform convergence on X (defined in Exercise 9).

9. Since X is compact, each continuous function $f : X \to \mathbf{R}$ is bounded. The conclusions now follow from Problem 3.14.

10. Use the hint for the reverse implication. For the forward implication, suppose that F is a compact subset of $C(X)$. The nested family of open sets $U_M = \{f \in C(X) : d(f, 0) < M\}$ cover $C(X)$. By compactness, F is contained in one of these sets, and so F is

bounded. The compact set F is closed in $C(X)$, by Corollary 6.4, since $C(X)$ is a Hausdorff space. To see equicontinuity, suppose $\varepsilon > 0$ is given. Since F is totally bounded (Theorem I.5.1), there exists a finite subset $\{f_1, \dots, f_m\}$ of F such that each $f \in F$ has distance less that ε from some f_j. Let $x \in X$. Choose a neighborhood U of x such that $|f_j(x) - f_j(y)| < \varepsilon$ for all $y \in U$ and $1 \leq j \leq m$. If $f \in F$, choose k such that $d(f, f_k) < \varepsilon$, and estimate $|f(x) - f(y)| \leq |f(x) - f_k(x)| + |f_k(x) - f_k(y)| + |f_k(y) - f(y)| < 3\varepsilon$ for $y \in U$. This establishes equicontinuity.

11. (a) Given a fixed stage of subdivision of the square, the curves eventually all belong to a given subsquare of that subdivision over the same time interval. Uniform convergence follows. (b) Given a point q, at the jth stage choose a parameter point t_j such that $f_j(t_j)$ lies in the same square as q. If the t_j's have t as an accumulation point, then the uniform convergence shows that $f(t) = q$. (c) Let M denote the sequence of moves for f_{k-1}. Let P denote the sequence of moves corresponding to rotating M counterclockwise by 90° and running it backwards. Let Q denote the sequence of moves corresponding to rotating M clockwise by 90° and running backwards. Then the sequence of moves for f_k is given by $P(R)M(D)M(L)Q$.

Section 7. Locally Compact Spaces

1. Suppose X is locally compact. Let $p \in X$ and let U be an open neighborhood of p. By the definition of local compactness, there is an open set W containing x such that $\overline{W}$ is compact. Then $V = W \cap U$ is an open subset of U containing p, and $\overline{V}$ is a closed subset of the compact set $\overline{W}$ hence compact. The reverse implication is trivial. Note that the hypothesis of being Hausdorff is superfluous.

2. Suppose Y is locally compact and $h : X \to Y$ is a homeomorphism. Let $p \in X$. Choose an open subset V of Y such that $h(p) \in V$ and $\overline{V}$ is compact. Set $W = h^{-1}(V)$. Then W is an open neighborhood of p. Since the image of $\overline{V}$ under the continuous function h^{-1} is $\overline{W}$, $\overline{W}$ is compact. Hence X is locally compact.

3. Let X be a locally compact Hausdorff space. Let C be a closed subset of X, and let $p \in X \backslash C$. Let W be an open set containing p such that $\overline{W}$ is compact. Replacing W by $W \backslash C$ (cf. Problem 1), we can assume that $W \cap C = \emptyset$. Now $\overline{W} \cap C$ is a closed subset of a compact Hausdorff space, hence compact. By Lemma 6.3, there are disjoint open sets U and V such that $p \in U$ and $\overline{W} \cap C \subset V$. Then $U \cap W$ and $V \cup (X \backslash \overline{W})$ are disjoint open sets containing p and C respectively.

4. Let p and q be distinct points of X, and let U be an open neighborhood of p such that $\overline{U}$ is a compact Hausdorff space. Since a compact Hausdorff space is regular, there are disjoint relatively open subsets V and W of $\overline{U}$ such that $p \in V$ and $(\overline{U} \backslash U) \cup \{q\} \subset W$. Since V is relatively open in $\overline{U}$ and $V \subset U$, V is open in X. Further, $q \notin \overline{V}$. Hence V and $X \backslash \overline{V}$ are disjoint open neighborhoods of p and q respectively, and X is Hausdorff.

5. The family $\mathcal{S}$ is still a topology on $Y = X \cup \{\infty\}$ for which Y is compact, since the argument given does not use the Hausdorff property of X. If X is not locally compact, then Y need not be a Hausdorff space even if X is Hausdorff. (Indeed, if Y is Hausdorff then X is locally compact.) If X is Hausdorff, the subspace topology for X as a subset of Y coincides with the given topology of X.

6. Suppose U is an open neighborhood of ∞ in Y. Then $X \backslash U$ is compact, by definition of the topology for Y. Since X is not compact, $X \backslash U \neq X$, and X meets U. Hence ∞ is adherent to X.

7. Use the hint.

8. The open n-ball is homeomorphic to $\mathbf{R^n}$. Now apply Exercise 7.

Section 8. Connectedness

1. If $h : X \to Y$ is a homeomorphism, and U and V are open, disjoint, nonempty subsets of Y whose union is Y, then $h^{-1}(U)$ and $h^{-1}(V)$ are open, disjoint, nonempty subsets of X whose union is X.

2. Let S be connected. If $a, b \in S$, $a < b$, then S includes every x such that $a < x < b$; indeed, if $x \notin S$, then $\{s \in S : s < x\}$ and $\{s \in S : s > x\}$ are open disjoint nonempty sets in S with union equal to S, contradicting connectedness. It follows that $S \supset (\inf S, \sup S)$, so S coincides with the interval $(\inf S, \sup S)$ together with possibly one or both of its endpoints.

3. The union of the open discs is not connected, since the discs themselves form a decomposition of the union into disjoint nonempty open subsets. Any open disc with all or part of its boundary is connected, since it is a union of radii issuing from the center, and Theorem 8.2 applies. Theorem 8.2 then also shows that the union of any two closed discs that meet at a point is connected. Similarly, if we adjoin the boundary point to the open disc and apply Theorem 8.2, we see that the union of one of the open discs together with the closure of the other is connected.

4. Let $h : X \to Y$ be a homeomorphism. If q is a cut point of Y, then $Y\backslash\{q\}$ is the disjoint union of nonempty open sets U and V. Then $X\backslash\{h^{-1}(q)\}$ is the disjoint union of the nonempty open sets $h^{-1}(U)$ and $h^{-1}(V)$, so that $h^{-1}(q)$ is a cut point for X.

5. Every point of the interval $(0, 1)$ is a cut point, while $[0, 1)$ has one non-cut point, and $[0, 1]$ has two non-cut points. The argument given in Exercise 4 shows that any homeomorphism caries cut points to cut points, so that the number of non-cut points is a topological invariant. Thus no two of the given spaces are homeomorphic.

6. No two of the spaces are homeomorphic. The square contains no cut points, while the other three have cut points. The second space has exactly three non-cut points, the third space has infinitely many, and the fourth space has exactly four.

7. First note that an open ball B in $\mathbf{R^n}$, $n \geq 2$, contains no cut points. Indeed, if $x \in B$, then $B\backslash\{x\}$ is the union of broken line segments starting at some fixed point, which are connected, so that by Theorem 8.2, $B\backslash\{x\}$ is connected. Now f is a one-to-one continuous map onto its image, and $[0, 1]$ is compact, so by Theorem 6.7, f is a homeomorphism onto its image. If the image of f contains an open ball B, then f maps $f^{-1}(B)$ homeomorphically onto B. But $f^{-1}(B)$ has at most two non-cut points, and this contradicts the fact that every point of B is a cut point.

8. Let X be a countable metric space with metric d. Fix $x \in X$. Since the distances from x to other points form a countable set, there is a decreasing sequence $r_n \to 0$ such that $d(x, y) \neq r_n$ for all $y \in X$ and $n \geq 1$. Then the ball $B(x, r_n)$ is both open and closed in X, so that the connected component of x is contained in this ball. Since the balls shrink to x, the connected component of x is the singleton $\{x\}$.

9. It suffices to show that the closure of a connected set is connected. Let E be a connected subset of X. Suppose U and V are disjoint relatively open subsets of $\overline{E}$ whose union is $\overline{E}$. Then $U \cap E$ and $V \cap E$ are disjoint relatively open subsets of E whose union is E. Since E is connected, one of these sets is empty, say $V \cap E = \emptyset$. Now $V = W \cap \overline{E}$ for

some open subset W of X. Since W is disjoint from E, no point of W lies in the closure of E. It follows that $V = \emptyset$, and we conclude that $\overline{E}$ is connected.

10. The connected components of $X = \{0\} \cup \{1/n : n = 1, 2, \dots\}$ are all singletons, by Exercise 8, though $\{0\}$ is not open.

11. Suppose C is the connected component containing a point p, and V is a connected open set containing p. Then $C \cup V$ is connected, by Theorem 8.2. Hence $C = C \cup V$, and C contains a neighborhood of p.

12. If $a, b \in \mathbf{R}$ and $a < b$, then $(-\infty, b)$ and $[b, +\infty)$ are disjoint $\mathcal{T}$-open subsets of $\mathbf{R}$ containing a and b respectively. So a and b do not belong to the same $\mathcal{T}$-component of $\mathbf{R}$, and the space is totally disconnected.

13. The space $(\mathbf{R}, \mathcal{T})$ is connected. To see this, suppose that $\mathbf{R} = \mathbf{U} \cup \mathbf{V}$ is a disjoint decomposition of $\mathbf{R}$ into $\mathcal{T}$-open subsets. Let $p \in U$. If p is irrational, then there is an open interval containing p that is contained in U. If p is rational, then there is an $\varepsilon > 0$ such that $(p - \varepsilon, p + \varepsilon) \cap \mathbf{R_0} \subset \mathbf{U}$. If q is any irrational in this interval, then there is an open interval containing q that is completely contained in either U or V, and since the rational points near q are contained in U, the interval is contained in U. In particular, $q \in U$, and $U \supset (p - \varepsilon, p + \varepsilon)$. In any event, U contains an open interval about each of its points, and U is open in the usual topology of $\mathbf{R}$. By the same token, V is open in the usual topology of $\mathbf{R}$. Since $\mathbf{R}$, with the usual topology, is connected, either U or V is empty. Thus $\mathbf{R}$ is $\mathcal{T}$-connected.

14. Suppose Y is disconnected. Let $Y = U \cup V$ be a disjoint decomposition of Y into nonempty open subsets. Since X is not compact, $X \cap U$ and $X \cap V$ are both nonempty, and they decompose X into disjoint open sets. Hence X is disconnected. Thus Y is connected whenever X is connected. But Y can be connected even if X is not. For example, take X to be the punctured interval $[-1, 1] \backslash \{0\}$. Then X is disconnected, though its one-point compactification is homeomorphic to $[-1, 1]$, which is connected.

Section 9. Path Connectedness

1. If a, b belong to an interval (of any kind), then $\gamma(t) = (1 - t)a + tb$, $0 \leq t \leq 1$, defines a path from a to b in the interval.

2. See Exercise 3 below.

3. If $p = f(x)$ and $q = f(y)$, and γ is a path in X from p to q, then $f \circ \gamma$ is a path in $f(X)$ from x to y.

4. If a path component contains a point x, then it contains the neighborhood U of x appearing in the definition of local path-connectedness. Hence path components are open. The complement of a path component is a union of path components, so the complement of a path component is also open. It follows that each connected component is contained in a path component. Since path components are connected, the path components coincide with the connected components.

5. For (a), see Exercise 4 above. For (b), take U to be the path component of V containing x, and apply (a) to V.

6. An open subset of $\mathbf{R^n}$ is locally path-connected, so Exercise 4 applies.

7. It is easy to check that the path components of X are E and F. Each connected

component of X is closed, by Exercise 8.9, and it is a union of path components. Since F is not closed, the connected component of X containing F must be $E \cup F = X$. Thus X is connected.

8. In Exercise 7, F is not closed and E is not open.

Section 10. Finite Product Spaces

1. The complement of $E_1 \times \cdots \times E_n$ is the union of open sets $X_1 \times \cdots \times X_{k-1} \times (X_k \backslash E_k) \times X_{k+1} \times \cdots \times X_n$, hence is open.

2. It suffices to show that $X_1 \times X_2$ is regular whenever X_1 and X_2 are regular. Suppose E is a closed subset of $X = X_1 \times X_2$, and let $p = (p_1, p_2) \in X \backslash E$. Since $X \backslash E$ is open, there exist open sets $U_1 \subset X_1$ and $U_2 \subset X_2$ such that $(p_1, p_2) \in U_1 \times U_2$ and $U_1 \times U_2 \subset X \backslash E$. By regularity of the X_i's, there are open sets V_1 and V_2 such that $p_i \in V_i$ and $\overline{V}_i \subset U_i$. Then $V = V_1 \times V_2$ is an open set containing p, whose closure $\overline{V}_1 \times \overline{V}_2$ is contained in $U_1 \times U_2$ hence disjoint from E.

3. Let E_j be a connected component of X_j. By Theorem 10.6, $E_1 \times \cdots \times E_n$ is connected. Let S be any connected set containing $E_1 \times \cdots \times E_n$. By Theorem 8.1, each $\pi_j(S)$ is connected, and it contains E_j. Hence $\pi_j(S) = E_j$, and $S \subset E_1 \times \cdots \times E_n$.

4. We can assume $X = X_1 \times X_2$. For (a), suppose $p, q \in X_1$, $p \neq q$. Fix a point y in X_2. Then there are disjoint open subsets U and V of X containing (p, y) and (q, y) respectively. The slices $\{x \in X_1 : (x, y) \in U\}$ and $\{x \in X_1 : (x, y) \in V\}$ are then disjoint open neighborhoods of p and q in X_1. Hence X_1 is Hausdorff. Parts (b) and (c) are similar. For (d), observe that the coordinate projection is continuous and apply Theorem 8.1. Similarly, part (e) follows from Exercise 9.3, and part (f) from Theorem 6.6.

5. Since $\mathcal{S}$ contains the usual metric topology of $\mathbf{R}^2$, any set that is closed in the usual topology is also $\mathcal{S}$-closed. Since each $\mathcal{S}$-open set $J_\varepsilon(x) = [x, x + \varepsilon) \times [-x, -x + \varepsilon)$ meets L in the singleton set $\{(x, -x)\}$, the singletons are relatively $\mathcal{S}$-open, and the relative toplogy of L is discrete. Hence any subset of L is relatively closed and open in L, with respect to $\mathcal{S}$. Since L is $\mathcal{S}$-closed, any subset of L is $\mathcal{S}$-closed in $\mathbf{R}^2$. Let F_n be the set of points $x \in \mathbf{R}$ such that $J_\varepsilon(x) \subset V$ for $\varepsilon = 1/n$. By the Baire Category Theorem, there is N such that the closure $\overline{F}_N$ of F_N in the usual metric topology has nonempty interior. If $s \in S$ is in the interior of $\overline{F}_N$, then $b = (s, -s) \in E$ has the property given in (d). Let U and V be $\mathcal{S}$-open sets containing E and F respectively, and choose $\delta > 0$ such that $J_\delta(s) \subset U$. If the x_k's are as in (d), then the sets $J_\varepsilon(x_k)$ eventually meet $J_\delta(s)$, so that U and V are not disjoint.

Section 11. Set Theory and Zorn's Lemma

1. We take the ordering to be set inclusion. If $\mathcal{T}$ is a totally ordered subset of $\mathcal{S}$, then the union U of all $T \in \mathcal{T}$ is in $\mathcal{S}$. Indeed, since $\mathcal{T}$ is totally ordered, any finite subset of U belongs to some $T \in \mathcal{T}$, hence to $\mathcal{S}$. By Zorn's Lemma, $\mathcal{S}$ has a maximal element.

2. Let e be the identity of R. An ideal I is proper if and only if $e \notin I$. We order the proper ideals by set inclusion. If $\mathcal{I}$ is a totally ordered set of proper ideals, then the union of $I \in \mathcal{I}$ is an ideal. Since e does not belong to any of the ideals in $\mathcal{I}$, e does not belong to their union, and the union is proper. Now apply Zorn's Lemma.

3. Consider the family $\mathcal{S}$ of all linearly independent subsets S of V that contain A, and follow the proof given in the text.

4. Follow the hint. The values of L on basis elements not in W can be assigned

arbitrarily, and then L is extended by linearity.

5. Choose a vector-space basis for $\mathcal{X}$. Without loss of generality, we can assume that the basis elements have unit norm. Since $\mathcal{X}$ is infinite dimensional, the basis contains a sequence of distinct elements $v_1, v_2, v_3, \dots$. Define L on the v_n's by $L(v_n) = n$, define L to be 0 on the remaining basis elements, and extend L to $\mathcal{X}$ by linearity. Then L is unbounded, hence discontinuous.

6. Suppose there is a sequence $\{x_1, x_2, \dots\}$ that forms a basis for $\mathcal{X}$. For fixed $n \geq 1$, let E_n be the linear span of the first n basis elements. The map $(t_1, \dots, t_n) \to \sum t_k x_k$ is a continuous function from $\mathbf{R^n}$ onto E_n. Since $\mathbf{R^n}$ is a union of a sequence of compact sets, and the image of a compact set under a continuous map is compact, E_n is the union of a sequence of compact sets. By hypothesis, the union of the E_n's is all of $\mathcal{X}$. Consequently $\mathcal{X}$ is the union of a sequence of compact subsets of the various E_n's. However, each $x \in E_n$ is a limit of elements of the form $x + \varepsilon x_m$, which do not belong to E_n for $m > n$, so each E_n has empty interior. In view of the Baire Category Theorem, we have a contradiction.

Section 12. Infinite Product Spaces

1. The complement of $\prod E_\alpha$ is the union of the open sets $\pi_\alpha^{-1}(X_\alpha \backslash E_\alpha)$, hence is open.

2. If U is a basic open set given by (12.2), then $\pi_\beta(U)$ coincides with U_β if β coincides with α_k, otherwise $\pi_\beta(U)$ coincides with X_β. (We are assuming, of course, that no X_α is empty.) In any event, $\pi_\beta(U)$ is open. Since any open set is a union of basic open sets, and since functions preserve unions, the image of any open set under π_β is open.

3. As in Exercise 2, one checks that the restriction is open. Since it is continuous and one-to-one, it is a homeomorphism.

4. Suppose x and y are distinct points of the product space. Choose β such that $x_\beta \neq y_\beta$. Since X_β is Hausdorff, there are disjoint open subsets U and V of X_β such that $x_\beta \in U$ and $y_\beta \in V$. Then $\pi_\beta^{-1}(U)$ and $\pi_\beta^{-1}(V)$ are disjoint open subsets of the product space that contain x and y respectively.

5. This is similar to Exercise 4.

6. Suppose the X_α's are connected. Fix a point $x \in \prod X_\alpha$, and let E be the connected component containing x. For any fixed finite set of indices $\alpha_1, \dots, \alpha_m$, the set of $z \in \prod X_\alpha$ satisfying $z_\beta = x_\beta$ for all indices β distinct from the α_j's is homeomorphic to the (finite) product of the X_{α_j}'s, hence is connected, by Theorem 10.6. Thus each such set is contained in E. It is easy to check that the union of these sets is dense in the product space, so since E is closed (Exercise 8.9), E coincides with the entire product space.

7. The connected components of X are the sets of the form $C = \prod C_\alpha$, where each C_α is a connected component of X_α. That each such set C is connected follows from Exercise 6. On the other hand, if S is any connected subset of X, then each $\pi_\alpha(S)$ is a connected subset of X_α, by Theorem 8.1. Hence each $\pi_\alpha(S)$ is contained in a component C_α of X_α, and S is contained in the product of the C_α's.

8. If $x, y \in \prod X_\alpha$, and if $\gamma_\alpha : [0,1] \to X_\alpha$ is a continuous function with $\gamma_\alpha(0) = x_\alpha$ and $\gamma_\alpha(1) = y_\alpha$, then $\gamma(t)_\alpha = \gamma_\alpha(t)$ defines a function $\gamma : [0,1] \to \prod X_\alpha$ that is continuous, by Theorem 12.2, hence a path from x to y.

9. Follow the hint.

10. For (a), identify $x \in X$ with the set of $s \in S$ such that $x(s) = 1$. For (b), apply Tychonoff's Theorem and Exercise 12.4. For each finite subset $T = \{t_1, \ldots, t_m\}$ of S and each sequence $q = (q_1, \ldots, q_m)$ of m zeros and ones, define U_{Tq} to be the set of all $x \in X$ such that $x_{t_j} = q_j$ for $1 \le j \le m$. By definition, the sets U_{Tq} form a base for the topology of X. If S is finite or countable, there are at most countably many sets in this base, so X is second-countable. On the other hand, if S is uncountable, then no point of X can obtained as the intersection of a sequence of sets in the base, and consequently X is not first-countable.

11. (a) Note that $\mathcal{B}$ is closed under finite intersections. (b) If each X_α is discrete, then $\mathcal{B}$ includes the one-point sets, and the box topology is discrete. (c) Follow the hint and apply (b). (d) The product of Hausdorff spaces, endowed with the usual product topology, is Hausdorff. Since the box topology is larger than the usual product topology, the product is also Hausdorff when endowed with the box topology. The product of regular spaces, with the box topology, is also regular. To see this, let $U = \Pi U_\alpha$ be a basic open set containing x. For each α, choose an open neighborhood V_α of x_α with closure contained in U_α. Then ΠV_α is a basic open set whose closure is contained in U. By Exercise 10.5, not even the product of two normal spaces is necessarily normal.

12. Since E is the union of basic open sets of the form $\{y : |x(\alpha) - y(\alpha)|w(\alpha) < C\}$, E is open. If $z \notin E$, then $|x(\alpha) - z(\alpha)|w(\alpha)$ is not bounded. Thus the set of y such that $|y(\alpha) - z(\alpha)|w(\alpha) < 1$ is a basic open set containing z that is disjoint from E. Hence the complement of E is open, and E is closed. For part (b), note that in fact the connected component containing a point $x \in X$ consists of all $y \in X$ such that $y(\alpha) = x(\alpha)$ for all but at most finitely many indices α. Indeed, if $y(\alpha) \neq x(\alpha)$ for infinitely many α's, we can find a weight function $w(\alpha) > 0$ such that the set E of part (a) does not contain y.

Section 13. Quotient Spaces

1. The quotient projection maps X continuously onto the quotient space. Thus (a) follows from Theorem 6.6, that the image of a compact space under a continuous map is compact. Similarly, (b) and (c) follow from Theorems 8.1 and Exercise 9.3 respectively.

2. A subset S of $X/\sim$ is closed if and only if $\pi^{-1}(S)$ is closed, and in particular the points in the quotient space are closed if and only if equivalence classes in X are closed sets. For the example, let $X = \mathbf{R}$, and define an equivalence relation by declaring any two points of the open interval $(0, 1)$ to be equivalent. In the quotient space, the point correspoinding to $(0, 1)$ is not closed, and in fact its closure contains exactly three points.

3. As a point set, $X/\sim$ can be viewed as the disjoint union of a closed interval I and a semi-open interval J. Points of J have the usual neighborhood base, but points s of I have a neighborhood base of sets that are unions of open subintervals of I containing s and open intervals at the open end of J. Thus neighborhoods of any two points of I contain in common a subinterval at the open end of J.

4. We declare $x \sim y$ if $f(x) = f(y)$. The induced map g from $X/\sim$ to Y is then one-to-one and onto, and by Theorem 13.3, it is continuous. If U is open in $X/\sim$, then $\pi^{-1}(U)$ is open in X, so by the hypothesis, $f(\pi^{-1}(U)) = g(U)$ is open in Y. Thus g is an open mapping, hence a homeomorphism.

5. The function g that maps each point of $X \backslash E$ to the corresponding one-point equivalence class and maps ∞ to the equivalence class E is a one-to-one correspondence of the one-point compactification of $X \backslash E$ and $X/\sim$. One checks directly from the definitions

that the open sets for the one-point compactification correspond exactly to those for the quotient topology. The open subsets of $X\backslash E$ correspond to the open subsets in the quotient space that do not contain the equivalence class E, and also to the open subsets of the one-point compactification that do not contain ∞. The open subsets of the quotient space that contain the equivalence class E are exactly the complements of the compact subsets of $X\backslash E$, and these correspond to the open subsets of the one-point compactification that contain ∞.

6. Let $N = (0, \dots, 0, 1)$ denote the north pole of the sphere $S^n \subset \mathbf{R}^{n+1}$. The punctured sphere $S^n\backslash\{N\}$ is homeomorphic to $\mathbf{R}^n$, and in fact the usual stereographic projection maps the punctured sphere homeomorphically onto the subspace $\{x_{n+1} = 0\}$ of $\mathbf{R}^{n+1}$. Now $\mathbf{R}^n$ is homeomorphic to its open unit ball, via for instance the map $x \to x||x||/(1 + ||x||)$. Thus the one-point compactification of $S^n\backslash\{N\}$, which is S^n, is homeomorphic to the one-point compactification of the open unit ball of $\mathbf{R}^n$, which by Exercise 5 is obtained from the closed unit ball by identifying the boundary sphere to a point.

7. The map $f(x_1, \dots, x_n) = x_1$ is a continuous map of the product onto X_1. Since the map is constant on each equivalence class, it determines by Theorem 13.3 a continuous quotient map g so that $f = g \circ \pi$. Evidently g is one-to-one and onto. Since f is open, also g is open, and consequently g is a homeomorphism.

8. The image of of a compact space under a continuous map is compact, so P^n is compact. Let $x \in S^n$, and let U be an open set containing x with small diameter. Then $U \cup -U$ is an open subset of S^n that is a union of equivalence classes, and the quotient map π is one-to-one on U. Since $V \cup -V$ is open for each open $V \subset U$, each $\pi(V)$ is open in P^n, and π maps U homeomorphically onto the open neighborhood $\pi(U)$ of $\pi(x)$ in P^n. To see that P^n is Hausdorff, consider small neighborhoods of two non-antipodal points of S^n. For (d), define the map f from B^n to P^n by $f(x_1, \dots, x_n) = \pi(x_0, x_1, \dots, x_n)$, where $x_0 \geq 0$ satisfies $x_0^2 + \cdots + x_n^2 = 1$, and apply Theorem 13.4. Part (c) is a special case of (d), since an interval with its endpoints identified is homeomorphic to a circle.

9. Let W_j denote the set of points $(z_0, \dots, z_n) \in \mathbf{C}^{n+1}$ satisfying $z_j \neq 0$, and let $U_j = \pi(W_j)$ be the corresponding set of points in $\mathbf{CP}^n$. The map from $(z_0, \dots, z_n) \in W_j$ to $(z_0/z_j, \dots, z_{j-1}/z_j, z_{j+1}/z_j, \dots, z_n/z_j)$ is constant on equivalence classes, so by Theorem 13.3 it determines a continuous map $g : U_j \to \mathbf{C}^n$. One checks that g is one-to-one and onto, and that g is a homeomorphism. Thus distinct points of U_j have disjoint open neighborhoods, and further any point in U_j has a closed neighborhood (corresponding to a closed ball in $\mathbf{C}^n$) that is disjoint from $\mathbf{CP}^n\backslash U_j$, so that $\mathbf{CP}^n$ is Hausdorff. The projection π maps the unit sphere of $\mathbf{C}^n$ onto $\mathbf{CP}^n$, so that $\mathbf{CP}^n$ is compact. One checks that the map $w \to (\pi(w), w_j/|w_j|)$ is a homeomorphism of $W_j \cap S_{2n+1}$ and $U_j \times S^1$. In particular, each fiber of this map is a circle. In the case $n = 1$, there is only one point in $\mathbf{CP}^1\backslash U_1$, which is the equivalence class of $(1, 0)$. Since U_1 is homeomorphic to $\mathbf{C}$, $\mathbf{CP}^1$ is the one-point compactification of $\mathbf{C}$, which is homeomorphic to S^2.

Chapter III. Homotopy Theory

Section 1. Groups

1. Suppose a group has three elements e, a, b. Since $a \neq e$ we have $ab \neq a$, and similarly $ab \neq b$, so $ab = e$. Also $ba = e$, $a^2 = b$, and $b^2 = a$. This determines completely the multiplication in the group, and the group is isomorphic to Z_3.

2. Follow the hint.

3. If the group has an element of order four, it is cyclic, isomorphic to Z_4. The group cannot have an element of order three; if a, a^2, a^3, and b are the group elements, and $a^3 = e$, application of the cancellation law shows that there is no possibility for ab. (More generally, the order of any element of a finite group divides the order of the group.) The only remaining possibility is that the group consists of e, a, b, c satisfying $a^2 = b^2 = c^2 = e$. In this case the cancellation law shows that the only possibility for multiplication is $ab = c$, $ac = b$, $bc = a$, and the group is isomorphic to $Z_2 \oplus Z_2$, the Klein four-group.

4. If $x = f(a)$ and $y = f(b)$, then $xy = f(ab)$ and $x^{-1} = f(a^{-1})$. Thus $f(G)$ is closed under multiplication and inverses, and that is sufficient to guarantee that it is a subgroup. It is easy to check also that $f^{-1}(e)$ is closed under multiplication and inverses, hence a subgroup.

5. This is a straightforward verification of the group axioms.

Section 2. Homotopic Paths

1. Suppose x and y are in the same path component. Let α be a path from x to y, and let γ be the constant path at x. Then $((\alpha\alpha^{-1})\gamma)(1/2) = a$, while $(\alpha(\alpha^{-1}\gamma))(1/2) = b$. If the operation is associative, then $a = b$, and path components reduce to points.

2. Let γ be a path from x to y. If α is any path from x to y, then the path $\alpha(\alpha^{-1}\gamma)$ passes through b, and it is homotopic to γ with endpoints fixed.

3. Define $\alpha_t(s) = (1-t)\alpha(s) + t\beta(s)$, so that $\alpha_t(s)$ moves along the straight line segment from $\alpha_0(s) = \alpha(s)$ to $\alpha_1(s) = \beta(s)$.

4. Let d be the distance from α to ∂D, as in the preceding exercise, and choose $0 = s_0 < s_1 < \cdots < s_n = 1$ such that $|\alpha(s) - \alpha(s_j)| < d$ for $s_j \le s \le s_{j+1}$. Define a path β by $\beta(s) = (s - s_{j+1})\alpha(s_j) + (s - s_j)\alpha(s_{j+1})$ for $s_j \le s \le s_{j+1}$. Then β is polygonal, and the formula from the preceding exercise provides a homotopy of α and β.

5. Proceed in the same way as in Exercise 4.

6. The formula $\gamma(s,t) = \gamma_t(s)$ determines a correspondence between homotopies $\gamma(s,t)$ of paths from a to b with endpoints fixed and paths $t \to \gamma_t$ in path-space.

Section 3. The Fundamental Group

1. If γ is a loop based at the south pole that does not pass through the north pole, we can homotopy γ to a point by pulling points continuously to the south pole along circles of longitude.

2. Fix a base point $b \in U \cap V$. Let γ be a loop based at b. Choose $0 = s_0 < s_1 < \cdots < s_n = 1$ such that each subarc $\gamma([s_j, s_{j+1}])$ is either contained in U or contained in V. Let β_j be a path in $U \cap V$ from b to $\gamma(s_j)$, $0 \le j \le n$. Let α_j be the loop, either in U or in V, obtained by following β_j from b to $\gamma(s_j)$, the subarc of γ from parameter value s_j to s_{j+1}, and then β_{j+1}^{-1} from $\gamma(s_{j+1})$ to b. Then γ is homotopic to $\alpha_0 \cdots \alpha_{n-1}$, and each α_j is homotopic to a point in U or in V, so γ is homotopic to a point.

3. Any loop γ based at x_0 is homotopied to a point by $\gamma_t(s) = F(\gamma(s), t)$, $0 \le t \le 1$.

4. Suppose $F(x,t)$ is a contraction of X to the point $x_0 = (0,1)$ with x_0 fixed. For each fixed point $x_n = (1/n, 1)$, the map $t \to F(x_n, t)$ is a path in X from x_n to x_0, which by connectivity must cross $(1/n, 0)$, say $F(x_n, t_n) = (1/n, 0)$. If t^* is a limit point of the t_n's as $n \to \infty$, then from the continuity of F we obtain $F(x_0, t^*) = (0,0)$, and this contradicts

$F(x_0, t) = x_0$.

5. Parts (a) and (b) are straightforward. For (c), define $F(x,t) = (1-t)w + tx$. For (d), combine Exercises 5(c) and 3.

6. Any loop in $X \times Y$ based at (x_0, y_0) has the form $\gamma(s) = (\alpha(s), \beta(s))$, where α and β are loops in X and Y based at x_0 and y_0 respectively. It is straightforward to verify that this determines a one-to-one correspondence between homotopy classes, and that the correspondence is a group isomorphism.

7. Apply Exercise 6.

8. Any loop γ in $\mathbf{R^n}\backslash\{0\}$ can be homotopied to a path on the unit sphere S^{n-1} by $\gamma_t(s) = \gamma(s)/||\gamma(s)||^t$. According to Exercise 1, paths on the sphere S^{n-1} can be homotopied to a point providing that $n > 2$.

9. That (b) implies (a) is trivial. To show (a) imples (c), let $f : S^1 \to X$. Then $\alpha_1(s) = f(e^{2\pi is})$ is a loop based at $b = f(1)$. If $\alpha_t(s)$ is a homotopy of α_1 to the point b, then $F(te^{2\pi is}) = \alpha_t(s)$ is a continuous extension of f to B^2. Finally, assume that (c) holds, and let α and β be paths from b to c. Define a continuous function f from S^1 to X by $f(e^{\pi is}) = \alpha(s)$ and $f(e^{-\pi is}) = \beta(s)$, for $0 \le s \le 1$. Let F be a continuous extension of f to B^2. A homotopy of α and β with endpoints fixed is then given by $\alpha_t(s) = (1-t)F(e^{\pi is}) + tF(e^{-\pi is})$, which is the composition of F and a homotopy of the top semicircle to the bottom along vertical lines.

Section 4. Induced Homomorphisms

1. Apply Corollary 4.4.

2. Consider $f(x) = x/||x||$.

3. Since $f \circ j$ is the identity map of A, $f_* \circ j_*$ is the identity homomorphism. Hence j_* is one-to-one and f_* is onto. If X is simply connected, then $\pi_1(X, x_0) = 0$, so since f_* is onto, $\pi_1(A, x_0) = 0$, and A is simply connected. (More directly, if α_t is a homotopy of a loop α in A to a point $x_0 \in A$ in X, then $f \circ \alpha_t$ is a homotopy of α to a point in A.)

Section 5. Covering Spaces

1. Since the index of α_m is m, the result follows from Theorem 5.6.

2. The proof follows the same idea as the proof of Theorem 5.6. It must be shown that the correspondence from the homotopy class of the loop α to the terminal point of the lift of α starting at e is a group homomorphism. For this, let α_1 and α_2 be loops in X based at b, and let β_1 and β_2 be lifts starting at e. Using the continuity hypothesis, we see that the lift of $\alpha_1\alpha_2$ is given by $\beta_1(2s)$ for $0 \le s \le 1/2$, and $\beta_1(1)\beta_2(2s-1)$ for $1/2 \le s \le 1$. Since this terminates at the product $\beta_1(1)\beta_2(1)$ of the terminal points, the correspondence is a homomorphism.

3. The first statement follows from the remark that products of covering spaces are covering spaces. The isomorphism statement follows from Exercise 2. A loop corresponding to the n-tuple $(m_1, \ldots, m_n)$ is given by $\alpha(s) = (e^{2\pi m_1 s}, \ldots, e^{2\pi m_n s})$.

4. If D is any disk in the complex plane that does not contain 0, then $p^{-1}(D)$ is a disjoint union of domains, all translates of each other by integral multiples of $2\pi i$, and each mapped homeomorphically by p onto D. In this case, $p^{-1}(\{1\}) = 2\pi iZ$, so the fundamental group of $C\backslash\{0\}$ is isomorphic to Z.

5. The fundamental groups of both the open and closed annuli are isomorphic to Z.

6. Let γ be a loop based at 0. Break the parameter interval $[0,1]$ into subintervals so that if $\gamma(s)=0$ at any point of a subinterval, then γ is near 0 on the entire subinterval. Note that the image under γ of any subinterval on which $\gamma \neq 0$ is completely contained in one of the two circles of the figure-eight. Now by deforming γ by a homotopy, we can assume that on any subinterval, either $\gamma = 0$, or else γ vanishes only at the endpoints. Then the restriction of γ to any of the subintervals is homotopic to a constant or to one of α, β, α^{-1}, β^{-1}. Thus γ is homotopic to a product of these, and this yields a product representation for $[\gamma]$. The uniqueness of the decomposition follows by considering the appropriate covering space, in analogy with the discussion after Theorem 5.3.

7. Let $y \in Y$. Since Y is simply connected, in particular Y is path connected. Let γ be any path from c to y. Then $f \circ \gamma$ is a path in X from b to $f(y)$, which by Theorem 5.2 lifts to a path in E from e to some point z. Since any two paths from c to y are homotopic, Theorem 5.3 shows that the terminal point is independent of the path γ, and we may define $g(y)$ unambiguously to be the terminal point z. Evidently $p(g(y)) = f(y)$, so that g is a lift of f. Towards showing that g is continuous, let $y_0 \in Y$ and let W be an open neighborhood of $g(y_0)$. Shrinking W if necessary, we can assume that W is mapped homeomorphically by p onto an open neighborhood W_0 of $p(g(y_0)) = f(y_0)$. Let $U = f^{-1}(W_0)$, and let V be a neighborhood of y_0 such that any point of V can be joined to y_0 by a path in U. If $y \in V$, let γ be a path that proceeds from c to y_0 in Y, and then from y_0 to y in U. The lift of $f \circ \gamma$ to E then proceeds from e to $g(y_0)$ and subsequently remains in W. In particular, we obtain $g(y) \in W$, and consequently g is continuous at y_0. The uniqueness of the lift g follows from Lemma 1.1, and (a) is proved. For (b), apply the construction used in part (a) to lift the covering map $p : E \to X$ uniquely to a continuous map $h : E \to Y$ satisfying $h(e) = c$. One checks that h provides an inverse for g.

8. The verification of the group properties is straightforward. For (b), let $c \in p^{-1}(b)$. By Exercise 7, the projection p lifts uniquely to a homeomorphism f_c of E that satisfies $f_c(e) = c$. The isomorphism is now obtained from Theorem 5.4, upon showing that under the one-to-one correspondence of that theorem, the composition of covering maps corresponds to path products in the fundamental group.

9. If $f : S^2 \to S^2$ is a covering map over P^2, then f maps each pair of antipodal points onto itself, that is, $f(x) = \pm x$ for each x. Since f is continuous, either $f(x) = x$ for all x, or $f(x) = -x$ for all x. Thus there are only two covering maps, and this corresponds to the fact that $\pi_1(P^2) \cong Z_2$ has only two elements. If $f : \mathbf{R} \to \mathbf{R}$ is a covering transformation with respect to the covering map $t \to e^{2\pi i t}$ of $\mathbf{R}$ over S^1, then $f(0) = m$ for some integer m, and f is the transformation $f(t) = t + m$.

10. This problem requires a stronger hypothesis, that every $x \in X$ has a neighborhood base of simply-connected open sets.

11. The example given does not work. It should be replaced by the union X of the circles in the plane centered at $(0, 1/n)$ and of radius $1/n$, $n \geq 1$. For this example, any evenly covered neighborhood of the origin contains a circle that lifts to circular paths in the covering space that are not homotopic to points.

Section 6. Some Applications of the Index

1. Write $g(z) = (z - f(z))/|z - f(z)|$.

2. Let B be a large ball containing U. Fix $\varepsilon > 0$ and w_0. For w near w_0, the parallel

planes through $(1 \pm \varepsilon)g_0(w_0)w$ and perpendicular to w do not meet the corresponding plane through $g_0(w_0)w_0$ within B. Consequently the plane perpendicular to w and cutting U in half lies between the parallel planes, that is, $g_0(w_0) - \varepsilon < g_0(w) < g_0(w_0) + \varepsilon$. Thus g_0 is continuous at w_0, and similarly for g_1 and g_2.

3. A plane divides the three balls in half by volume if and only if it passes through the centers of the balls. Thus the plane is unique if and only if the three centers are distinct.

4. Apply the Ham Sandwich Theorem to U, V, and any ball centered at w.

5. Suppose $h : X \to Y$ is a homeomorphism, and X has the fixed-point property. If f is any map from Y to Y, and if x is a fixed point of $h^{-1} \circ f \circ h$, then $h(x)$ is a fixed point of f.

6. If f is a retraction of B^n onto S^{n-1}, then $g = -f$ is a self-map of B^n with no fixed points. In the reverse direction, any self-map f of B^n with no fixed point can be used to define a retraction g of B^n onto S^{n-1} as in the figure in the proof of Theorem 6.3.

7. The function $g(e^{i\theta}) = f(e^{i\theta}) - f(-e^{i\theta})$ is continuous and satisfies $g(-e^{i\theta}) = -g(e^{i\theta})$. Thus f assumes the same value at antipodal points if and only if $g = 0$ at the points. Since g is odd, and the image $g(S^1)$ is connected, the image is an interval (or point) symmetric about 0, hence the image contains 0.

8. For each $e^{i\theta} \in S^1$ there is a unique oriented line that is parallel to the ray from 0 to $e^{i\theta}$ and that bisects the set U. Further, this line moves continuously with θ. Let $f(e^{i\theta})$ be the area of the part of V that lies to the left of the line when we traverse the line in the positive direction. Then f depends continuously on θ. By Exercise 7, there are antipodal points of S^1 at which f has the same value, and these correspond to a line that cuts V in half by area.

9. If g is the retraction described in the hint, then $h = g \circ f$ is a retraction of B^2 onto S^1, and $-h$ is a self-map of B^2 with no fixed point, in contradiction to Theorem 6.3.

Section 7. Homotopic Maps

1. Let $f : Y \to X$ and $g : Z \to Y$ be homotopy equivalences, and let $F : X \to Y$ and $G : Y \to Z$ be homotopy inverses for f and g respectively. Then $F \circ f$ is homotopic to the identity map of Y. Pre- and post-composing this homotopy with g and G respectively, we see that $G \circ F \circ f \circ g$ is homotopic to $G \circ g$, which is in turn homotopic to the identity map of Z. Similarly, $f \circ g \circ G \circ F$ is homotopic to the identity map of X. Thus $f \circ g : Z \to X$ is a homotopy equivalence, with homotopy inverse $G \circ F$, and the relation of being homotopy equivalent is transitive.

2. Let $F(y,t)$ be a contraction of Y to the point y_0. We claim that the map $f(x) = (x, y_0)$ of X into $X \times Y$ and the projection $g(x,y) = x$ of $X \times Y$ onto X are homotopy inverses. In fact, $g \circ f$ is the identity map of X, and $(f \circ g)(x,y) = (x, y_0)$ is homotopic to the identity of $X \times Y$ via the homotopy $(x,y,t) \to (x, F(y,t))$.

3-6. Exercises 3,4,5, and 6 are all similar to the worked example in the text. In the case of Exercise 5, for instance, one checks that the mapping $g(x) = x/||x||$ of $\mathbf{R}^{n+1} \backslash \{0\}$ onto S^n and the inclusion mapping of S^n into $\mathbf{R}^{n+1} \backslash \{0\}$ are homotopy inverses.

7. If $F(y,t)$ is a homotopy of Y to a point y_0, then any map $f : X \to Y$ is homotopic to the constant map y_0 via the homotopy $f_t(x) = F(f(x), t)$.

8. Let $F : X \times [0,1] \to X$ be a contraction of X to a point x_0. Any map $f : X \to Y$ is homotopic to the constant map $y_0 = f(x_0)$ via the homotopy $f_t(x) = F(f(x), t)$. Since Y is path-connected, any constant map to y_0 is homotopic to the constant map to any other point y_1, through constant maps determined by a path from y_0 to y_1.

9. Suppose that f_t is a homotopy of f_0 and f_1. Define a loop β based at b by $\beta(t) = f_t(1)$. Then α_1 is homotopic to $(\beta^{-1}\alpha_0)\beta$ with b fixed. One way to construct a homotopy is to define α_s to follow β^{-1} from parameter value $t = 1$ to $t = s$, then follow $f_s(t)$ from parameter value $t = 0$ to $t = 1$, then return along β from parameter value $t = s$ to $t = 1$. For the converse, we can replace α_1 by a homotopic loop based at b and assume that α_1 is actually homotopic to $(\beta^{-1}\alpha_0)\beta$ with b fixed. Using the homotopy, it is straightforward to find a continuous family $t \to \gamma_t$ of loops deforming α_0 to α_1, with γ_t based at $\beta(t)$. The homotopy of f_0 and f_1 is then given by $f_t(e^{2\pi i s}) = \gamma_t(s)$.

10. Apply Exercise 9.

Section 8. Maps into the Punctured Plane

1. Let ε be the infimum of $|f(x)|$ for $x \in X$. If $|g(x) - f(x)| < \varepsilon$, then the straight line segment from $g(x)$ to $f(x)$ does not meet the origin, so $g_t(x) = (1-t)g(x) + tf(x)$, $0 \le t \le 1$, defines a homotopy of $g_0 = g$ and $g_1 = f$ through maps of X into $C\backslash\{0\}$. By Theorem 8.2, g is an exponential.

2. It suffices to show that f is homotopic to the constant function 1 if and only if f is an exponential. One direction is clear; if $f = e^h$, then $f_t = e^{th}$ defines a homotopy of f to 1. For the converse, suppose that $F(x,t)$ is a homotopy with $F(x,1) = f(x)$ and $F(x,0) = 1$. Since the unit interval is contractible, for each fixed t we can write $F(x,t) = e^{H(x,t)}$, where $H(x,t)$ is continuous in t, and $H(x,0) = 0$. The function H is unique. Since f is the exponential of $H(x,1)$, it suffices to show that H is continuous on $X \times [0,1]$. Since X is locally compact, it suffices to show that H is continuous on $E \times [0,1]$ for any compact subset E of X. Thus we can assume that X is compact. Then we have $|F| \ge \delta > 0$ on $X \times [0,1]$. Fix $x_0 \in X$. On account of the compactness of $[0,1]$ and the continuity of F, there is a neighborhood U of x_0 such that $|F(x,t) - F(x_0,t)| < \delta$ for all $x \in U$ and $0 \le t \le 1$. Then $|1 - F(x,t)/F(x_0,t)| < 1$, so that $F(x,t)/F(x_0,t)$ lies in the right half-plane, and there is a unique $G(x,t)$ satisfying $e^{G(x,t)} = F(x,t)/F(x_0,t)$ for which the imaginary part of G lies strictly between $-\pi$ and π (the principal value of the logarithm). Further, $G(x,t)$ depends continuously on x and t, and $G(x,0) = 0$. Since $e^{G(x,t)+H(x_0,t)} = F(x,t)$, we obtain from the uniqueness of H that $G(x,t) + H(x_0,t) = H(x,t)$, and consequently H is continuous.

3. If X is contractible, then any map from X to $C\backslash\{0\}$ is homotopic to a point, and Exercise 2 applies.

4. Let $\varepsilon = \inf|f(x)|$. Any g satisfying $|g - f| < \varepsilon$ is homotopic to f, as in Exercise 1, hence has the same index.

5. Since $f/g \ne -1$, we can express f/g uniquely in the form e^{ih} where $-\pi < h < \pi$. The function h is continuous, and $f = e^{ih}g$, so f and g have the same index.

6. Let H_+ and H_- denote the closed upper and lower hemispheres of S^n respectively. These spaces are compact and contractible, so that Corollary 8.5 applies, and there are continuous functions g_+ and g_- on the respective hemispheres whose exponentials coincide with f. Then the exponentials of g_+ and g_- coincide on the equator $E = H_+ \cap H_-$. Thus the values of $g_+ - g_-$ on E are integral multiples of $2\pi i$. Since $n \ge 2$, the equator E is connected, and hence $g_+ - g_-$ is constant on E, say $g_+ - g_- = 2\pi i m$ on E. We then obtain

a continuous function g on S^n by setting $g = g_+$ on H_+ and $g = g_- + 2\pi im$ on H_-, and $f = e^g$.

7. Let $\pi : S^n \to P^n$ denote the projection. By the preceding exercise, there is a continuous function g on S^n such that $f \circ \pi = e^g$ on S^n. Since $f \circ \pi$ attains equal values at antipodal points of S^n, $g(x) - g(-x)$ is an integral multiple of $2\pi i$ for each $x \in S^n$. Since S^n is connected, this function is constant on S^n, say $g(x) - g(-x) = 2\pi im$ on S^n. Replacing x by $-x$, we obtain $g(-x) - g(x) = 2\pi im$, and consequently $m = 0$. Thus $g(-x) = g(x)$, and g determines a continuous function h on P^n, which satisfies $f = e^h$. Towards explaining this state of affairs, note that the group of maps to the punctured plane modulo exponentials has no elements of finite order, whereas the elements of the fundamental group of P^n have finite order.

8. The homotopy classes are uniquely determined by the winding number around each loop of the figure-eight, hence are classified by $Z \oplus Z$.

Section 9. Vector Fields

1. In the complex plane, define $V(z) = z^2$ if $|z| \leq 1$, and $V(z) = z^2/|z|^2$ if $|z| \geq 1$. Regarded as a vector field on the sphere, $V(z)$ extends continuously to the point at infinity and is represented near ∞ by the figure before the proof of Theorem 9.2. The only zero of $V(z)$ is at $z = 0$, and $V(z)$ has index 2 there (as required).

2. If V is a constant vector field tangent to the sphere, then V is orthogonal to itself, hence $V = 0$.

3. Any vector field on the sphere S^2 can be decomposed in the form $F(p) = T(p) + N(p)$, where $T(p)$ is tangent to the sphere at p and $N(p)$ is normal to the sphere. Apply Theorem 9.2 to the tangent vector field $T(p)$.

4. Compute the partial derivatives of $\sigma_1 = x/(1-z)$ and $\sigma_2 = y/(1-z)$, and substitute into the formula, to obtain $\sigma^* F(u,v) = G(u,v)/(1-z)^2$.

5. Let $p_0 \in \mathbf{C}\backslash\gamma(S^1)$, and let ρ denote the distance from p_0 to $\gamma(S^1)$. If $|p - p_0| < \rho$, then by Theorem 8.2, $(\gamma - p)/(\gamma - p_0)$ is an exponential, so the index of $\gamma - p$ is the same as the index of $\gamma - p_0$. Thus the winding number of γ around p is locally constant, consequently it is constant on each connected component of $\mathbf{C}\backslash\gamma(S^1)$. If $|\gamma(S^1)| < M$, then Theorem 8.2 shows that $\gamma - M$ is an exponential, so its index is zero, and consequently the winding number of γ is zero around any point in the unbounded component of $\mathbf{C}\backslash\gamma(S^1)$.

6. If both γ_0 and γ_1 have winding number 1 about p, then $\gamma_0 - p$ and $\gamma_1 - p$ have the same index, so they are homotopic as maps to $B(p,\varepsilon)\backslash\{p\}$. Consequently $F \circ \gamma_0$ is homotopic to $F \circ \gamma_1$, regarded as maps to $\mathbf{C}\backslash\{0\}$, and F has the same index around γ_0 and γ_1. In (b), the index of F around γ is m times the index of F at p. To see this, show that γ is homotopic to the path $p + \varepsilon e^{2\pi imt}$, $0 \leq t \leq 1$, and unravel the definitions. For(c), both the vector fields $F(x,y) = (2x, 2y)$ and $G(x,y) = (-y, x)$ have index $+1$ at the origin. In (d), the index of $-F$ is also m, and in fact the index of any constant rotate of F is the same as the index of F.

7. First treat the special case described in the hint, covering two zeros, by writing the function as an exponential on each semicircle and noting the cancellation of the contributions along the common segment. Then combine the basic idea with an induction argument on the number of zeros to treat the general case.

8. For the idea, see the figure before the proof of Theorem 9.2.

Section 10. The Jordan Curve Theorem

1. Let W_0 and W_1 be the two connected components of $U \cap V$. Then W_0 and W_1 are both path-connected, and their closures are disjoint. Fix $b \in W_0$. Let γ be a loop in X based at b. Since $\gamma(\overline{W_0})$ and $\gamma(\overline{W_1})$ are disjoint closed sets, it is easy to find parameter values $0 = s_0 < s_1 < \cdots < s_{2n} = 1$ such that $\gamma(s_j)$ belongs to W_0 if j is even and to W_1 if j is odd, and $\Gamma_j = \gamma([s_j, s_{j+1}])$ is contained in either U or V. We assign an integer m to this decomposition as follows. For each even integer j we count 0 if $\Gamma_j \cup \Gamma_{j+1}$ is a subset of U or V, otherwise we count $+1$ if $\Gamma_j \subset U$ and $\Gamma_{j+1} \subset V$ and we count -1 if $\Gamma_j \subset V$ and $\Gamma_{j+1} \subset U$. The integer m is the sum of the ± 1's obtained in this manner. One shows that the integer m does not depend on the choice of s_j's satisfying the above conditions. (Hint: Argue that if $0 = t_0 < t_1 < \cdots < t_{2k} = 1$ is another such choice, and if $t_2 \leq s_2$, we can replace t_1 by s_1 and get another such partition with the same m.) The integer m is the same for any path close to γ, hence the integer m depends only on the homotopy class of γ. The path product evidently corresponds to adding the m's. One checks easily that the correspondence from paths to Z is onto, and that a path with $m = 0$ is homotopic to a point.

2. Follow the hint.

3. The case of an arc in $\mathbf{R}^2$ can be reduced to the case of an arc in S^2 by adjoining the point at ∞. For a simple arc in S^2, place one endpoint of the arc at ∞ and apply the preceding exercise.

4. Let U be a connected component of $\mathbf{C}\backslash\Gamma$, and let $z \in U$. Let $w \in \Gamma$, and let Γ_ε denote an open interval on Γ containing w and contained in an ε-neighborhood of w. Then $\Gamma\backslash\Gamma_\varepsilon$ is a simple arc, so by either Exercise 3 or 6, its complement is connected. Hence we can connect z to w by an arc disjoint from $\Gamma\backslash\Gamma_\varepsilon$. This arc meets Γ only within the ε-neighborhood of w, and consequently there are points inside the ε-neighborhood of w that belong to U. The closure of U then contains each $w \in \Gamma$, so the boundary of U is Γ.

5. If p lies in the unbounded component of $\mathbf{C}\backslash E$, we can connect p to ∞ by a path $p(t)$ in $\mathbf{C}\backslash E$. This induces a homotopy $t \to -(1 - z/p(t))$ of $z - p$ and a constant function in the space of functions from E to $\mathbf{C}\backslash\{0\}$. By Theorem 8.3, $z - p$ is an exponential. For the converse, suppose p lies in a bounded component U of $\mathbf{C}\backslash E$, and suppose for purposes of obtaining a contradiction that $z - p$ is an exponential on E. Write $z - p = e^h$ on E, and extend h continuously to the complex plane so that $h(z) \to 0$ as $z \to \infty$. Define $g(z) = z - p$ for $z \in U$ and $g(z) = e^{h(z)}$ for $z \in \mathbf{C}\backslash U$. Then g maps $\mathbf{C}\backslash\{p\}$ continuously into $\mathbf{C}\backslash\{0\}$, the index of g around a small circle centered at p is 1, and the index of g around a large circle centered at p is 0. Since the circle loops are homotopic in $\mathbf{C}\backslash\{p\}$, this contradicts Theorem 8.7.

6. Since E is contractible and compact, this follows from Corollary 8.5 and Exercise 5.

7. The proof is virtually the same as that of Exercise 5.

8. Suppose p_1 and p_2 are in different components of the complement of a simple closed curve Γ. By Exercise 5, $z - p_j$ is not an exponential on Γ, so the index of $z - p_j$ around Γ is an integer $m_j \neq 0$. Then the index of $(z - p_1)^{m_2}(z - p_2)^{-m_1}$ around Γ is 0. Thus $(z - p_1)^{m_2}(z - p_2)^{-m_1}$ is an exponential on Γ, contradicting Exercise 7.

9. Show that if L is a straight line segment in Γ, then the index of $z - p$ increases by

+1 as p crosses over L from right to left, with respect to the orientation of L induced by Γ.

10. Let γ be a simple closed curve in $\mathbf{C}$, parametrized for convenience by the interval $-1/2 \leq t \leq 1/2$. Assume $\gamma(0) = 0$, and denote $\Gamma = \gamma([-1/2, 1/2])$. Let $\varepsilon > 0$ be small. Choose $\eta > 0$ so that if $|s| < \eta$, then $|\gamma(s)| < \varepsilon$. Choose $\delta > 0$ so that if $|\gamma(s)| < \delta$, then $|s| < \eta$. Let L be a straight line segment from 0 to another point $q \in \Gamma$ satisfying $|q| < \delta$. If there is an interval on L contained in Γ, then we are done, by the preceding exercise. If not, then there is a nonempty open interval in $L\backslash\Gamma$, with endpoints in Γ, say the endpoints are $\gamma(a)$ and $\gamma(b)$, where $a < b$. Note that $|a|, |b| < \eta$. Let α be the loop that coincides with γ outside (a, b) and that runs linearly along L from $\gamma(a)$ to $\gamma(b)$ inside (a, b). Then $\Gamma_0 = \alpha([-1/2, 1/2])$ is a simple closed curve that contains a straight line segment. By the preceding exercise, $\mathbf{C}\backslash\Gamma_0$ has a bounded component, which by Exercise 4 includes Γ_0 in its closure. Thus if p_0 is any point of Γ such that $|p_0| > \varepsilon$, there are points p near p_0, $p \notin \Gamma$, for which the index of $\alpha - p$ is not zero. For such p, the function $\alpha - p$ can be homotopied to $\gamma - p$, by moving the values of α along straight line segments inside the ε-disk centered at 0 to the values of γ. Thus the index of $\gamma - p$ is also nonzero, and p belongs to a bounded component of $\mathbf{C}\backslash\Gamma$. In particular, $\mathbf{C}\backslash\Gamma$ is not connected, and the alternative proof of the Jordan Curve Theorem (laid out in Exercises 4 through 10) is complete.

Chapter IV. Higher Dimensional Homotopy

Section 1. Higher Homotopy Groups

1. Use the identification of $\pi_k(S^1, 1)$ with the set of homotopy classes of maps $S^k \to S^1$, where the maps and homotopies take a fixed $b \in S^k$ to $1 \in S^1$. Apply Exercise III.5.7 to lift $F : S^k \to S^1$ to $\hat{F} : S^k \to \mathbf{R}$. Since $\hat{F}$ is homotopic to a constant map with the image of b fixed during the homotopy, so is F.

2. A map $F : S^k \to X \times Y$ is canonically identified with a pair (F_1, F_2) of maps $F_1 : S^k \to X$ and $F_2 : S^k \to Y$. The conclusion follows by checking behavior of homotopies and base points.

3. This is analogous to the proof for π_1 given in Lemma III.2.5.

4. Follow the hint.

5. See Exercise III.3.9 and proceed analogously.

6. The deformation of the identity $X \to X$ to the constant map $X \to b$ gives immediately (by composition) a homotopy of an arbitrary map $S^k \to X$ to the constant map $S^k \to b$. Exercise 5 then applies.

7. If $[F] \in \pi_k(Y, c)$ then $[g \circ F]$ is a well-defined element of $\pi_k(X, b)$. The required verifications are now routine.

8. If $h : (X, b) \to (Y, c)$ is a homotopy inverse of g, then $g_* \circ h_* = (g \circ h)_*$ by Exercise 7. Also $(h \circ g)_* = h_* \circ g_*$. Since $g \circ h$ and $h \circ g$ are homotopic to identity maps on X and Y respectively, $(g \circ h)_*$ and $(h \circ g)_*$ are the identities, so h_* is an inverse for g_*.

Section 2. Noncontractibility of S^n

1. Set $r(t) = x + t(y - x)$ and compute $|r(t)|^2 - 1 = t^2(|x|^2 + |y|^2 - 2x \cdot y) - 2t(|x|^2 + x \cdot y) + (|x|^2 - 1)$. This is a quadratic equation in t, which has two real roots, one positive and one negative, corresponding to the points where there line through x and y hits S^{n-1}.

The positive root is the parameter value for which the ray from x to y meets S^{n-1}. This root is expressed in terms of the quadratic formula, and this yields the expression for $h(x, y)$.

2. The point $t = 0$ is the only potentially troublesome point for existence and continuity of derivatives. Derivatives of all orders of $\exp(-1/t^2)$ are sums of terms of the form $ct^{-m}\exp(-1/t^2)$. Such terms tend to 0 as $t \to 0$. For (b) and (c), follow the hints.

3. A retraction h of Λ onto the slice $X \times \{0\}$ determines a homotopy of X to a point by simply composing h with the quotient map from $X \times [0,1]$ onto Λ. Conversely, since homotopies to a point are constant on the end slice, they determine maps from the quotient space Λ to X. The condition that the homotopy be the identity map at the other end slice corresponds to the condition that the quotient map h be a retraction onto the end slice.

4. For Theorem 2.1, note that S^n is homeomorphic to $S^{n-1} \times [-1,1]$ with the slices $S^{n-1} \times \{-1\}$ and $S^{n-1} \times \{1\}$ each identified to a point. A contraction of S^{n-1} to a point would thus give a deformation of S^n to a closed interval, and hence to a point. The argument for Theorem 2.2 is similar, using the identification of a ball as a double cone on a ball of one lower dimension. For Theorem 2.3, note that B^n is homeomorphic to $B^{n-1} \times [0,1]$. If $F : B^{n-1} \to B^{n-1}$ had no fixed point, then $(x,t) \to (F(x), t)$ would have none.

5. Follow the given outline.

6. Follow the hint.

Section 3. Simplexes and Barycentric Subdivision

1. If u is the barycenter of the simplex and v belongs to a face, map $u + t(v-u)$ to $t(v-u)/||v-u||$, $0 \le t \le 1$.

2. Let S have vertices $v_0, \ldots, v_k$, and let c be the maximum of $f(v_0), \ldots, f(v_k)$. Suppose that $t_j \ge 0$, $\sum t_j = 1$. Then $f(\sum t_j v_j) = \sum t_j f(v_j) \le c \sum t_j = c$, with equality if and only if $t_j = 0$ whenever $f(v_j) < c$. Thus $f = c$ precisely on the face of the simplex generated by the vertices at which $f = c$. For the second statement, if F is a face, define f on the simplex by $f(\sum t_j v_j) = \sum \{t_j : v_j \in F\}$. Then f attains its maximum value 1 over the simplex precisely on F.

3. The barycentric subdivision of a k simplex has $k+1$ k-simplexes, one opposite each vertex. Proceeding by induction, we obtain $(k+1)^N$ for the number of k-simplexes in the N-th barycentric subdivision.

4. This follows by repeated application of Lemma 3.2.

5. Use Lemma 3.2.

6. For the nontrivial implication, suppose S is convex. By definition, convex combinations of any two vectors in S belong to S. We proceed by induction and suppose that convex combinations of n vectors in S are in S. Suppose that $v = \sum t_j v_j$ is a convex combination of the $n+1$ vectors $v_0, \ldots, v_n$ in S. We can assume $t_0 \ne 1$. Set $r = 1 - t_0 = t_1 + \cdots + t_n$. The vector $u = (t_1/r)v_1 + \cdots + (t_n/r)v_n$ is then a convex combination of n vectors in S. Hence $u \in S$, and since $v = t_0 v_0 + (1-t_0)u$, also $v \in S$.

7. Use the formula $\sum_i s_i(\sum_j t_j v_j) = \sum_j (\sum_i s_i t_j) v_j$ to show that any convex combination of convex combinations is again a convex combination.

8. Suppose S is not affinely independent. Then there are $w_j \in S$ and $t_j \ne 0$ such that $\sum t_j = 0$ and $\sum t_j w_j = 0$. Let $\sum'$ denote the sum over the indices j for which $t_j > 0$, and let

$\sum''$ denote the sum over the indices j for which $t_j < 0$. Multiplying the t_j's by a constant, we can assume that $\sum' t_j = 1$. Then $\sum''(-t_j) = 1$, so that $\sum' t_j w_j = \sum''(-t_j)w_j$ gives two different representations of an element of the convex hull of S as a convex combination of elements of S. The simplex statement follows from Exercise 7.

Section 4. Approximation by Piecewise Linear Maps

1. If h is affine, then by definition $h(v) = u + L(v)$, where u is fixed and L is linear. If $\sum t_j = 1$, we then have $h(\sum t_j v_j) = u + L(\sum t_j v_j) = (\sum t_j)u + \sum t_j L(v_j) = \sum t_j h(v_j)$. In particular, h preserves convex combinations. For the converse, set $u = h(0)$, and define $L(v) = h(v) - u$. Then L also preserves convex combinations, and $L(0) = 0$. It is then straightforward to check that L is linear. (First observe that L(-w)=-L(w), since 0 is the midpoint of the line seqment from $-w$ to w. Next check that $L(tw) = tL(w)$ for $0 \leq t \leq 1$, then for $t \geq 1$, then for all t. Finally check that $L(v + w) = L(v) + L(w)$.)

2. If the vertex sets are $v_0, \ldots, v_k$ and $w_0, \ldots, w_k$, send $\sum t_j v_j$ to $\sum t_j w_j$.

3. For (a), let $x_1, \ldots, x_{k+2} \in x_0 + W$. Since W is k-dimensional, there is a nontrivial linear relation $c_2(x_2 - x_1) + \cdots + c_{k+2}(x_{k+2} - x_1) = 0$. Setting $c_1 = -(c_2 + \cdots + c_{k+2})$, we obtain a nontrivial affine relation $\sum c_j x_j = 0$, $\sum c_j = 0$, so that $x_1, \ldots, x_{k+2}$ are affinely dependent. For (b), it is easy to check that the indicated set E is a k-flat, since the vectors $w_j - w_0$, $1 \leq j \leq k$, are linearly independent. Evidently $w_0, \ldots, w_k \in E$, and any k-flat containing the w_j's also contains E. Any strictly larger flat corresponds to a subspace of strictly larger dimension, so that E is unique.

4. If g is linear, then g maps subspaces to subspaces, and the linearity shows that g maps flats to flats. Since an affine map is a linear map plus a constant, and since translates of flats are flats, any affine map sends flats to flats. This establishes (a), and (b) is similar.

5. We make an induction hypothesis, and assume that the statement is true for simplexes of dimension less than k. We can assume then that Q meets the interior of S; otherwise $Q \cap S$ is contained in a face of S, and we can apply the induction hypothesis to the face. If Q is a point, we take the $(k-q)$-face to be S itself. If Q is not a point, then Q meets some $(k-1)$-face F of S. In this case, there is a subface T of F of minimal dimension, say t, such that T contains $Q \cap F$. Evidently Q contains internal points of T (points not on a boundary face). Let P be the intersection of Q and the flat generated by T. Since Q contains internal points of T, the dimension p of P satisfies $p \geq q - (k-t)$, that is, $t - p \leq k - q$. Further, P is generated by points in $P \cap T$. By the induction hypothesis, P meets some $(t-p)$-face of T, and consequently Q meets a $(k-q)$-face of S.

6. We are assuming that $k < n$. The intersection is a closed set. If the intersection had nonempty interior in the k-face, then it would contain $k+1$ affinely independent vectors, and the flat would be the unique k-flat containing the face. However, this flat does not intersect the interior of the simplex.

7. Use $F(x,t) = (tf(x) + (1-t)g(x))/||tf(x) + (1-t)g(x)||$.

8. Follow the hint.

9. Suppose $m < n$. By Exercise 8, $\pi_m(S^n) = 0$. On the other hand, it was established in Section 2 (Theorem 2.1) that $\pi_m(S^m) \neq 0$. Hence S^n cannot be homeomorphic to S^m.

10. Recall that S^n is homeomorphic to the one-point compactification of $\mathbf{R}^n$ and follow the hint.

11. If $F(x,t)$ were a homotopy of the identity map of $\mathbf{R}^{n+1}\backslash\{0\}$ and a constant, then $F(x,t)/||F(x,t)||$ would be a homotopy of the identity map of S^n and a constant, contradicting Theorem 2.1.

12. Apply the procedure used in Exercise 11 to a small n-sphere centered at p.

13. We can assume that $p = 0$ and that the radius of B is greater than 1, so that B contains S^n. Any map f from S^k to $B\backslash\{0\}$ is then homotopic to a map from S^k to S^n, with homotopy $F(x,t) = f(x)/||f(x)||^t$. Now apply Exercise 8.

14. Suppose for some $n > m$ there is a homeomorphism h from a nonempty open subset V of $\mathbf{R}^n$ onto an open subset U of $\mathbf{R}^m$. By restricting h, we can assume that V is a ball, say with center p. By Exercise 13, $\pi_{m-1}(B\backslash\{p\}) = 0$. Hence $\pi_{m-1}(h(U)\backslash\{h(p)\}) = 0$, and this contradicts Exercise 12.

Section 5. Degrees of Maps

1. For the homotopy of f and a constant, define $F(e^{i\theta}, t)$ to be $e^{2ki\theta t}$ if $0 \leq \theta \leq \pi$, and to be $e^{2ki(2\pi-\theta)t}$ if $\pi \leq \theta \leq 2\pi$. Each point of the circle is hit exactly $2k$ times by f, that is, each inverse image has $2k$ points.

2-5. Follow the hints.

6. This exercise should have the additional hypothesis that the vector field V be nowhere zero. Then we can replace V by $V/||V||$ and assume that $||V|| = 1$ everywhere. Now follow the hint, and use the orthogonality of p and $V(p)$ to get $||H|| = 1$.

7. Since the Jacobian of f is continuous, the set of singular points of f is closed, hence compact. Since the image of a compact set under f is compact, the critical points of f also form a compact set.

8. The Jacobian determinant formula follows by direct calculation. Then the singular points of $\hat{f}$ are the points on the rays from the origin passing through the singular points of f, and similarly for the critical points. The conclusion follows.

9. Follow the hint.

10. If σ were homotopic to a constant map in $B(p; 2\varepsilon)\backslash\{p\}$, then the composition of the homotopy with f would give a homotopy of the map $x \to f(p) + \delta x$ to a constant in $f(B(p; 2\varepsilon))\backslash\{f(p)\}$. But this map is not even homotopic to a constant in $\mathbf{R}^n\backslash\{f(p)\}$, much less in $f(B(p; 2\varepsilon)\backslash\{f(p)\}$. This establishes (a). For (b), note that the maps $(\sigma - p)/||\sigma - p||$ corresponding to different δ's are homotopic. For (c), note that the maps $(\sigma - q)/||\sigma - q||$ are homotopic to each other for q near p. Hence the set of points at which f is orientation-preserving is open, as is the set of points at which f is orientation-reversing. Since U is connected, one of these sets is U, the other empty. The statements in (d) follow from the fact that the degree of the composition of two maps of S^n is the product of the degrees. For (e), use the Taylor approximation $\sigma(x) \approx p + J^{-1}x$, where J^{-1} is the inverse of the Jacobian matrix of f at p. From Exercise 4.7, for instance, we see that $(\sigma - p)/||\sigma - p||$ is homotopic to $J^{-1}x/||J^{-1}x||$. The degree of this latter map is positive if and only if $\det J^{-1}$ is positive, which occurs if and only if $\det J$ is.

11. We cover X by two coordinate sets U and V, where U corresponds to the set where $0 < x < 1$, and V to the set where $x \neq 1/2$. For coordinatizing maps we take $\varphi(x,y) = (x,y)$ on U, and $\psi(x,y) = (x,y)$ for $(x,y) \in V$ with $0 \leq x < 1/2$, $\psi(x,y) = (1-x, 1-y)$ for $(x,y) \in V$ with $1/2 < x \leq 1$. The coordinate functions are well-defined on X, and the

overlap maps are C^∞, so X is a differentiable manifold. If X were oriented, then the U coordinate would be either everywhere positively oriented or everywhere negatively oriented relative to the orientation of X, since U is connected. Similarly, since V is connected, the V coordinate would also be either everywhere positively or everywhere negatively oriented relative to the orientation of X. But then the overlap map $\psi \circ \varphi^{-1}$ on $U \cap V$ would be everywhere orientation-preserving or everywhere orientation-reversing. However, $U \cap V$ has two components, on one of which the overlap map preserves orientation, while on the other it reverses orientation. We conclude that X is not orientable.

12. Compute the overlap maps explicitly to see that they are C^∞. Define an orientation by declaring that $(x_1, \dots, x_{n+1}) \to (x_1, \dots, x_{k-1}, x_{k+1}, \dots, x_{n+1})$ preserves orientation if either $x_k > 0$ and k is odd or if $x_k < 0$ and k is even, otherwise reverses orientation. Then check that the overlap maps all preserve orientation.

13. This is a direct verification.

14. For n odd, $(x_1, x_2, \dots, x_n, x_{n+1}) \to (-x_2, x_1, -x_4, x_3 \cdots)$ is a tangent vector field without zeros. For n even, define V as in the hint, and check that V is a tangent vector field with two zeros, at the north and south poles $(0, \dots, 0, 1)$ and $(0, \dots, 0, -1)$. In terms of the coordinates $(x_1, \dots, x_n)$ around $(0, \dots, 0, 1)$, the vector field has the form $(-x_1(1 - \sum x_j^2)^{1/2}, \dots, -x_n(1 - \sum x_j^2)^{1/2})$, which has the same index as $(-x_1, \dots, -x_n)$, namely 1. Similarly, the index at $(0, \dots, 0, -1)$ is also 1, so the sum of the indices is 2.

Index

A CATALOG OF SELECTED

DOVER BOOKS

IN SCIENCE AND MATHEMATICS

ASYMPTOTIC METHODS IN ANALYSIS, N.G. de Bruijn. An inexpensive, comprehensive guide to asymptotic methods–the pioneering work that teaches by explaining worked examples in detail. Index. 224pp. 5⅜ x 8½. 64221-6 Pa. $7.95

OPTICAL RESONANCE AND TWO-LEVEL ATOMS, L. Allen and J. H. Eberly. Clear, comprehensive introduction to basic principles behind all quantum optical resonance phenomena. 53 illustrations. Preface. Index. 256pp. 5⅜ x 8½. 65533-4 Pa. $8.95

COMPLEX VARIABLES, Francis J. Flanigan. Unusual approach, delaying complex algebra till harmonic functions have been analyzed from real variable viewpoint. Includes problems with answers. 364pp. 5⅜ x 8½. 61388-7 Pa. $9.95

ATOMIC SPECTRA AND ATOMIC STRUCTURE, Gerhard Herzberg. One of best introductions; especially for specialist in other fields. Treatment is physical rather than mathematical. 80 illustrations. 257pp. 5⅜ x 8½. 60115-3 Pa. $7.95

APPLIED COMPLEX VARIABLES, John W. Dettman. Step-by-step coverage of fundamentals of analytic function theory–plus lucid exposition of five important applications: Potential Theory; Ordinary Differential Equations; Fourier Transforms; Laplace Transforms; Asymptotic Expansions. 66 figures. Exercises at chapter ends. 512pp. 5⅜ x 8½. 64670-X Pa. $12.95

ULTRASONIC ABSORPTION: An Introduction to the Theory of Sound Absorption and Dispersion in Gases, Liquids and Solids, A.B. Bhatia. Standard reference in the field provides a clear, systematically organized introductory review of fundamental concepts for advanced graduate students, research workers. Numerous diagrams. Bibliography. 440pp. 5⅜ x 8½. 64917-2 Pa. $11.95

UNBOUNDED LINEAR OPERATORS: Theory and Applications, Seymour Goldberg. Classic presents systematic treatment of the theory of unbounded linear operators in normed linear spaces with applications to differential equations. Bibliography. 199pp. 5⅜ x 8½. 64830-3 Pa. $7.95

LIGHT SCATTERING BY SMALL PARTICLES, H.C. van de Hulst. Comprehensive treatment including full range of useful approximation methods for researchers in chemistry, meteorology and astronomy. 44 illustrations. 470pp. 5⅜ x 8½. 64228-3 Pa. $12.95

CONFORMAL MAPPING ON RIEMANN SURFACES, Harvey Cohn. Lucid, insightful book presents ideal coverage of subject. 334 exercises make book perfect for self-study. 55 figures. 352pp. 5⅝ x 8¼. 64025-6 Pa. $11.95

OPTICKS, Sir Isaac Newton. Newton's own experiments with spectroscopy, colors, lenses, reflection, refraction, etc., in language the layman can follow. Foreword by Albert Einstein. 532pp. 5⅜ x 8½. 60205-2 Pa. $12.95

GENERALIZED INTEGRAL TRANSFORMATIONS, A.H. Zemanian. Graduate-level study of recent generalizations of the Laplace, Mellin, Hankel, K. Weierstrass, convolution and other simple transformations. Bibliography. 320pp. 5⅜ x 8½. 65375-7 Pa. $8.95